The Comprehensive
Handbook of
SCHOOL
SAFETY

Occupational Safety and Health Guide Series

Series Editor

Thomas D. Schneid
Eastern Kentucky University
Richmond, Kentucky

Published Titles

Forthcoming Titles

The Comprehensive
Handbook of
SCHOOL
SAFETY

Edited by
E. Scott Dunlap

CRC Press
Taylor & Francis Group
Boca Raton London New York

CRC Press is an imprint of the
Taylor & Francis Group, an **informa** business

CRC Press
Taylor & Francis Group
6000 Broken Sound Parkway NW, Suite 300
Boca Raton, FL 33487-2742

© 2013 by Taylor & Francis Group, LLC
CRC Press is an imprint of Taylor & Francis Group, an Informa business

No claim to original U.S. Government works

Version Date: 20120522

International Standard Book Number: 978-1-4398-7407-3 (Hardback)

This book contains information obtained from authentic and highly regarded sources. Reasonable efforts have been made to publish reliable data and information, but the author and publisher cannot assume responsibility for the validity of all materials or the consequences of their use. The authors and publishers have attempted to trace the copyright holders of all material reproduced in this publication and apologize to copyright holders if permission to publish in this form has not been obtained. If any copyright material has not been acknowledged please write and let us know so we may rectify in any future reprint.

Except as permitted under U.S. Copyright Law, no part of this book may be reprinted, reproduced, transmitted, or utilized in any form by any electronic, mechanical, or other means, now known or hereafter invented, including photocopying, microfilming, and recording, or in any information storage or retrieval system, without written permission from the publishers.

For permission to photocopy or use material electronically from this work, please access www.copyright.com (http://www.copyright.com/) or contact the Copyright Clearance Center, Inc. (CCC), 222 Rosewood Drive, Danvers, MA 01923, 978-750-8400. CCC is a not-for-profit organization that provides licenses and registration for a variety of users. For organizations that have been granted a photocopy license by the CCC, a separate system of payment has been arranged.

Trademark Notice: Product or corporate names may be trademarks or registered trademarks, and are used only for identification and explanation without intent to infringe.

Library of Congress Cataloging-in-Publication Data

The comprehensive handbook of school safety / edited by E. Scott Dunlap.
    p. cm. -- (Occupational safety & health guide series)
  Includes bibliographical references and index.
  ISBN 978-1-4398-7407-3 (hardback)
    1. Schools--Safety measures--Handbooks, manuals, etc. 2. School violence--Prevention--Handbooks, manuals, etc. I. Dunlap, E. Scott (Erik Scott)

LB2864.5.C65 2012
363.11'9371--dc23   2012016135

Visit the Taylor & Francis Web site at
http://www.taylorandfrancis.com

and the CRC Press Web site at
http://www.crcpress.com

Contents

SECTION I School Security

SECTION II   School Safety

SECTION III   *Emergency Management*

SECTION IV   *Program Development and Execution*

Preface

"Do you believe in God?" the shooter asked young Cassie Bernall. Standing among the chaos of what had previously been a normal day at school, she replied, "Yes." And moments later her life was brought to an end by a gunshot wound. Cassie was one of many students who suffered the unexpected tragedy of active shooters at Columbine High School (Bernall, 1999). The tragedy of Columbine took our nation by surprise. Though we had been accustomed to security and safety issues in other areas of our culture, Columbine brought the issue to our schools.

In the modern school, it may not be unusual to see

- Metal detectors to prevent handguns and other weapons from being brought onto school property
- Students in standardized uniforms to prevent the appearance of gang affiliations
- Police officers patrolling the property to deter violent activity as well as being on hand in the event that an incident were to occur

Such evolutions have forever changed how we view the safety of our students. However, the phrase "school safety" goes beyond issues of security that are in place to protect students, faculty, and staff. There are also environmental safety risks that need to be discussed in order to address school safety from a comprehensive perspective. As each summer passes, children continue to be injured and lose their lives as a result of heat stress from extracurricular activity or of being abandoned in a vehicle. Faculty members are significantly injured from simple activities such as climbing on a chair instead of a ladder to hang classroom decorations. Staff members are exposed to risks and suffer injuries from activities such as failing to lockout an energy source prior to working on energized equipment.

This text seeks to expand the dialogue on school safety to comprehensively address the spectrum of safety risks to students, faculty, and staff. The material contained in this text can be used by school administrators to develop appropriate programs to protect these individuals from harm.

REFERENCE

Bernall, M. (1999). *She Said Yes*. Pocket Books, New York.

Editor

E. Scott Dunlap, EdD, CSP, holds a doctorate of education from the University of Memphis and an MS in loss prevention and safety from Eastern Kentucky University. Dunlap is a professional member of the American Society of Safety Engineers (ASSE) and is a board certified safety professional (CSP). Prior to beginning his career in workplace safety, he taught fire prevention education in the public school system in eastern Kentucky.

Dunlap began his career in workplace safety as a safety specialist at a state psychiatric hospital in a suburb of Baltimore, Maryland. He then moved to Zanesville, Ohio, where he served as the safety manager at an AutoZone distribution center. After three years with AutoZone, Dunlap went on to spend six years in the grain industry, working for Cargill in Minneapolis, Minnesota, and Archer Daniels Midland (ADM) in Decatur, Illinois. He served as corporate health and safety director for both organizations in their respective grain divisions. He then led environment safety and health initiatives for Nike in Memphis, Tennessee, where the majority of equipment, footwear, and apparel are distributed in the United States and where Nike's dot-com business is operated. In his most recent role, Dunlap is an assistant professor in Eastern Kentucky University's graduate safety security and emergency management program.

In addition to his work responsibilities, Dunlap served in a number of leadership roles in two grain industry trade organizations, the Grain Elevator and Processing Society (GEAPS) and the National Grain and Feed Association (NGFA). He was a featured grain industry speaker at numerous national and regional conferences throughout the United States and Canada. Dunlap has also been a speaker at the National Safety Congress and ASSE national professional development conferences.

Dunlap is a published writer on such topics as Occupational Safety and Health Association (OSHA) compliance, Department of Transportation compliance, behavior-based safety, and organizational safety culture. His work has appeared in *Professional Safety,* the *Online Journal of Distance Learning Administration,* the *Journal of Emergency Medical Services,* and *World Grain Magazine.* His most substantial work has been the publication of textbooks on motor carrier safety and auditing. His current research is in the area of industry leader development and involvement in workplace safety leadership.

Contributors

Ryan K. Baggett, MS, is an assistant professor of homeland security within the Department of Safety, Security, and Emergency Management at Eastern Kentucky University (EKU). Prior to this position, Ryan served as the director of homeland security programs within the Justice and Safety Center at EKU. In this capacity, he provided oversight for two U.S. Department of Homeland Security (DHS) and Federal Emergency Management Agency (FEMA)-funded programs. Baggett started at EKU in the fall of 1999, after working three years at a university police department. During his tenure at the center, he has served as a program manager for various public safety and security projects from both the Department of Justice and the Department of Homeland Security. Topical areas within these programs include, but are not limited to, the national incident management system/national response framework, technology evaluations and assessments, and training and technical assistance to rural and remote agencies. Prior to the fall of 2011, Baggett taught as adjunct faculty in the safety, security, and emergency management and the criminal justice and police studies departments within the college. He has coauthored several book chapters and recently coauthored the book *Homeland Security and Critical Infrastructure Protection* (Praeger International, 2009). Baggett holds an MS in criminal justice (with a concentration in police administration) from Eastern Kentucky University and a BS in criminal justice from Murray State University. He is currently at all-but-dissertation status on his doctoral degree in educational leadership and policy studies.

Pamela A. Collins, EdD, is the principal investigator and senior program manager for Department of Homeland Security projects that are federally funded in the Justice and Safety Center (JSC) housed in the College of Justice and Safety at Eastern Kentucky University (EKU) in Richmond, Kentucky. She has also held the positions of acting dean of the graduate school and department chair of loss prevention and safety, and she is a professor of both undergraduate and graduate security studies in the Department of Safety, Security, and Emergency Management. Collins was instrumental in the development and implementation of the Justice and Safety Center (JSC) beginning in 1998. When she first organized the center, there were three federally funded projects totaling about $2 million. The JSC grew in size from three projects and $2 million to a center managing more than seventy federal, state, and local programs and projects totaling more than $125 million over ten years.

Collins holds a BS in security and public safety, an MS in criminal justice from EKU, and a doctorate in educational policy studies from the University of Kentucky. She is a certified fraud examiner. Collins is completing postdoctoral education at both the University of California, East Bay, and Stanford University.

Before coming to EKU, Collins worked as an industrial security specialist for General Electric, Aircraft Engine Business Group Division, and as a fire and safety engineer for Industrial Risk Insurers. She also worked as a consultant for Toyota in

Georgetown, Kentucky, where she spent approximately a year setting up their security program. She has also provided security consulting to various companies both nationally and internationally.

Collins served for several years on the American Society for Industrial Security International Guidelines Commission and spearheaded the development of a private security officer guideline that may evolve into a standard. Part of the early work on the commission was going through the process of becoming an standards development organization, which was granted in the second year of the commission. She has also chaired various committees for ASIS International and the Academy of Criminal Justice Sciences, all relating to security and crime prevention.

Ronald Dotson, MS, is a professor of occupational safety and health (OSHA) at Eastern Kentucky University and serves on the board of directors for the Kentucky Safety and Health Network. He is an OSHA Training Institute construction trainer, and he operates an excavation business. Dotson is a certified safety and health manager with ISHM, a construction health and safety technologist through BCSP, and an active member of the American Society of Safety Engineers, the Institute for Safety and Health Management, and the National Association of Safety Professionals. His current research interests include playground safety and occupational injuries of educational service employees.

Dotson's safety background includes a variety of technical skills and management environments, including military construction project operations with the U.S. Marine Corps Reserve, several small excavation contractors, and his personal excavation business. Recently, Dotson has provided safety training to residential contractors in Kentucky. He works as an instructor and head football coach for a public school district in Kentucky, various security and personal protection projects, heavy equipment operations training, commercial vehicle driving, diesel mechanics, and law enforcement.

A highly decorated police officer serving in Ashland, Kentucky, Dotson earned several awards, including a Medal of Honor. After performing patrol and investigation duties, he became an instructor at the Department of Criminal Justice Training (DOCJT) in Richmond, Kentucky. He performed duties as an instructor in defensive tactics and physical fitness for recruits as well as for veteran officers. Shortly before leaving DOCJT, he developed training for the Homeland Security Department of DOCJT for chemical awareness and readiness, personal protective equipment, and suspicious package-handling procedures.

While working as a safety manager with KI USA Corporation, Dotson led the company in reducing injuries by 46% and became an inaugural member of Kentucky EXCEL. KI saw its lowest worker's compensation expenditure in its history during his tenure.

Dotson has served on a curriculum advisory committee for heavy equipment operations for the Kentucky Community and Technical College in Maysville. He also serves on several committees for Eastern Kentucky University and the scholarship committee for the Kentucky Safety and Health Network.

Paul English, CSP, is an assistant professor at Eastern Kentucky University in the safety, security, and emergency management department. He has worked for Fortune

100 companies, including Nestlé and Ford Motor Company, in different facets of occupational safety, security, and emergency response operations. While at the Ford Motor Company, he received the President's Health and Safety Award for Innovation representing the Americas in reducing injuries and illnesses while launching three new vehicles. His most recent position is with E-ONE, Inc. in Ocala, Florida, as director of environmental health and safety.

Gary D. Folckemer, BA, is a police corporal specializing in emergency management with the Eastern Kentucky University Division of Public Safety. He began his career as a U.S. Army military policeman and went on to work as a security manager for the Ritz-Carlton Hotel Company, a correctional officer for Corrections Corporation of America, and a university police officer for American University and Eastern Kentucky University. He earned an AS in criminal justice from Gulf Coast Community College in Panama City, Florida, and a BA in psychology from American University in Washington, D.C. He is pursuing an MS in safety, security, and emergency management in the College of Justice and Safety at Eastern Kentucky University. He lives in Richmond, Kentucky, with his wife, Nancy.

Greg Gorbett, MS, is an assistant professor in the Fire and Safety Engineering Technology Program at Eastern Kentucky University in Richmond, Kentucky. He currently serves as a member of the National Fire Protection Association (NFPA) Technical Committee, which writes the "Standard for Professional Qualifications for Fire Investigators," serves as the vice chairperson for the NFPA Fire Science and Technology Educators' section, and is a director for the National Association of Fire Investigators. For the past eleven years he has worked as a fire and explosion expert with Kennedy & Associates, Madison County Fire Investigation Task Force, and Kodiak Fire & Safety. He operates his own consulting firm. Gorbett holds two BS degrees, one in fire science and the other in forensic science. He also holds two MS degrees, one in executive fire service leadership and the other in fire protection engineering. He is a PhD candidate in fire protection engineering at Worcester Polytechnic Institute. He is a certified fire and explosion investigator (CFEI), certified fire investigator (IAAI-CFI), certified fire protection specialist (CFPS), certified vehicle fire investigator (CVFI), and certified fire investigation instructor (CFII). Gorbett has been elected to full member status in the Institute of Fire Engineers (MIFireE).

Kelly Gorbett, PhD, is a nationally certified school psychologist and works as a full-time school psychologist for Fayette County public schools in Lexington, Kentucky. Gorbett graduated with a BA in psychology and criminal justice from Tri-State University in Angola, Indiana. She attended graduate school at Ball State University in Muncie, Indiana, where she earned an MA in counseling and a PhD in school psychology. She is a member of the National Association of School Psychologists and the Kentucky Association for Psychology in the Schools.

William D. Hicks, MS, is an assistant professor for the Department of Safety, Security and Emergency Management at Eastern Kentucky University in Richmond, Kentucky. Hicks has a BA in fire and safety engineering technology and an MS in loss prevention and safety from Eastern Kentucky University. He serves as the fire chief of the Whitehall Fire Department. He holds the executive fire officer designation from the National Fire Academy and chief fire officer from the International Association of Fire Chiefs. He is National Association of Fire Investigators– and International Association of Arson Investigators–certified as a fire, arson, and explosive investigator and a National Fire Protection Association–certified fire protection specialist. He is an industry consultant in several technical areas, including fire protection detection and suppression systems, emergency response, and emergency management and preparedness.

Amy C. Hughes, MS, serves as executive director of the Rural Domestic Preparedness Consortium, a federal grant project within Eastern Kentucky University's (EKU) Justice and Safety Center. Hughes oversees all aspects of the multimillion-dollar national project to develop and deliver high-quality, all-hazards training tailored for small and rural emergency response communities.

Hughes has a BS in business communications from the University of Kentucky and an MS in safety, security and emergency management (SSEM) from EKU. She has extensive experience in state government policy, including homeland security, emergency management, and disaster management issues. Hughes served as senior policy analyst for the National Emergency Management Association and as the Emergency Management Assistance Compact administrator, supporting the mobilization of more than $30 million of human and equipment assets during the mutual aid response to Hurricanes Charlie, Frances, Ivan, and Jeanne in 2004. Hughes is an adjunct faculty member within the EKU SSEM department, a published researcher, and an author. She has been involved in grants management and administration for the past fifteen years.

Terry Kline, EdD, is a professor at Eastern Kentucky University, where he has served since August 1997, and the program coordinator for the EKU Traffic Safety Institute, where he has served since July 2007. He is project director for Kentucky Transportation Cabinet Contracts for State Traffic School (classroom) and the Graduated Licensing Program for Novice Drivers (classroom), which provides $1.4 million in funding for Traffic Safety Institute–contracted programs. He also serves as project director for a Kentucky Justice and Public Safety Cabinet contract for the EKU Kentucky Motorcycle Rider Education Program, which provides $1.1 million in funding for novice and experienced rider training through twenty-four regional site providers. He serves the American Driver and Traffic Safety Education Association (ADTSEA) as the ADTSEA Curriculum Standards chairperson. He holds degrees from Millersville State University (BS in education), Central Missouri State University (MS in safety), and Texas A&M University (EdD in industrial education/curriculum specialist).

Kline has experience as an adjunct professor at Indiana University of Pennsylvania, a research associate and lecturer at Texas A&M University, an adjunct professor at Central Washington University, a regional traffic safety specialist in Washington State, a graduate assistant at Central Missouri State University, and a high school driver education instructor in Pennsylvania. He has lectured on driver skill enhancement techniques in more than eighty state and national association conferences. Kline has developed curriculum for highway-railroad grade crossings, elementary bicycle education, junior high school traffic safety, secondary alcohol and driver education, adult commercial driver skill enhancement, and adult alcohol education programs. He has been the editor of *The Chronicle of ADTSEA* and *ADTSEA News and Views* and a refereed national journal for traffic safety education, and he has served as editor for state associations in Kentucky, Pennsylvania, Washington, and Texas. Publications include more than forty articles on related driver and traffic safety issues and contributions to several national magazines. He is the textbook author for the Prentice-Hall Driver Right Support Materials and the *Encyclopedia of Education* concerning driver and traffic safety education.

He has worked with local and national news media in supporting traffic safety education concerns and was honored with the coveted ADTSEA Richard Kaywood Award at the 2004 Portland Conference. Kline is working with statewide curriculum development and in-car instructor training programs for Kentucky Driver and Traffic Safety Education Endorsement.

Michael Land, EdD, is the director of labs and networks for Eastern Kentucky University, College of Justice and Safety. In addition to his twenty years at EKU, he has consulted for state and federal agencies in different facets of technology design and implementation. Land has taught approximately thirty courses in security and loss prevention throughout his career at EKU. Land has also owned and consulted with several private businesses relating to workplace safety, compliance, and security. He holds a BS in security and loss prevention and an MS in loss prevention administration, both from Eastern Kentucky University, as well as a doctorate in educational leadership from Lincoln Memorial University.

Lynn McCoy-Simandle, PhD, is a graduate of the University of Kentucky and, as a school psychologist, has worked with numerous children and youth who bully or have been bullied. Since retiring from Fayette County public schools in Lexington, Kentucky, McCoy-Simandle is a consultant for the Kentucky Center for School Safety and has presented countless workshops on bullying in schools for teachers, administrators, parents, and students.

Sheila Pressley, DrPH, is an associate professor in the Department of Environmental Health Science at Eastern Kentucky University (EKU) in Richmond, Kentucky. Her research interests include children's environmental health issues as they relate to toxic substance exposures from lead and illegal drug lab environments, as well as food safety and security. In addition to her book publications, she has also published articles in various newsletters, magazines, and journals. Pressley received a BS in environmental health science from Western Carolina University. She received an MS from the Civil Engineering School at Tufts University and a DrPH from the University of Kentucky's College of Public Health. She also holds a number of professional certifications, such as the Registered Environmental Health Specialist credential from the National Environmental Health Association and the Certified in Public Health credential from the National Board of Public Health Examiners. Prior to teaching at EKU, she worked for a company in Chicago, where she directed a series of environmental health and safety training programs in the Midwest and on the East Coast that were funded by the National Institute of Environmental Health Sciences and the United States Environmental Protection Agency.

Rebbecca Schramm, MS, holds an MS in safety, security, and emergency management from Eastern Kentucky University and a BA in secondary education (English major, communications minor) from Western Michigan University. She works as a part-time facilitator at Eastern Kentucky University and has experience in human resources, regulatory affairs, and facilities engineering. Schramm is the environmental health and safety supervisor at Stryker Instruments in Kalamazoo, Michigan, where she lives with her husband, Alex, and her two children, Max and Mackenzie.

James P. Stephens, MS, is a two-time graduate from Eastern Kentucky University with a BA in police administration in 1995 and a master's in 2010. He is a part-time instructor for the Eastern Kentucky University College of Justice and Safety. James has over sixteen years of experience in law enforcement. His career in law enforcement began in 1995 with the Ashland, Kentucky, police department, graduating from the Department of Criminal Justice Training Academy in Richmond, Kentucky, in 1996.

His career with the Kentucky State Police began in 1999, when he graduated from the Kentucky State Police Training Academy. During his tenure with the Kentucky State Police, he has held the rank of trooper, trooper first class, sergeant, lieutenant, and captain.

Stephens is active in the improvement and implementation of response to and preparation for critical incidents in school facilities, churches, hospitals, and other organizations. This includes meetings with multiple school and law enforcement jurisdictions for a review and assessment of their emergency response policies.

Stephens has presented to various groups and conventions, including the Kentucky Center for School Safety Convention in 2008, 2009, 2010, and 2011, the Kentucky Association of School Resource Officers, and the Kentucky Healthcare Coalition Conference in 2008, 2009, 2010, and 2011. Stephens serves as a post commander for the Kentucky State Police.

Section I

School Security

1 School Vulnerability Assessments

Ryan K. Baggett and Pamela A. Collins

CONTENTS

During the 2009 school year in the United States, approximately 56 million students were enrolled in prekindergarten through grade twelve with approximately 3.6 million full-time equivalent teachers providing them with instruction (Snyder and Dillow, 2011). Unfortunately, these students, faculty, and staff are not immune from critical incidents that can occur in the school environment. These incidents can range from natural hazards (tornadoes, high-rising water, hurricanes, and earthquakes) to pandemic and infectious disease, threats of terrorism, and other criminal acts. Although the threat of these events will never be eliminated from a school, stakeholders can better prepare and recover from them by conducting a comprehensive vulnerability assessment. Assessments can help prevent certain incidents, as well as mitigate and lessen the impact of incidents if they do occur. To facilitate the assessment process, this chapter will identify the components of a school vulnerability assessment while describing the overall advantages to the school environment.

    The chapter will begin with a presentation of information on the purpose and advantages of implementing a team approach in conducting assessments. This section will include information on potential group composition, roles and responsibilities, and potential tasks for the group. Next, three assessment components will be highlighted to include assets and infrastructure identification, threat and hazard assessment, and the vulnerability assessment.

The information on assets and infrastructure identification will help stakeholders begin to identify and prioritize a school's critical assets and infrastructure. This component must be undertaken in an effort to identify the appropriate mitigation strategies of the facility or complex. In relation to this, the threat and hazard assessment will focus on potential school threats and hazards while emphasizing the need for situational awareness of incidents both in the complex and the surrounding community. Both assessments will prepare the team for the execution of the vulnerability assessment.

As noted in the chapter, the vulnerability-assessment process will assist in identifying vulnerabilities and risks that are apparent within individual schools or school districts. After identification, stakeholders can develop plans to address and help remedy those problematic areas. In addition to describing the assessment, the chapter will also provide information on reporting and the prioritization of findings. It is important to understand that even if the completion of the assessment is successful, the report and associated findings must be developed into a usable format that will give stakeholders information on how to improve the safety and security of the complex.

The last component of the chapter is a fictional case study that will help readers critically apply the information they have obtained. The case study can be completed individually, in small groups, or as a classroom discussion tool. Collectively, the chapter sections will introduce general information regarding vulnerability assessments in a school environment; however, before moving forward with the chapter, there are a few points that should be considered.

First, this chapter is not intended to be a step-by-step or a "how-to" guide for conducting assessments. Rather, it is intended to provide school stakeholders with an overview of the process to encourage additional discussions with others in the community who each have diverse knowledge, skills, and abilities that will be necessary during the assessment process in the hope that these discussions will lead to the development of an assessment team and a subsequent assessment.

Next, it is important to understand that there are multiple ways in which to conduct an assessment. In fact, some methodologies may use varying terminology to describe the process. This chapter will provide a general overview of the assessment process by describing common terminology and approaches. Factors such as available resources, willingness of community stakeholders, and school demographics and logistics will all contribute to the assessment. However, it is essential to note that assessments are not a "one size fits all" approach; in order to prove useful, assessments must be customized to the school or school district to inform subsequent activities.

THE ASSESSMENT TEAM

Communities, whether large or small, are comprised of individuals from varying backgrounds, education, and experiences. Because it is not expected that school personnel will have all of the required knowledge and skills to conduct a vulnerability assessment, the inclusion of community stakeholders with varying perspectives is essential. Not only will a diverse assessment team be helpful during the execution of the assessment, but the various members can also be beneficial during the planning and report-development phases. The first step in forming the assessment team is to determine the type of team, as well as the types of individuals that should be involved.

From an internal perspective, administrators must decide whether the team will serve the school district or be limited to a single school. It may be useful to form a district-wide team that can perform multiple assessments for the school board (U.S. Department of Education, 2008). Although a district team may help to avoid duplication of efforts and maximize resources, caution should be taken to ensure that customization at each school remains a fundamental premise of the assessment process. Following the determination of the type of team, the composition of the team must be established.

A tendency when forming an assessment team may be inclusiveness of all individuals who may have some usefulness to the process. However, it is important to ensure that the team is manageable in size and that each member is selected based on a specific skill set that will be essential to the assessment. The size of the team will depend on the size and complexity of the school or district. Next, before seeking assistance from the outside community, school leaders should determine what expertise resides within the organization. School personnel that should be considered for inclusion on an assessment team are listed in Table 1.1.

The inclusion of administrators may be a wise decision because they typically have the authority for funding allocations as well as policy changes. Both attributes may be useful in implementing changes based on the vulnerability-assessment results. Next, the remaining positions listed in Table 1.1 will have indispensable input into the daily occurrences within a school. In an effort to ensure that the team

---

TABLE 1.1

School Personnel Positions for Possible Inclusion on Assessment Team

Administrators
  Principal/vice-principal
  District/central office representative
Educators
  General
  Special
School resource officer/security officer
School nurse
Building and grounds staff
  Cafeteria staff
  Custodians
  Bus drivers

---

is manageable in size as previously noted, these individuals may not be acting members of the assessment team but rather individuals who are interviewed to gather additional information. Following the identification of the internal team, outside community stakeholders should be identified.

The diversity of individual skill sets and experiences are fundamental in the assessment process. Table 1.2 lists several potential members of the team. First, public safety officials have experience in responding to crises and will be beneficial in identifying potential hazards and threats to the school (U.S. Department of Education, 2008). Second, representatives from local or county government will not only be helpful in their specific areas of expertise (as they apply to the school), but they may also have experience in conducting assessments at the city or county level. Related, some private sector entities are well-versed in vulnerability assessments; it would benefit the school team to include this level of experience on the team. Next, in an effort to garner the support and skill sets of the school's parents, a representative of the school's site-based council or parent-teacher organization may be a wise decision. It is possible that this representative may also fill another role that is presented in Table 1.2. Finally, administrators should consult with the state's department of education. It is likely that they have been involved in other school-assessment processes and could provide a representative to help guide the district's assessment team. Further, they may be able to refer the team to other resources in the state such as a center for school safety or other subject-matter experts. After the establishment of team members, the group should convene to discuss potential tasks and associated timelines.

It is important for assessment-team members to understand the commitment of joining such a team. Assessments are not a one-time event but involve a systematic process of continual evaluation. It is important to maintain continuity of membership on the team during the planning, execution, and report-development phases. During the team's initial meeting, time should be spent on introductions where individuals

---

TABLE 1.2

Community Stakeholder Positions for Possible Inclusion on Assessment Team

Public safety officials
    Law enforcement officers
    Fire service workers
    Emergency manager
    Public health officials
Site-based council/parent-teacher organization representative
Local government officials
    Public utilities official
    Transportation official
    Code enforcement/building inspector
Private sector representative
State department of education representative

can speak to their knowledge and skills, as well as to what they believe they will contribute to the assessment team. The team should also develop clear goals with associated objectives and realistic timelines. These items will ensure that the team stays on task and completes the assessment in a timely manner.

ASSETS AND INFRASTRUCTURE IDENTIFICATION

An asset is any resource considered to have some value to the school that requires protection. An asset can be tangible (e.g., students, faculty, staff, school buildings, facilities, or equipment) or intangible (e.g., processes, information, or a school's reputation). Identifying and prioritizing a school's critical assets is a vital first step in the process of assets and infrastructure identification. Only by identifying a school's assets can mitigation steps be implemented to establish an acceptable level of protection prior to an act of school violence. Clearly, most if not all would agree that the people (students and staff) are a school's most critical asset. The process begins with these individuals and then moves to identifying and prioritizing the school infrastructure.

The process of identifying people and the value of a school's assets begins with interviews of the people who are most familiar with the schools policies and facilities. Feedback from school personnel such as those listed previously in Table 1.1, as well as any others who can help identify the most valuable assets, provides the necessary information to conduct a thorough assessment. One recommended approach to structuring these interviews is to develop a list of questions or topics that need to be addressed prior to the interviews. Although any member of the assessment team could potentially conduct the assessment, it is recommended that the tasks are assigned to team members (such as the school resource officer) who can then report back to the team and inform the final report.

When the process for identifying the core functions and processes is completed, an evaluation of school infrastructure can begin. Below are some potential questions to ask in order to better understand and assign value to the identified infrastructure:

- How many people may be injured or killed during a terrorist attack that directly affects the infrastructure?
- What happens to school functions, services, or student satisfaction if a specific asset is lost or degraded? (Can primary services continue?)
- What is the impact on other organizational assets if the component is lost or cannot function?
- Is critical or sensitive information stored or handled at the school?
- Do backups exist for the school's assets?
- What is the availability of replacements?

Next, the team should make a determination of the potential for injuries or deaths that could possibly occur from an "all-hazards" approach to any catastrophic event at the school. The all-hazards planning approach focuses on developing capacities and capabilities that are critical to preparedness for a full spectrum of emergencies or disasters. This process can begin by addressing the following issues:

- Identify any critical faculty, staff, or administrators whose loss would degrade or seriously complicate the safety of students, faculty, and staff during an emergency
- Determine whether the school's assets can be replaced and identify replacement costs if the school building is lost
- Identify the locations of key equipment
- Determine the locations of personnel work areas and systems within a school
- Identify the locations of any personnel operating "outside" a school's controlled areas
- Determine, in detail, the physical locations of critical support architectures:
  1. Communications and information technology (IT, the flow of critical information)
  2. Utilities (e.g., facility power, water, air conditioning, etc.)
  3. Lines of communication that provide access to external resources and provide movement of students and faculty (e.g., road, rail, air transportation)
- Determine the location, availability, and readiness condition of emergency response assets and the state of training of school staff in their use (Federal Emergency Management Agency, 2003).

The assets and infrastructure identification assessment can be a very time-consuming process but is necessary to adequately identify the school's infrastructure. This information is used to target efforts on those systems and assets that are necessary to best protect the students, staff, and other individuals located within the school. When this information has been collected, the assessment team should focus on the execution of an inclusive threat and hazard assessment.

THREAT AND HAZARD ASSESSMENT

A comprehensive school-security review is an essential step when addressing the security needs of your school. The review should involve a tour of the school grounds (inside and outside), which would ideally be led by the person responsible for school security, accompanied by members of the assessment team. During the tour, key hazards should be noted, photographs should be taken, and a fundamental review of all security issues and measures should be undertaken. In health and safety terms, a "hazard" is any situation with the potential to cause harm or damage. A hazard can describe unsafe conditions (e.g., a physical condition that can make the workplace unsafe), such as a slippery floor, or unsafe acts, such as horseplay. Unsafe omissions such as the failure to follow safe systems or wear protective equipment can also be termed hazards. During this step, it is important to imagine worst-case scenarios.

During the 1990s, twenty-eight instances of targeted violence occurred in America's public school systems, culminating with the largest mass homicide in history on April 20, 1999, at Columbine High School in Littleton, Colorado (Omoike, 2000; Vossekuil et al., 2002). In the wake of this horrific attack, states and school districts across the nation revisited their outdated crisis-management plans, turning to corporate public-relations research to inform changes in the mostly reactive characteristics of gun-violence prevention, investigating strategic and anticipatory models of crisis intervention. On the collective recommendations of the Department of Education, the U.S. Secret Service, and the Department of Justice, school administrators began to train themselves and their personnel in threat-assessment techniques in hopes to prevent their school from becoming the next Columbine (Vossekuil et al., 2002).

A seminal work on school safety was the Safe School Initiative (2002), implemented through the Secret Service's National Threat Assessment Center and the Department of Education's Safe and Drug-Free Schools Program. The Safe School Initiative began with a study of the thinking, planning, and other preattack behaviors engaged in by students who carried out school shootings. That study examined thirty-seven incidents of targeted school violence that occurred in the United States from December 1974 through May 2000 (Vossekuil et al., 2002).

The result of this study was the development of *Threat Assessment in Schools: A Guide to Managing Threatening Situations and to Creating Safe School Climates.* The guide provides school administrators with basic information on how to incorporate the threat assessments into strategies to prevent school violence. The purpose of the guide was to assist school administrators and teachers with the information and tools necessary to create a safe and secure school environment. The guide provided insights into the process for investigating, evaluating, and managing targeted violence, as well as how school officials should respond responsibly, prudently, and effectively to threats and other behaviors that raise concern about potential violence (Fein et al., 2002).

Some of the key findings of the Safe School Initiative suggest that most attackers did not threaten their targets directly but did engage in preattack behaviors that would have indicated an inclination toward or the potential for targeted violence had they been identified. Findings about the preattack behaviors of perpetrators of targeted violence validated the "fact-based" approach of the threat-assessment process. This process relies primarily on an appraisal of behaviors, rather than on stated threats or traits, as the basis for determining whether there is cause for concern.

Students who engaged in school-based attacks typically did not "just snap" and engage in impulsive or random acts of targeted school violence. Instead, the attacks examined under the Safe School Initiative appeared to be the end result of a comprehensible process of thinking and behavior that typically began with an idea, progressed to the development of a plan, moved on to securing the means to carry out the plan, and culminated in an attack. The Safe School Initiative found that the time span between the attacker's decision to mount an attack and the actual incident may be short. Consequently, when indications that a student may pose a threat to the school community arise in the form of information about a possible planned attack, school administrators and law enforcement officials will need to move quickly to inquire about and intervene in that possible plan.

The study concluded that it was not really possible to create a specific profile for an individual who would become violent. Also the use of profiles to determine whether a student is thinking about or planning a violent attack is not an effective approach to identifying students who may pose a risk for targeted violence at school. What the study did recommend is that the focus should be on students' behavior and the things they may say to other students or school personnel.

Interestingly, bullying was not a factor in many of the cases and is not a primary indicator that a student who is bullied poses a threat to the school as a form of retaliation. Bullying, however, is a real concern and has been a factor in some of the cases in which students who were the targets of taunts and bullying described this as one of the reasons for their actions. When bullying was a factor, it was often described by the victim as "torment," and if this same behavior was present in a workplace, it could be described as harassment or assault (Vossekuil et al., 2002).

The weapon of choice in the majority of school violence cases was either a handgun or rifle. These weapons were almost always taken from the student's own home or a friend's or relative's home. A good predictor of students who may use violence is that they have some sort of history with guns.

The duration of most of these incidents is very short, and law enforcement seldom actually stops the attack points. It is therefore imperative for schools, staff, and community stakeholders to develop preventive measures whenever possible in addition to maintaining current and well-exercised emergency plans. The preventive measures should include protocols and procedures for responding to and managing threats and other behavioral issues. Following acquisition of hazard and threat information, the team is ready to proceed to the vulnerability-assessment process.

VULNERABILITY ASSESSMENT

A vulnerability assessment evaluates weaknesses, which can be exploited by an aggressor, of critical assets across a broad range of identified threats and provides a basis for determining mitigation measures for protection of people and critical assets. At the conclusion of the chapter, several online references will be provided that compile best practices based upon technologies and scientific research to consider during the design of a new building or an assessment of an existing school building. These checklists can be used as a screening tool for an initial vulnerability assessment or be used by subject-matter experts for a comprehensive vulnerability assessment of existing school buildings.

The assessment of any vulnerability of a school building should be done within the context of the defined threats and the value of the school's assets, as previously identified in this chapter. For example, the school building should be analyzed for vulnerabilities to each threat, and a vulnerability rating should be assigned. Table 1.3, extracted from FEMA 428 (Federal Emergency Management Agency, 2008), provides an example of how the vulnerability-assessment process would include a value rating system for assets. These ratings are not set and can change based upon each school's unique circumstances.

A vulnerability assessment is a continual process of identifying existing and proposed assets and then making a determination of the value of those assets against possible threats and hazards. There are many techniques and methods to conducting

TABLE 1.3
Nominal High School Asset Value Assessment

Asset	Value	Numeric Value
Students	Very High	10
Faculty	Very High	10
Staff	Very High	10
Designated shelter (safe haven)	Very High	10
Main school building	High	9
Teaching functions	High	9
IT/communications	High	8
Utilities associated with shelter	Medium High	7
Nurses' station	Medium High	7
School/student records	Medium High	7
Transportation	Medium High	7
Physical security equipment	Medium High	7
Administrative functions	Medium	5
Temporary classrooms	Medium Low	4
Food service	Medium Low	4
Library	Low	3
Custodial functions	Low	3
Vocational shops	Low	3
Indoor sports facilities	Low	2
Outdoor sports facilities	Very Low	1

Source: From Federal Emergency Management Agency. (2008). *PRIMER: To Design Safe School Projects in Case of Terrorist Attacks–FEMA 428.* U.S. Department of Homeland Security, Washington, D.C.

these types of assessments, and at times the information can become overwhelming. Although every situation is different, the Federal Emergency Management Agency has developed an entire risk management series and provides a number of guides and tools that school administrators and staff may want to consult when developing their vulnerability assessments. A good place to start is to refer to the following site for a complete listing of those resources: http://fema.gov/plan/prevent/rms/.

REPORTING AND PRIORITIZING FINDINGS

Following the vulnerability assessment data-collection phase, the information must be presented in a usable format that will aid decision makers in implementing measures to increase the safety and security of the complex. If this phase is not conducted in a timely and effective manner, the previous steps will likely be negated. The primary purpose of the report is to prioritize vulnerabilities that pose the greatest risk to the school and/or school district. There are several helpful tips to consider when developing the final report (U.S. Department of Education, 2008).

TABLE 1.4
Risk Index Worksheet

Hazard	Frequency	Magnitude	Warning	Severity	Risk Priority
	4 – Highly likely	4 – Catastrophic	4 – Minimal	4 – Catastrophic	3 – High
_____	3 – Likely	3 – Critical	3 – 6–12 hours	3 – Critical	2 – Medium
	2 – Possible	2 – Limited	2 – 12–24 hours	2 – Limited	1 – Low
	1 – Unlikely	1 – Negligible	1 – 24 or more hours	1 – Negligible	

CONSIDER USING A RISK MATRIX TO DETERMINE WHICH HAZARDS AND VULNERABILITIES WOULD HAVE THE GREATEST CONSEQUENCES FOR EACH SCHOOL

As outlined on pages 26 and 27 of the 2008 U.S. Department of Education's Guide to School Vulnerability Assessments, a risk matrix is one way a school can determine change priorities. As displayed in Table 1.4, schools can determine which hazards should be higher or lower priority based on frequency, magnitude, warning, severity, and overall risk priority.

PROVIDE AN ACCURATE AND OBJECTIVE LISTING OF VULNERABILITIES PRIORITIZED BY RISK

A high level of objectivity must be maintained by the assessment team at all times. In doing so, it is important to give the report stakeholders an accurate and unabated picture of the complex. All vulnerabilities, no matter how large or small, should be included in the report. These vulnerabilities can be identified through the tour of school grounds, interviews with school stakeholders, or other community and hazards research that was conducted.

DO NOT JUST DWELL ON THE NEGATIVES, ALSO ACCENTUATE THE POSITIVES

If the school or school district has implemented successful measures that were identified during the assessment, they should also be noted within the report. Not only does this add to the situational awareness of the report among stakeholders, but it also provides best-practice suggestions for other schools that may be experiencing similar challenges.

CONSIDER ADDING PHOTOS OR OTHER DIAGRAMS AND VISUAL AIDS TO THE REPORT

The inclusion of photographs of hazards that were encountered in the school complex will provide the reader with a nearly firsthand review of what was encountered. Other charts and visual aids will benefit the visual learner with a graphical representation of the important facts of the assessment. Additionally, the assessment team may decide to collect the information in a presentation software tool for presentation at a school board or other meeting.

Allow All Team Members a Final Review of the Report Before It Is Submitted

Although all of the team members may not have a hand in the actual writing of the report, it is imperative that each individual be given an opportunity to provide a final review of the document. By providing final reviews, the team members can clarify any remaining points, request revisions on areas that are inaccurate or poorly written, or provide additional information that may not have made it into the final version of the report.

Following development and submission of the final assessment report, it is vital that follow-up be conducted. Execution of the assessment and subsequent development and submission of the report are very important steps; however, stakeholders should continue to work together to begin the development of plans to reduce the hazards and vulnerabilities noted in the report. Budgetary and time constraints will undoubtedly play a role in the follow-up discussions, but both short- and long-term plans should be implemented to improve the overall safety and security of the school complex.

CONCLUSION

The results of a systematic and comprehensive vulnerability assessment can be used to prioritize mitigation activities and can help inform disaster recovery, mitigation, and response planning. As evidenced in the chapter, the process is a multifaceted approach that requires dedicated and knowledgeable team members, a sound methodology, and clear goals and objectives with a realistic timeline. It should be noted that an assessment is not a one-time event but should be conducted annually as part of the district's emergency-planning activities (Department of Education, 2008). Each year, a continual improvement process with corresponding reviews should be implemented by team members identifying the pros and cons of the schedule, methodology and tools, and overall process. Although not a panacea, the assessment results can help stakeholders make better-informed decisions regarding the overall safety and security of the school or school district.

CASE STUDY

Based on increasing violence in the community and pressure from the community, the Samuel County School District announced today that it will conduct a comprehensive vulnerability assessment of the fifteen schools located in the district. Because of the number of schools, the district will take an incremental approach based on perceived need of each school. For purposes of this activity, the assessment team will conduct the first assessment at Benjamin Franklin High School in the southern part of Samuel County.

Benjamin Franklin High School has a student population of 2,200 students and is located on a twenty-acre complex in a major urban area. The complex has several buildings with athletic fields and parking lots. Based on intelligence obtained from local law enforcement, there has been a credible threat that the high school will be the site of an incident that could involve anywhere from a lone active shooter to a large scale firefight between neighboring gangs. The superintendent has convened a meeting of school administrators and security personnel to ensure that the district's vulnerability assessments are conducted in a timely matter, especially the one at Benjamin Franklin High School.

- What is your first step in the assessment process?
- Who would comprise your team based on the information you have been provided?
- What assets and infrastructure are critical in a traditional high school setting?
- How would you determine what assets and infrastructure exist within this complex?
- While one specific threat has been provided, what additional information would you like to know about the threat?
- How would you obtain additional data on threats and hazards for this particular school?
- What steps would you take in conducting a vulnerability assessment at the school?
- What types of information would you include in the final report, and why is the final report critical in this situation?

EXERCISES

1. Articulate the advantages and potential challenges of conducting a vulnerability assessment in a school or school district.
2. Although the case study in this chapter featured a large urban school, consider the assessment process in a small, rural school. What similarities and differences exist between the two processes?
3. School administrators have limited time given their multiple responsibilities. How might the assessment team garner the initial interest and support of administration for the assessment process?
4. Should students play a role in the assessment process? If so, how?
5. Caveat emptor is the Latin phrase for "Let the buyer beware." When news is released that a school or school district is preparing to conduct an assessment, it is likely that vendors and other private entities will contact the school in an effort to sell their products (guides, assessment software, etc.). What consideration should the assessment team make when evaluating the products that may be helpful to the assessment?

REFERENCES

Federal Emergency Management Agency. (2003). *Reference Manual to Mitigate Potential Terrorist Attacks against Buildings–FEMA 426.* U.S. Department of Homeland Security, Washington, D.C.

Federal Emergency Management Agency. (2008). *PRIMER: To Design Safe School Projects in Case of Terrorist Attacks–FEMA 428.* U.S. Department of Homeland Security, Washington, D.C.

Fein, R. A., Vossekuil, B., Pollack, W. S., Borum, R., Modzeleski, W., and Reedy, M. (2002). *Threat Assessment in Schools: A Guide to Managing Threatening Situations and to Creating Safe School Climates.* United States Secret Service and the U.S. Department of Education, Washington, D.C.

Omoike, I. I. (2000). *The Columbine High School Massacres: An Investigatory Analysis.* Omoike Publishing, Baton Rouge, LA.

Snyder, T. D., and Dillow, S. A. (2011). *Digest of Education Statistics, 2010.* NCES 2011-015. U.S. Department of Education, Washington, D.C.

U.S. Department of Education. (2008). *A Guide to School Vulnerability Assessments: Key Principles for Safe Schools.* Office of Safe and Drug-Free Schools, Washington, D.C.

Vossekuil, B., Fein, R. A., Reddy, M., Borum, R., and Modzeleski, W. (2002). *The Final Report and Findings of the "Safe School Initiative": Implications for the Prevention of School Attacks in the United States.* U.S. Department of Education and U.S. Secret Service, Washington, D.C.

USEFUL WEB SITES

Federal Emergency Management Agency, Security Risk Management Series, http://www.fema.gov/plan/prevent/rms

National Clearinghouse for Educational Facilities, School and University Safety and Security Assessment, http://www.ncef.org/rl/safety_assessment.cfm

National School Safety Center, http://www.schoolsafety.us

U.S. Department of Education, Office of Safe and Drug-Free Schools, http://www2.ed.gov/about/offices/list/osdfs/index.html

U.S. Secret Service, National Threat Assessment Center, Safe Schools Initiative, http://www.secretservice.gov/ntac.shtml

2 Access Control

Ronald Dotson

CONTENTS

Access to a public school is a combination of security and public relations. You want the parents to feel welcome, yet access must be limited and security must be observable by the public to have a deterrent effect. Understanding the community that the school is located in and handling diverse issues in an informative and open manner will ease the stern feeling of strict security. Understanding the underlying philosophy for access control is important. Access-control measures can be strategic, tactical, and protective. They have the goal of allowing timely access to those that need it and excluding access to those who may present a hazard.

As a general rule, begin with only allowing one access point for all to enter the school. Then assess the need for and risks of additional access points. Additional access points that might be reasonable may include an entry for teachers and other employees and access for vendors such as truck drivers delivering cafeteria supplies. General mail delivery should take place without formal entry to the school. The rule of one access point means that the grounds should be fenced on at least the sides of the building or grounds other than the front. This way doors that are needed for quick entry from playgrounds or other educational areas can be left unlocked during recess or physical activity. When an unplanned incident occurs and injury results, the stress will create an added barrier to access. When keys are needed for certain entry points or security identification cards are required to be swiped or passed by a sensor, it can impede rapid response. This does not mean that the door should be left uncontrolled or accessible to anyone other than the school personnel and the children participating in the activity. The fenced area with locked gates and observation protect the door until rapid access is no longer required. It should then be secured again. The goal is to be able to provide at least first aid response within four minutes and be able to summon more advanced aid quickly.

STP METHOD: STRATEGIC, TACTICAL, AND PROTECTIVE BARRIERS

In order to achieve one access point or to reasonably minimize the number of access points, barriers and features will be utilized from the strategic, tactical, and protective (STP) method. The first level of barrier and feature to control access is referred to as a strategic barrier. Strategic barriers may not appear to be barriers at all. They are features of landscape, signage, design of sidewalks, crossings, and building features. These features guide or influence the movement of any person to the main access point. During the design stage of school buildings and property, access control must be considered. The flow of the sidewalks, the placement of parking areas, the placement of fencing, and the direction of the face of the building, along with the design of access doors, must be considered with the initial architecture. For example, having one driveway for bus loading and unloading and a second driveway for visitor or parent drop-off should be considered. Having a third driveway for employees only or a fourth driveway for commercial vehicle access is ideal. An additional type of strategic barrier is view. The area that is visible or observable through means such as cameras should also be considered in initial design. The corners of buildings, placement of structural columns, and obscure doorways or cubbyholes within the structure of the school building should be considered in relation to how they impede viewing.

Tactical barriers are always physical in nature. Tactical barriers prevent certain views, prevent or limit approach, force physical movement, or prevent access altogether. They are generally perimeter devices. However, they are not always ominous features that portray a prison environment. Park-style benches anchored in reinforced concrete, for example, can have curb appeal and also protect from vehicle approach and unintended impacts or at least slow down access to allow response actions. Trash cans, shrubbery, trees, high curbs, speed bumps, direction stop sticks, statues, and many other creative features can be used as tactical barriers.

Protective barriers are features that are meant to provide protection from contact. Inevitably close contact will be made with entrants. Protective barriers can take the shape of bulletproof or impact-resistant glass, screens, tables, desks, doors, and many other items. Distance is foundational to protective barriers. When needing a protective measure, distance is always a friend of the defender. It is also assumed that most people with bad intent will migrate as close to the school building and as far inward or toward the person of contact as possible.

Desks, tables, counters, and other items that allow for close contact without a full physical barrier for separation should be wide enough to impede a quick and easy defeat of the barrier. Stay out of reach. Police officers are taught in defensive tactics to maintain a six-foot barrier for general approach and initial contact. This is a good rule of thumb to remember. Counters, for example, should be wide enough so that the distance between staff and visitor can be easily maintained at about six feet. Although counters and desks are not usually this wide, employees can stand back from them. The height of the counter in addition to its depth will allow for defensive avoidance of a lunge or grabbing motion across the counter. Without a tactical physical barrier, a person with ill intent can lunge for an individual or charge 18–21 feet in such a quick time that defensive strikes or counters can be difficult if the defender is off guard.

Access-control measures are useless if the personnel using them are not familiar with the STP strategy and not trained on how to recognize signs of aggression and how to use defensive countermeasures. All front-office personnel responsible for greeting and signing in visitors should be trained in these techniques, along with any personnel who might have contact with an upset member of the community. It is reasonable to expect a possible assault on teachers during parent-teacher meetings, on coaches and staff at events, or on principals, who generally deal as the frontline receiver for angry complaints. Access control must fit well with any response to aggressive behavior or emergency response planning and training.

POSITIONING

Pressure Point Control Tactics created by Bruce Siddle (2001) is a widely accepted system for defensive tactics for police as well as for teachers. The program offers instruction for teachers dealing with out of control students. At a minimum, teachers and staff should be familiar with positioning and should be able to couple positioning tactics with tactical and protective-barrier strategy.

When considering positioning tactics in Siddle's program, the positions are numbered from the perspective of the person to whom the teacher or staff member is speaking. Directly in front of the subject is considered the 0 position. Standing to the front and side of the person at a 45-degree angle is the 1 position. Straight out from either side is the 2 position, and directly behind the person is the 3 position. Off to the rear at a 45-degree angle is the 2½ position.

The 0 position should be avoided. The preferred position to speak with a person or make initial contact is the 1 position. Because teachers are usually unarmed, the 1 position away from the person's dominant hand is preferred. Therefore, if an unarmed person is making initial contact with an unfamiliar person, one that has a potential to be violent or appears upset, an attempt to determine the dominant hand is quite a challenge. Standing in the 1 position away from a person's dominant hand means that usually in order to throw a punch, the violent person has to reposition or cross his body, shortening the reach of his or her punch, slowing the action, and allowing the teacher or staff person to avoid the assault and use a barrier. When a second school person or official enters the situation, he should try to position himself in either the opposite 1 position or one of the 2½ positions (Siddle, 2001).

This tactic of positioning can also be more effective when barriers in the offices are arranged to the point that a visiting person must negotiate around multiple barriers. It may be possible for the barriers to facilitate the school employee naturally being in the 1 position or having a second person approach from the 2½ position. This can be done by positioning multiple personnel in the front office where visitors are initially greeted. Two personnel at a minimum on opposite sides with entrance in the middle is preferred.

ACCESS STRATEGIES

Main-entrance access is an obvious focus. The preferred entrance will funnel visitors to a front desk where an initial greeting is made. Access to the school should be made only after passing by the front-office administrators or from access granted from a visual observation and recognition of identity. For example, after entering the front door, the access to the main school should be a locked doorway. A side door may then lead to the front desk. Preferably, this side door is locked and allows for visual observation of the person requesting entry and allows for the greeter to allow them in or unlock the door from a distance. Once signed in, the visitor should be escorted unless the visitor is a regular trusted person that has had a background check from the school on file, but the general rule is to escort or bring the person or student requested for visit or pickup to the office.

Special events are regular occurrences at a school. For such events, after signing in, a visitor can be directed to the relevant areas of the school with either a designated tour guide or with personnel awaiting their arrival at the relevant area in the school. General patrol of the halls and secured doors to unauthorized areas should be standard practice. For example, when an open house is occurring, teachers and principals may be busy speaking with parents. School staff and educational assistants can help fill the roles of patrolling or checking hallways, locked doors, unauthorized areas, and directing visitor traffic.

Depending on school policy and practice, many special events may occur in the gymnasium or on school grounds. Some schools allow facilities to be rented or loaned to community organizations. In this case, doors and access to the school should be secured. Visitors for the event should have access to public restrooms, first aid kits, and egress that is sufficient for the number of people and the arrangement of the facility. However, the main offices and classrooms should not be accessible.

Outdoor grounds must have barriers and access points arranged in the same strategy as discussed previously for main buildings. However, with outdoor grounds such as athletic fields, barriers are used at outer perimeters and inner perimeters. Outer perimeters have much looser controls because of limited resources. Nevertheless, strategic barriers are the main tool to funnel different types of traffic to the access points of the inner perimeter.

Access points for outdoor events deviate from the single-access point rule. Here, keeping crowds to a minimum is the priority. Access points need to be convenient and sufficient in number to allow quick access. The areas in front of access points should again use barriers to organize and funnel to the access points. Personnel that can assist and signage can help keep crowds moving in an orderly manner and avoid crowd-control problems.

Outdoor grounds have some other special considerations. Animals must be kept out of playground areas. Stray dogs should be reported to animal control, but fences should extend to the ground and have stakes or cables to secure fencing to the ground and prevent access under the fence. Homes around schools should also be required to keep animals in fences, in kennels, or on leashes.

DOOR CONTROL

Implementing policy and training on not propping doors open is common, but it is not enough. If doors are propped open to allow access for recess needs or other situations, a problem with the management exists. Although some schools may have security staff, most do not have this resource available or have not experienced the need. Additionally, implementing a check of doors is impractical because it would interrupt the duties of teachers and staff.

The solution is to utilize door alarms. These local alarms should be integrated with the comprehensive alarm system for the school. An additional option is to have a monitor panel for the door alarms at the front desk, which should always be staffed. At a minimum, an audible buzzer on the door when open will alert teachers, students, and staff in the area of an open door.

This might seem impractical because many doors will be utilized to gain outside access for teachers and students on breaks, at recess, and at other times. The alarms can be delayed to get louder or to even begin after a certain amount of time. Because access from the outside is controlled to allow only for authorized entry, this can be achieved with minimal risk. At a minimum, a couple of doors could be left without an alarm for ease of outdoor access. This would minimize the number of doors needing to be monitored. This could be effective especially if the doors were within fencing around the school. Doors outside an inner perimeter would require alarms.

Preventing doors from being easily accessed by unauthorized persons is critical for access control. If cafeteria vendors enter or deliveries take place without checking in using the usual front-desk procedure, then the access point must be controlled properly by staff in the area in question. The door should have an intercom-style voice or audible system for the driver to request access. The staff person should have a method to visually identify the vendor before granting entrance. The request should match expected or scheduled delivery times. If not, the driver should check in at the front office.

PEDESTRIAN ACCESS TO SCHOOL

All schools have pedestrian traffic in and around the school. Some schools have "walkers," which are those students who walk to and from school each day. A challenge exists in getting the pedestrian traffic to and from school grounds in a safe way. The manner in which pedestrian traffic is handled while on school grounds can affect the safety of the pedestrians while off school grounds as well.

In 2009, approximately thirteen thousand school-aged children were injured in vehicle versus pedestrian accidents in the United States (Safe Kids USA, 2011). Many of these incidents occurred off of school-controlled grounds. However, the procedures used by individual schools may impact the behavior of children while off of school grounds. The habits and behavior patterns established at school may carry more credibility with children because schools are viewed as teaching correct ethical and safe behavior. It is therefore imperative that any procedure established at school be examined critically.

Critical examinations require explanation, practicality, and the establishment of norms (Stanford Encyclopedia of Philosophy, 2005). This translates to identifying a problem, identifying the actors involved, setting standards for critique, and setting measurable goals for success. This technique is not commonly used for the examination of safety practices in K–12 schools. However, an empathetic approach to examining common practices will shed light on the true impact on future behaviors of students.

Most educators are mistaken about the definition of school safety. School safety has three elements that are closely related. School safety is the management of the workplace safety of educational service employees, the safety and security of students, and the safety and security of school guests and of the community that comes in contact with the educational service employees while conducting school business. The failure to properly manage teacher and school employee safety sends an underlying message to all involved with the school. However, the impact that it has on the child impacts the whole of the community. Teachers lead the way in injured educational service employees in Kentucky according to Kentucky School Board Association Statistics from 2005 until 2009. The Kentucky School Board Association was the workers' compensation carrier and manager for 98 of 174 districts during that time (Isaacs, 2010). With that in mind, we must ask ourselves whether or not the example is seen and understood by children in a manner that transfers to potentially injurious behavior in other places than school and later when working in construction, general industry, or any other occupational field.

Let us examine a daily practice for parent drop-off at one regional elementary school. Parents pull up near the front door to a stop bar in front of a marked crosswalk in order to drop off their child. A teacher is waiting at the end of the crosswalk closest to the school. When other employees, children, or guests walk from the parking lot or city sidewalk and approach the school entrance, the teacher puts up his hand to instruct the driver to wait, steps in front of the vehicle, and allows the pedestrians to cross. For a vast majority of the times observed, the pedestrians do not stop at the crosswalk and confirm that the vehicle has stopped and that the driver recognizes their presence and intent to cross but rather without hesitation meet the teacher in the middle of the crosswalk. Of course, the teacher is not following established safety protocol for roadway crossing either. The teacher does not ensure that the driver acknowledges the stop order, and the teacher is not wearing a high-visibility vest or utilizing a stop paddle for the stop command. Perhaps the most intriguing aspect of the protocol is the unnecessary personal exposure of the teacher. The potentially injurious behavior established for the young child is that stopping before crossing a roadway, looking both directions, listening for oncoming traffic, and waiting to cross until the road is clear of all traffic or until traffic is stopped and awaiting their crossing is not established. Furthermore, this practice establishes that personal exposure to a hazard in order to perform a job is somehow proper and, because of the social standing of the teacher and of a school in general, is somehow "chivalry in practice."

What is the underlying message that your students get from established safety procedures? In the specific example above, the habits could impact the action of the child while away from school or later in life. In this examination, there is a problem with the number of children injured as pedestrians each year. Even one child is too many. Zero here is a standard. We will, as trusted educators, strive to improve safety continually. This does not necessarily mean that the measure of success is zero, but that we will continually examine each practice and policy in order to reduce hazards and incidents. So, thirteen thousand child pedestrian injuries are not acceptable.

The key actors from an educational perspective are most definitely the teachers and administrators who have control for establishing policy and practice. The best management practices for safety are well established for most circumstances either specifically or holistically. Here specifically, we have mentioned the standard practices and teachings for roadway crossings. Obviously, measures of success might include the number of child pedestrian injuries and near misses of students, but they might also be an observed measure of behaviors exhibited by the child while at school and in the community.

Pedestrian crossings from public walks to school property are a perfect environment for the responsible students in the most senior grade. Students can be supervised as crossing guards by school personnel in joint partnership with local law enforcement. Portable stop signs can be placed at crosswalks for traffic. A crossing guard is then utilized to stop pedestrian traffic and wait to ensure that vehicular traffic is stopped and that drivers acknowledge the crossing. The crossing guard will then signal the walkers to cross via a walk/do not walk paddle. Of course, high-visibility vests and student training is required.

The program ideally should be implemented in partnership with local law enforcement to be present as many mornings as possible to patrol or target traffic violations, such as speeding infractions and failures to stop. The police department can be utilized to provide training and help with equipment costs. This type of program is common in larger communities and is very successful.

It may first seem that students would be placed in harm's way. However, a commitment for adequate supervision and the stationing of the students on sidewalks, avoiding exposure to traffic, can minimize the risk. The benefit is an obvious one for the establishment of civic duty, personal safety, the safety of others, responsibility, and overall leadership. Rewards for a year of successful service can be given. Field trips to state capitals, museums, and Washington, D.C. have all been used by school districts as rewards for safe behavior.

The ideal protection for pedestrians to avoid drop-off and pickup traffic is to use pedestrian crossovers or bridges and tunnels to keep vehicles and foot traffic physically separated. This can be an added feature to existing buildings. However, proper planning in the early design stage of the school will be much cheaper and avoid construction interruptions later in the life of the school.

CONCLUSION

Access control is a necessity in today's unstable environment. Schools have experienced what were once unthinkable acts of violence and betrayal. Schools that once kept ball courts, playgrounds, and gymnasiums open for public use after hours have been defending themselves from expensive suits from those injured on their property. Today, access control not only means placing easily unnoticed barriers on school grounds, using them and tactical positioning to deter unwanted access, and managing perimeters on outdoor grounds, locked doors, and access gates, but it also means that grounds must be constantly audited for unsafe conditions to avoid lawsuits from trespassers or welcomed after-hours users. Cameras are another useful deterrent strategy for controlling access and limiting damage. They should be heavily announced to all entering the grounds. It is clear that today's environment has necessitated that schools can no longer be wide-open and welcoming institutions to the community at large. Principals, educators, and staff must now add "security officer" to their wide array of job duties.

CASE STUDY

At your school, many parents drop off their child in the morning between 7:30 a.m. and 8:00 a.m. Although the buses have a separate drive for entering and leaving the school, foot traffic from employees, visitors, and parents who walk their child to school must cross a drive for parent drop-offs. Two weeks ago a child was struck by a vehicle that had just dropped off a child. The child pedestrian was not hurt. The vehicle had just begun to pull forward. The parent of the pedestrian had pulled into the employee parking lot and dropped the child off there rather than waiting in the drop-off line because of being late for work. The parent who was pulling forward was in a large heavy-duty truck with towing mirrors and did not see the child enter the crosswalk. A parent witness behind the truck stated that the child was jogging across the driveway but did not know whether the child had stopped or even hesitated before crossing.

- In what ways could this incident have been prevented?
- What program components would you implement in your school to address this risk?

EXERCISES

1. Design a plan for guarding pedestrian traffic without a tunnel or crossover. What would your procedure encompass?
2. Perform a hazard analysis for school employees on traffic duty.
3. Should the school train or educate children on pedestrian safety? What if the school is in an urban setting and many children walk to school?
4. Design a lesson plan for educating your students on walking to and from school.

REFERENCES

Isaacs, J. (2010). School liability issues. 2010 Kentucky School Board Association Annual Meeting, Louisville, KY.

Safe Kids USA. (2011). Pedestrian safety fact sheet. Accessed April 30, 2012. http://www.safekids.org/our-work/research/fact-sheets/pedestrian-safety-fact-sheet.html.

Siddle, B. (2001). Pressure point control tactics management systems. Instructor certification course, Department of Criminal Justice Training, Richmond, KY, June 2001. Conducted by Tim Anderson.

Stanford Encyclopedia of Philosophy. (2005). Critical theory. Accessed March 3, 2012. http://plato.stanford.edu/entries/critical-theory.

3 Dress Codes

E. Scott Dunlap

CONTENTS

School administrators can utilize a standardized student dress code to address a number of issues that can lead to school safety risks. This can be done by implementing a basic dress code that takes into consideration personal clothing worn by students or may include a formal school uniform.

BASIC DRESS CODE

School administrators may choose to address school safety risks through the implementation of a basic dress code that allows students to wear personal clothing, but with delineated considerations and requirements. Following is an example of such a dress code:

- Holes in the blue jeans that show skin above the knees are NOT permitted.
- Students will NOT wear any clothing that exposes the mid part of the body.
- No tank tops.
- No spaghetti straps or even halters are permitted.
- The lengths of shorts, skirt, or dress must be extended to at least the student's mid-thigh.
- No head wear or hats.
- No clothing or articles depicting violence will be permitted.
- No baggy jeans where the undergarments are expose[d].
- No clothing, jewelry, or accessories which may be considered derogatory, towards a race, culture, or religion.
- Pierced body ornaments are restricted to the ears.
- 1st offense - parent conference/1 day of lunch detention/change of clothing
- 2nd offense - parent conference/1 to 3 days of ISS/change of clothing
- 3rd offense - parent conference/1 to 3 days of OSS/change of clothing (Scott High School, 2011)

Although much of this dress code addresses matters of appearance, the dress code also addresses issues that can present risks to school safety. For example, the dress code prohibits "clothing, jewelry, or accessories which may be considered derogatory, towards a race, culture, or religion." Students wearing such attire could provoke an act of violence because of the bias depicted on an article of clothing. A hostile environment could be created by the presence of derogatory communications that could result in a negative emotional impact on targeted students or escalate to physical altercations. A basic dress code provides school administrators the opportunity to evaluate school safety risks that exist in relation to student dress and design an acceptable standard of dress that seeks to eliminate the potential for incidents to occur.

SCHOOL UNIFORMS

School administrators may choose to go beyond a basic dress code and require the use of school uniforms. School uniforms can range from simple slacks/skirts and a golf shirt to the more formal appearance of a coat/sweater and tie. Research (Brunsma and Rockquemore, 2003; Firmin et al., 2006; Boutelle, 2008) traces a common argument for widespread implementation of school uniform policies to the 1996 State of the Union address in which President Bill Clinton said, "And if [teaching character, values, and citizenship] means teenagers will stop killing each other over designer jackets, then our public schools should be able to require their students to wear school uniforms." Research has evolved since that statement was made to assess the benefit of implementing the use of school uniform policies.

Konheim-Kalkstein (2006) evaluated the argument as to whether the implementation of school uniforms was truly responsible for improving school climate as manifested in such things as decreased violence. It appears that "gang violence is one of the most influential reasons for adopting uniform policies" (p. 25). Although her review of literature indicated that there is a positive correlation between the implementation of school uniforms and student achievement, she found that one area of conflict has been the potential infringement of school uniforms on constitutional rights. However, the Supreme Court upheld the right for schools to implement uniform policies in *Canady vs. Bossier Parrish School Board* given:

- First, that the school board has the power to make such a policy;
- Second, that the policy promotes a substantial interest of the board;
- Third, that the board does not adopt the policy to censor student expression; and
- Fourth, that the policy's "incidental" restrictions on student expression are not greater than necessary to promote the board's interest (Konheim-Kalkstein, 2006, p. 27).

Konheim-Kalkstein concluded that although the research is not conclusive in that additional variables need to be considered, the current research and testimonies of those who have been involved in the implementation of school uniforms is of importance.

Brunsma and Rockquemore (2003) sought to determine through quantitative research whether the implementation of school uniforms was effective in improving the school environment. In a response to a critique of earlier research that they had conducted, they sought to clarify the statistical applications they used to research the differences between students with uniforms and those without uniforms. A unique issue that they addressed was the need to explore the effectiveness of school uniforms beyond sound bites taken by the media or agendas that individuals or groups might have in promoting the use of school uniforms. The research discussed was an early attempt to quantify the effectiveness of school uniforms apart from rhetoric on the subject, which at the time was only supported by anecdotal information. Their findings indicated the need to consider additional variables that impact the school environment other than the sole variable of implementing a school uniform policy.

Challenges to implementing a school uniform policy include wording that is used in regulations that gives school districts the ability to create uniform policies and perceptions held by parents. Boutelle (2008) drew attention to a California regulation that gave school districts the ability to enact dress codes but also included wording that gave parents the ability "not to have their children comply with an adopted school uniform policy" (p. 34). Implementing a school uniform dress code under such a law can create a great deal of confusion among faculty who are responsible for managing the dress code in their classes. It can be a challenge to ensure that a student who is seen as not being in compliance is simply in violation of the uniform dress code or has been formally excused from having to comply with the policy because of the objection of a parent. Outreach and training may be necessary on the part of a school district to inform parents of the justification for implementing a dress code that requires uniforms. The dissemination of such information could assist parents in overcoming initial reservations regarding the transition to school uniforms.

The California law was tested soon after the state's Long Beach district enacted a school uniform policy among its elementary and middle schools (Portner, 1996). The American Civil Liberties Union (ACLU) filed suit against the district because of its failure to fully inform parents of their ability to opt their children out of the policy. The ACLU focused on the financial hardship that the implementation of a school uniform dress code placed on impoverished families. The dispute was productively settled through a number of process improvements:

- Additional communications regarding the school uniform policy were distributed to parent via the mail.
- Charities were identified that provided free uniforms, and this information was communicated to parents.
- Liaisons were created who facilitated communication between the schools and parents.

Walmsley (2011) presented a case study exploring the impact of school uniforms in a public school in England. Admittedly, she was "never an advocate of school uniforms" (p. 64), but the level of behavior exhibited by students helped to change her perception. She looked beyond the issue of school violence to benefits such as lowered cost for uniforms compared to costly individual outfits and the culture that is created in a school through the use of school uniforms.

TABLE 3.1

Implementing School Uniform Programs

Walmsley (2011, p. 65)	Konheim-Kalkstein (2006, p. 26)
Seek parental and community input	Get parents involved from the beginning
Research what has been done before	Protect students' religious expression
List goals for implementing the policy	Protect students' other rights of expression
Arrange for vouchers or seek grant funding to assist low-income families to buy items	Determine whether the policy should be voluntary or mandatory
Start with elementary grades and move upward as students move through grades	Determine whether to have an "opt out" provision
	Do not require students to wear a message (opinion or viewpoint of the school)
	Assist families that need financial help
	Treat school uniforms as part of an overall safety program

Source: Adapted from Konheim-Kalkstein, Y. (2006), *American School Board Journal*, August, 25–27; and Walmsley (2011), *Phi Delta Kappan, 92*, 65.

With research resulting in various findings, it is clear that instituting a school uniform dress code can have a positive impact on reducing school violence, but additional variables need to be considered. A key element in the literature regarding school uniform dress codes is the increased level of school safety through such things as:

- Decreased violence and theft
- The identification of nonstudent intruders
- Prevention of gang attire (Daugherty, 2002, p. 391)

If the decision is made to pursue these benefits, Table 3.1 identifies issues to consider when implementing a school uniform dress code.

A theme in these two strategies for implementation is to integrate parental considerations into the process. This is manifested in both strategies addressing the need to engage parents in the developmental phase of the policy, as well as to provide financial assistance for those who are in need. Although enforcement of a school uniform dress code policy will come with its own challenges, the two strategies in Table 3.1 present the need to address parents as a key element to set the process up for success before the first student enters the school building wearing the desired uniform.

CASE STUDY

East School District has experienced a consistent increase in school safety concerns over the past three years. Acts of violence among students have risen in addition to periodic incidents of nonstudent intruders accessing school campuses. Administrators have convened to discuss the implementation of a school uniform dress code based on reports they have seen in the media. Some of the administrators are in support of the decision to move forward with requiring students to wear uniforms, whereas others have reservations.

- What variables should be considered in making this decision?
- What methodology should be used to implement a school uniform dress code if that is the final decision of the school administrators?

EXERCISES

For the following questions, identify a single school environment in which you would like to situate your responses and answer each question accordingly.

1. What items would you include when designing a basic dress code for a school?
2. What criteria did the Supreme Court use in upholding the right of a school district to implement the use of a school uniform dress code?
3. How would you integrate parents into the development and implementation of a school uniform dress code?
4. What challenges are presented by an "opt out" provision of a state law?
5. Should a school uniform dress code be mandatory or voluntary? Why?

REFERENCES

Boutelle, M. (2008). Uniforms: Are they a good fit? *The Education Digest*, February, 34–37.

Brunsma, D., and Rockquemore, K. A. (2003). Statistics, sound bites, and school uniforms: A reply to Bodine. *The Journal of Educational Research*, 97(2), 72–77.

Daugherty, R. (2002). Leadership in action: Piloting a school uniform program. *Education,* 123(2), 390–393.

Firmin, M., Smith, S., and Perry, L. (2006). School uniforms: Analysis of aims and accomplishments at two Christian schools. *Journal of Research on Christian Education,* 15(2), 143–168.

Konheim-Kalkstein, Y. (2006). A uniform look. *American School Board Journal*, August, 25–27, 3.

Portner, J. (1996). Suit challenging Long Beach uniform policy dropped. *Education Week*, 15(23).

Scott High School. (2011). Dress code. Accessed November 2, 2011. http://www.scotthighs-kyhawks.com/dresscode.htm.

Walmsley, A. (2011). What the United Kingdom can teach the United States about school uniforms. *Phi Delta Kappan*, 92(6), 63–66.

4 School Resource Officer

James P. Stephens

CONTENTS

Violent acts within school facilities create a fear of personal safety and are a growing concern. During the 2007–2008 school year, twenty-one homicides and five suicides of school-aged children occurred on school property in America. Accompanying the violent crimes were over 1.5 million nonviolent crimes on school grounds during that same period (National Center for Education Statistics, 2009).

School safety and security is a great concern and one that requires the attention of all. Therefore, multiple measures are implemented in school districts throughout this nation for mitigating the various safety and security risks present in schools each day. One method of addressing the safety and security within the school districts is through a School Resource Officer (SRO) program. This program weaves a member or members of the local law enforcement community into the fabric of the school setting. Many instinctively believe that placing a law enforcement officer into the school setting is primarily for the enforcement of the law. However, research shows that although this is one of the functions and roles of the SRO, an effective program focuses their presence in a multifaceted manner.

Although the deterrence and response aspect of a SRO is important, a properly instituted program will also address the overall culture of safety within the institution through the relationships built between the officers, students, and educators (Kennedy, 2001). The SRO should undertake three primary roles: 1) problem solver and liaison to community resources, 2) educator, and 3) safety expert and law enforcer (Raymond, 2010). These roles are important for maximizing the resource of a law enforcement officer assigned to any school district. In order to reduce the amount of criminal activity, a culture of safety must be implemented, which is possible through a well-designed and properly implemented SRO program.

Although a nationwide study on the effectiveness of SRO programs is nonexistent, multiple studies on individual districts, states, and one study of nineteen different SRO programs are available. One such survey conducted on nineteen large SRO programs revealed substantially positive attitudes toward the program (Finn and McDevitt, 2005). This research reveals many aspects of an effective SRO program and specifically identifies a direct correlation between those individuals with an increased perception of safety within the school and their personal approval and satisfaction of the SRO.

INTRODUCTION

The beginning of each school year in America brings forth many uncertainties in the lives of the students and educators alike. Normal uncertainties stem from various elements of the school and learning environment including unfamiliar teachers, students, and atmosphere. Unfortunately, the students and educators of today must face many other worries and concerns as they enter their sanctuary of learning that are certainly not ordinary in nature. The ongoing trend of violent acts within school facilities creates fears regarding personal safety and is a growing concern.

With advances in school safety efforts including video surveillance, access control, emergency-management planning, and electronic screening, one important element remains missing. The human element offered by a properly designed and implemented SRO program provides more than an effectual deterrent or emergency response to criminal activity in the school. The implementation of an SRO program is a valuable and practical method to enhance the overall safety and security within a school district, including the perceptions of a safe environment by those within.

HISTORY OF SCHOOL RESOURCE OFFICERS

Although statistics show that the chances of becoming a victim of a crime within the school setting are less than if you are in the general public, the perception and fear of criminal behavior in the school can disrupt the learning environment. The Omnibus Crime Control and Safe Streets Act of 1968 developed strategic operatives for law enforcement to combat criminal activity by becoming more involved in the community and provided funding for various programs to implement this strategy. This act, as amended in 1998 in section 1709, defines a school resource officer and the multifaceted roles of the SRO (National Institute of Justice, 2000). The SRO is defined as "a career law enforcement officer, with sworn authority, deployed in community-oriented policing, and assigned by the employing police department or agency to work in collaboration with the school and community-based organizations" (National Institute of Justice, 2000, p. 1). This does not necessarily define all SROs across the nation; however, it does give us a generalized understanding of the SRO.

The first official School Resource Officer program began in Flint, Michigan, in the early 1950s as a way to integrate law enforcement into the schools in a community-oriented manner. The idea of the Flint program was creating better relationships between the students and the officers and assisting the students to become better citizens. This model of character building and seeking an improved safety environment in the schools became the model for several other SRO programs; however, some instituted a traditional law enforcement approach (Burke, 2001). Today, the Flint, Michigan, model is widely accepted as the best approach for a successful SRO program.

The concept of SROs spread in the 1960s and 1970s and then observed resurgence following the unfortunate school tragedies of the late 1990s and early 2000s. A renewed focus on police officers within the schools appeared. In 2000, the Department of Justice awarded $68 million in grants to fund the hiring of 599 SROs in 289 communities across the nation. The intention of this federal assistance was to enhance community-oriented policing through the schools. This enhancement is achievable through the creation of better working relationships between the police, students, parents, and school administrators (Girouard, 2001).

North Carolina experienced a continual growth from 243 school resource officers in 1995 to 849 in 2009. Over the past fourteen years, the number of SROs in North Carolina has increased over 249% from 243 in 1996 to 849 as identified in the 2008–2009 School Resource Officer Census for that state. Also noteworthy from this census is the fact that of the 375 high schools in North Carolina, 330 have a SRO primarily assigned to them, meaning they do not share their time with other schools. Figure 4.1 illustrates the continued growth of the SRO programs in North Carolina from 1995 to 2008 (*North Carolina School Resource Officer Census 2008–2009*, 2009). This is demonstrative of their continued belief in the benefits of a broad SRO program.

State statutes often define and characterize what the school resource officer is in a particular state. In 1998, Kentucky defined an SRO within Kentucky Revised Statute 158.441 as

> a sworn law enforcement officer who has specialized training to work with youth at a school site. The school resource officer shall be employed through a contract between a local law enforcement agency and a school district (Kentucky Revised Statutes, 1998).

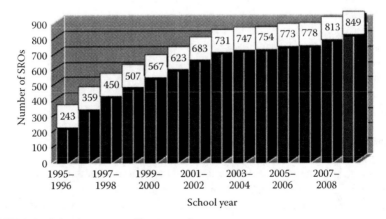

FIGURE 4.1   School resource officer growth.

This definition provides a basis for the SRO within Kentucky, with some important elements and requirements within. One requirement is that the SRO is a sworn law enforcement officer. Therefore, this provides the SRO with provisions given to a law enforcement officer within their given jurisdiction.

In order to be a qualified peace officer, they must receive basic police training from an accredited training agency within the Commonwealth of Kentucky. This training is provided by organizations designated by the Kentucky Law Enforcement Council, as identified in Kentucky Revised Statute Chapter 15, which also specifies the general qualifications and basic training requirements for obtaining certified peace officer status in Kentucky.

Also, the law states that the SRO must have specialized training to work with youth at a school site. Ambiguity regarding the type and amount of specialized training creates a wide spectrum of qualifications held by SROs and the training they receive. Although the statute does not establish or mandate what specialized training is required, there are areas of training for consideration to enable the SRO to effectively fulfill their duties. Those areas of training are discussed further in this chapter.

ROLE OF THE SCHOOL RESOURCE OFFICER

In 1991, the National Association of School Resource Officers (NASRO) was established as a nonprofit organization to bring forth a more collective and coordinated establishment of the SRO roles and responsibilities throughout the nation. One of the first priorities of the newly formed NASRO was the establishment of the triad model, which defined the three primary roles of a SRO. These three roles are law enforcement officer, teacher, and counselor (Burke, 2001). These primary roles are different in various districts yet remain as the predominant roles of SROs across the nation.

The role of the school resource officer extends far beyond traditional "enforcement" activities. Although the SRO does carry the legal authority to carry out an arrest, they also have the expertise to provide educational advancement in the area of legal matters for teachers and students. Developing a mentoring relationship through positive interaction with students is a positive attribute and result of an SRO program. The SROs become liaisons between students and school officials as they provide insight to the current trends and items of concern to both parties (Mulqueen, 1999). With recent budget cuts negatively affecting law enforcement, many departments still view the SRO programs as too valuable to lose (Black, 2009). Their ability to resolve situations before they become violent or escalate further is indeed a benefit that is difficult to measure monetarily.

RETURN ON INVESTMENT

As with most investments, school administrators are concerned with their return on investment from implementing an SRO program. This return is sometimes difficult to measure, because the measure is not monetary gain. For those school districts with SRO programs, they look for a different return: an improved safety culture and a reduction in criminal activity within their district. Therefore, returns from a properly managed SRO program far exceed monetary gain.

Some of these returns include an increased perception of safety, character building of students through mentoring programs, improved relationships between juveniles and law enforcement, and reduced workload for the administrators and other law enforcement agencies (Finn, 2006). According to Finn (2006), the establishment of SRO programs produces multiple benefits for the agency and the school district. One such benefit is a reduction in workload for the sponsoring agency. With the officer already within the school, the SRO, providing they are on duty at the time of the call for service, can handle most calls of service emanating from the school. This can be a substantial burden lifted from the agency, because it prevents a patrol unit from responding to the schools on a daily basis.

Another benefit of the SRO program identified by Finn (2006) is an improved image of law enforcement as perceived by the students, resulting in improved trust between the two. When trust is established because of positive daily interaction between the students and the SRO, an increased level of crime reporting can result. This increased level of reporting leads to an amplification of crime prevention through received knowledge of criminal activity from the students.

A properly established SRO program can also see a positive return with a better relationship between the students, faculty, parents, and the community. This improved relationship fosters a heightened level of trust between all parties involved. Having a positive relationship between the law enforcement agency and the community is vital for effective crime-reduction efforts. Finally, Finn (2006) posits that a SRO program can enhance the image of the sponsoring law enforcement agency. A positive image of the law enforcement agency is undoubtedly a result that any sheriff or chief will enjoy.

EFFECTIVENESS

Although the overall effectiveness of the SRO programs throughout America is difficult to determine, various individual studies are available that provide insight to specific SRO programs already implemented. One study examined the effectiveness of nineteen SRO programs representing a wide array of jurisdictions. This study found that the schools with SROs that they profiled had overwhelming positive results in many areas of improving the safety culture within their district and substantially positive attitudes toward the program (Finn and McDevitt, 2005).

This research also discusses many aspects of an effective SRO program, specifically identifying a direct correlation between those individuals with an increased perception of safety within the school and their personal approval and satisfaction with the SRO. A 2005 study on three newly implemented SRO programs discovered 92% of the students with a positive opinion of the officers feel safe in the school (McDevitt and Panniello, 2005). However, 76% of the students with a negative opinion of the SRO still report feeling safe in the school. Therefore, we see a direct correlation between the students' opinion of the SRO and an increase in the students' feelings of safety in the school.

In this report, the research conducted through surveys, interviews, and focus groups identified multiple elements present in successful SRO programs. Further identified are areas of consideration when developing and sustaining an SRO program. The research of McDevitt and Panniello (2005) states that when implementing an SRO program, the focus must be upon the following areas:

- Choosing a program model
- Defining specific SRO roles and responsibilities
- Recruiting SROs
- Training and supervising SROs
- Collaborating with school administrators and teachers
- Working with students and parents
- Evaluating SRO programs

This study also revealed that in one of the sites surveyed, the officer spends 10% of their time on law enforcement activities; 30% advising students, faculty, and administrators; 40% teaching in a classroom; and 20% on other activities such as paperwork (Finn and McDevitt, 2005). This is in keeping with the NASRO triad model regarding the roles of SROs within the school environment. Finn and McDevitt opine that although they did not attain empirical evidence on the effectiveness of the SRO programs, a change in the overall culture of the school regarding positive behavior was evident. Reductions in tobacco usage and gang activity and an increase of perceived safety of the students by the parents are some of the recognized improvements.

There are differing views on how to measure the success of an SRO program. With little empirical data available on actual reduction of criminal activity, measuring success is most often based on the perceptions of the individuals affected by the program. If the arrest rates increase after implementation of an SRO program, does this mean that crime has increased and therefore the SRO program is ineffective? Alternatively, does this mean that the arrest rates have increased because there is a law enforcement officer now at the facility? These are a few of the questions raised when determining the value of the program (Brown, 2006). Certainly, judging the success or failure of the program cannot lie solely upon the rise or fall of criminal arrests within the facility.

This is a brief representation of the importance of not relying totally upon empirical data and an explanation of why judging the effectiveness of an SRO program cannot derive from a generalized statement regarding the reduction or increase in criminal behavior or disciplinary actions taken from year to year.

CONCERNS

Concerns over SRO programs must receive notice in order to gain a total perspective of an established SRO program. One concern is the selection and training process for the SRO program. An article published by the American Civil Liberties Union discusses concerns over nonregulated standards for the SRO programs nationwide. This article does not discuss whether the programs should exist; instead it posits that if such programs do exist, then the officers should ensure a safe environment while respecting the students' rights and overall school climate. In order to ensure that the SRO respects the students' rights, it is important that the school administrators, law enforcement officer, and the community they serve clearly understand the goal of the SRO program and receive proper training before deployment into the schools (Kim and Geronimo, 2010). Kim and Geronimo (2010) further posit that disciplinary actions on school misconduct and ordinary disciplinary issues should not result in formal law enforcement intervention. This will allow the school administrators to intervene in these noncriminal or minimal incidents without formal interjection by the law enforcement officer. The school administrators must not feel the necessity of law enforcement involvement in every school disciplinary measure.

Other basic measures facilitate the respect of the students' rights during interaction with the SRO. One such measure is model language in the memorandum of understanding between the school and the law enforcement agency that specifies that SRO involvement in disciplinary misconduct is limited until the situation or incident becomes criminal behavior. The SRO is to consider the application of the students' protections such as their Fifth Amendment right against self-incrimination and their Miranda rights as relating to interviews and interrogations within the school setting (Kim and Geronimo, 2010).

Furthermore, protection of the students' Fourth Amendment rights against unreasonable searches and seizures is essential with any safety program or endeavor. During the development of the SRO program, the administration and law enforcement officer should discuss and understand that the officer will not take over "all" disciplinary problems in the school. Those actions of misconduct such as skipping class and wandering in the halls are not the concern of the officer in relation to disciplinary action further than notifying an administrator of the actions. The 1995 case of *Tinker v. T.L.O.* did provide a certain amount of flexibility to the school administrators in formulating necessary evidentiary standard for conducting a search of student property. In this ruling, the court held that the school needs only "reasonable suspicion" to conduct the warrantless search of student property. In order for the search to be legal under this ruling, reasonable suspicion based on articulable evidence and a limited scope is necessary so as not to become overly intrusive to the student (Beger, 2002).

The U.S. Supreme Court held in *Tinker v. Des Moines Independent Community School District* that the student does not shed their constitutional rights when they enter the school (Pickrell and Wheeler, 2005). Protecting student rights against self-incrimination and allowing students the provisions under *Miranda v. Arizona*, the SRO walks a very fine line when engaging in interviews or interrogations of the student. An individual's Miranda rights are intended to notify the individual of their right to remain silent, their right to an attorney, and that their statements may be used against them in a court proceeding if necessary. When determining whether one is required to read the student their Miranda rights, bifurcated elements must coexist in order for the Miranda rights to attach. First, the student must be in custody and not free to leave at that time. Second, the process of questioning must be interrogatory in manner and not simply a question of personal information such as name and date of birth (Pickrell and Wheeler, 2005).

Although some SROs get around these fundamental processes normally taken in custodial situations by stating they were acting as "an agent of the school," this can become a very dubious situation. Utilizing precautions against violating the students' fundamental rights and ensuring proper actions by all parties involved, including the school administration and law enforcement officer, are a paramount concern. This is possible through the proper selection and training of the SRO within the school district.

Some state that the SRO programs across this nation would benefit from standardized legislation that defines the ambiguous roles they assume. With standardized requirements and training for SROs, unresolved and often unintentional legal implications of the SRO program can negatively affect the efforts of the program. These issues pose a litigious threat to the school district, the SRO, and sponsoring law enforcement agency. Abating these issues is possible through various steps including standardized legislation defining the roles of SROs in this nation. Furthermore, the SRO must be viewed as not only a law enforcement officer but also a partner with other school personnel for a more timely interjection of their actions into school problems. Through the role of teacher and counselor, the SRO becomes a team member with the school administrators and acts as a preventative countermeasure to disciplinary problems and criminal actions (Maranzano, 2001).

There also needs to be a balance between safety and education in the school environment. In order to ensure that security within the schools does not become overly onerous, a focus on the physical security and educational aspect to improve the safety culture is necessary (Kennedy, 2002). If the SRO is engaged in the teaching and counseling roles, this balance is achievable. By interacting with the students in a nonenforcement manner, the SRO within the school becomes more than a typical police officer with the school as his assigned beat.

SCHOOL RESOURCE OFFICER SELECTION

A critical part of an SRO program is the selection of the personnel best suited to take on the roles previously discussed in this chapter. Assignment to the position of SRO must not be a disciplinary measure, and it cannot be a position only for an officer who was not able to fulfill his duties on "the street." Although this may seem like a good idea to some administrators, it quickly diminishes the credibility of the SRO program and establishes a foundation for a program that is certain to fail.

In actuality, the selection process begins before posting the position. The first step in this process is the development of a selection committee that includes equal representation from the school administration and the law enforcement department. An equal voice in the selection process establishes the cooperative efforts that will continue through the tenure of the SRO program. Choosing the correct personnel to serve on the selection committee is a vitally important step in the entire selection process for the SRO.

The school administrators and the law enforcement department working in the partnership must then determine the role their SRO will play. Furthermore, they must determine what characteristics that officer must have to fulfill that role effectively. After determining what characteristics they deem most important for this vital role, the selection committee must develop a list of interview questions they will use in the selection process. The interview questions must be carefully constructed and reviewed to elicit the information desired from the candidates and based upon the established role of the SRO.

The complete list of interview questions developed by the selection committee will vary in relation to the determined role of the SRO and the needs of the school administrators. However, a few questions should be included in any selection interview process in order to build the foundation for the supporting questions. The following is a brief list of suggested questions for inclusion in the selection interview process:

- Give us an overview of your career and training, including any specialized training you have received.
- Why do you want to become a school resource officer?
- Why do you believe you are the best candidate for this position?
- What is your definition of a school resource officer?
- What do you believe is the most important role for a school resource officer?
- Do you believe you can work in a collaborative manner with your sponsoring agency and the school administration?
- Can you provide us an example of how you have worked in a collaborative effort with an outside administration on a project?
- Will you be available for call and response if necessary during times outside of the normal school hours?

As previously stated, the complete list of interview questions will vary depending on the needs of the administration. The interview is only a portion of the overall selection process. Through proper screening of interested applicants' work performance, training, personality, and interview, the school administration and the law enforcement agency can choose the officer best suited for the role of SRO.

TRAINING

In order for SROs to carry out their assigned duties, they must possess or obtain skills through training that are conducive to the school and law enforcement community. The skills obtained during their basic training period and time on the street is the base for which all other specialized training is built. Furthermore, with national associations such as NASRO, a standard for basic training is established.

The NASRO basic SRO training program is a forty-hour block of instruction providing the officers with lessons in the following areas (National Association of School Resource Officers, 2011):

- History of school-based policing
- Roles and responsibilities of the SRO
- Being a positive role model
- Public speaking
- Law-related education
- Classroom instruction techniques and lesson plan development
- Counseling for child abuse, dysfunctional families, and special education
- School safety procedures
- Emergency management
- Crime prevention

- Substance abuse
- School law

This basic course curriculum blends educational areas covering the various duties that the SRO will carry out. Providing a basic understanding of the SRO program, public speaking, classroom presentation, positive role model development, and strengthening the understanding of the law enforcement function of the SRO are accomplished in this basic training course. NASRO also offers a twenty-four-hour instructional block for advanced SRO training. This course builds upon the basic course, focusing on the identification and mitigation of potentially dangerous situations in the school environment.

Certainly, other state and local organizations and associations offer great instruction in the area of school safety. All SROs and their administrations should seek opportunities for ongoing career development training to enhance their abilities to perform their obligations. Other recommended areas of training for the SRO are:

- Active shooter response
- Social media
- Situational awareness
- Crime prevention through environmental design
- Risk, vulnerability, and threat assessment
- Advanced interview techniques
- Crisis intervention techniques

CONCLUSION

With little empirical data to substantiate or refute the effectiveness of SRO programs, measuring the effectiveness of the program largely relies upon the perception of safety by those affected by the program. However, the perception of increased safety and improvements in community relations with law enforcement is a valuable resource in the ongoing efforts to create a safer community.

Various steps must be taken in order for the SRO program to be successful, including the following (Finn and McDevitt, 2005):

- Clearly established roles and responsibilities of the SRO
- Careful screening and selection of the SRO
- Ongoing collaboration between the SRO and school officials
- Continued training of the SRO
- An ongoing assessment of the program by the school administration and supporting law enforcement agency

SROs must remain mindful of the delicate balance they must keep between their roles as law enforcement officer, teacher, and counselor. As previously discussed, the fundamental rights of the students are not eliminated with the presence of an SRO within the school. The SRO is expected to honor those fundamental, protected rights of the students in each interaction, which will further improve their relationship and image with the students and school administrators.

REFERENCES

Beger, R. (2002). Expansion of police power in public schools and the vanishing rights of students. *Social Justice*, 29(1/2), 119–130.

Black, S. (2009). Security and the SRO. *American School Board Journal*, 196(6), 30–31. Accessed April 11, 2010. http://www.ebscohost.com/academic/academic-search-premier.

Brown, B. (2006). Understanding and assessing school police officers: A conceptual and methodological comment. *Journal of Criminal Justice*, 34(6), 591–604. DOI:10.1016/j.jcrimjus.2006.09.013.

Burke, S. (2001). The advantages of a school resource officer. *Law & Order*, 49(9), 73–75. Accessed April 18, 2010. Career and Technical Education (Document ID: 82188017).

Finn, P. (2006). School resource officer programs. *FBI Law Enforcement Bulletin*, 5(8), 1–7.

Finn, P., and McDevitt, J. (2005). National assessment of school resource officer programs final project report, document number 209273. Accessed April 13, 2008. http://www.ncjrs.gov/pdffiles1/nij/grants/209273.pdf.

Girouard, C. (2001). School resource officer training program. U.S. Department of Justice, Office of Justice Programs, Office of Juvenile Justice and Delinquency Prevention, Washington, D.C.

Kennedy, M. (2001). Teachers with a badge. *American School & University*, 73(6), 36. Accessed April 15, 2010. http://www.ebscohost.com/academic/academic-search-premier.

Kennedy, M. (2002). Balancing security and learning. *American School & University*, 74(6), SS8. Accessed April 16, 2010. http://www.ebscohost.com/academic/academic-search-premier.

Kentucky Revised Statutes. (1998). Accessed March 10, 2011. http://www.lrc.ky.gov/KRS/158-00/441.pdf.

Kim, C.Y., and Geronimo, I. (2010). Policing in schools. *Education Digest*, 75(5), 28–35.

Maranzano, C. (2001). The legal implications of school resource officers in public schools. *NASSP Bulletin*, 85(621), 76. Accessed April 26, 2012. http://www.ebscohost.com/academic/academic-search-premier.

McDevitt, J., and Panniello, J. (2005). National assessment of school resource officer programs: Survey of students in three large new SRO programs. Document number 209270. Accessed April 15, 2008. http://www.worldcat.org/title/national-assessment-of-school-resource-officer-programs-survey-of-students-in-three-large-new-sro-programs-document-number-209270/oclc/425027484&referer=brief_results.

Mulqueen, C. (1999). School resource officers more than security guards. *American School & University*, 71(11), SS17.

National Association of School Resource Officers. (2011). Basic SRO. Accessed April 10, 2011. http://www.nasro.org/mc/page.do?sitePageId=114186&orgId=naasro.

National Center for Education Statistics. (2009). Indicators of school crime and safety. Accessed April 17, 2010. http://nces.ed.gov/programs/crimeindicators/crimeindicators2009/key.asp.

National Institute of Justice. (2000). A national assessment of school resource officer programs. Accessed April 25, 2012. http://www.ncjrs.gov/pdffiles1/nij/s/000394.pdf

North Carolina School Resource Officer Census 2008–2009. (2009). North Carolina Department of Juvenile Justice and Delinquency Prevention, Center for the Prevention of School Violence. Accessed March 3, 2012. http://www.ncdjjdp.org/cpsv/pdf_files/SRO_Census_08_09.pdf.

Pickrell, T., and Wheeler, T., II. (2005). Schools and the police. *American School Board Journal*, 192(12), 18–21. Accessed April 14, 2010. http://www.ebscohost.com/academic/academic-search-premier.

Raymond, B. (2010). Assigning police officers to schools. Washington, D.C.: Department of Justice, Office of Community Oriented Policing Services.

5 Bullying

Lynn McCoy-Simandle

CONTENTS

Imagine you wake up each morning with an overwhelming sense of dread. Imagine you search for any justification for pulling the covers over your head and staying in bed. Imagine slowly rising and wondering whether your tormentors have discovered new ways to make your life miserable. Now, imagine that you are twelve years old. Such is the life of too many students as they prepare for another day in the hallowed halls of education. Is it surprising that a student who is bullied repeatedly shows little interest in how the three branches of the American system of government function?

This chapter is not a comprehensive exploration of bullying in schools, but an effort to highlight the salient features surrounding bullying and possibly whet the appetite of the reader to delve more deeply into the subject. The case studies featured in the topic boxes are actual student stories with names changed to protect confidentiality. The purpose of the case studies is twofold: 1) to illustrate the varying range of behaviors that constitute bullying and 2) to allow the reader to generate possible responses to each scenario.

Although the victims of bullying are the topics of television and newspaper reports, special features, public service announcements, and even fodder for talk shows, long before our current interest in bullying, examples of bullying behavior were found in some of our most memorable pieces of literature. *Oliver Twist*, written by Charles Dickens and published in 1838, was one of the first novels in the English language to focus on bullying (Carpenter and Ferguson, 2009). William Golding's *Lord of the Flies* in 1954, S. E. Hinton's *The Outsiders* in 1967, and Judy Blume's *Blubber* in 1974 are just a few more literary examples of the impact of bullying on a child's life. This intensive interest in bullying is not because it is new, but because the landscape of bullying has changed.

The outbreak of school shootings in the 1990s, especially the horrific events at Columbine, focused America's attention on school safety unlike at any other time in history. School safety was foremost in the minds of all parents. Organizations not typically involved in the safety of students in general weighed in on the issue. One such organization, the Secret Service, sought to develop a profile of a school shooter and discovered that the one common thread among the school shooters was that more than two-thirds of the school shooters had been the targets of prolonged bullying (U.S. Secret Service, 2002). Following Columbine and the connection to bullying, the American Psychological Association passed a resolution to integrate bullying prevention into violence prevention activities to encourage public and private funding agencies to support research on bullying behaviors and anti-bullying interventions (American Psychological Association, 2004). The National Association of School Psychologists (National Association of School Psychologists, 2006), the American Medical Association, the World Health Organization, and the Center for Disease Control are a few of the national organizations who have made a concerted effort to stem the tide of bullying in American schools. Further, the U.S. Department of Education has threatened cuts to federal funding if bullying complaints are not addressed (Ali, 2010). Even the president of the United States has weighed in on the issue, labeling bullying an "epidemic" in a White House Conference in January 2011. Truly, the consensus in America is that bullying in schools deserves the massive amount of attention it is now receiving.

UNDERSTAND BULLYING BEHAVIORS

Bullying is direct or indirect behavior that conveys dominance over another and is motivated by the desire to cause fear and distress. Although definitions of bullying differ somewhat, most experts agree that bullying occurs over a period of time, is marked by an imbalance of power, and is intended to harm the victim (Farrington, 1993; Olweus, 1993; Smith et al., 1999; Mayo Clinic, 2001; Limber, 2002).

Although bullying is manifested in many behaviors, including hitting; shoving; pushing; having possessions taken away; name-calling; teasing; ostracizing; excluding; spreading rumors; threatening; harassment about race, ethnicity, religion, disability, sexual orientation, gender identity; and sexual harassment; those behaviors can be categorized into three main types: physical, emotional, and social, also called relational aggression. Physical bullying involves threats to a person's body or possessions and is the most direct type of bullying and the easiest to observe. Emotional bullying attacks one's self-esteem and can be

---

TOPIC BOX 5.1

Jonathan (all names have been changed) was a tenth-grader who never participated in school functions and appeared to have no friends at school. He shared little personal information with school staff and was generally thought to be unmotivated. After Jonathan committed suicide, administrators discovered that he had been physically and emotionally bullied since elementary school, but he had not reported the bullying to anyone since sixth grade.

direct or indirect. Social bullying, the most indirect type, damages someone's ability to establish or maintain social relationships (Olweus, 1993; Simmons, 2002; Wiseman, 2002). Males are more likely to report being physically bullied by their peers, whereas females are more often the targets of social bullying (Harris et al., 2002; Nansel et al., 2001). Factors that place children and youth at high risk of being bullied by their peers include not fitting in with a peer group, being shy, being anxious, being obese, having developmental disabilities, and being gay, lesbian, or transgendered or perceived to be so (Hooever et al., 1993; Nabuzka and Smith, 1993; Herschberger and D'Augelli, 1995; Olweus, 1995; Marini et al., 2001; Rigby, 2002; Janssen et al., 2004).

For most, the mental image of a bully involves a big arrogant loudmouth with a gang of followers who terrorizes students at will. Even though this kind of bully does exist, far more common is the kind of bullying that takes place within the context of a group where one or several children are rejected by the group and become targets for the group's mistreatment. The "face" of this bully can be any child's face and is often at odds with the general perception of a bully. Psychologist Michael Thompson (Thompson and Grace, 2001) notes that inclusion and exclusion are powerful instruments of group cohesion. Groups exclude or maltreat others to maintain their identity and to create the boundary between "us" and "them." Most children

TOPIC BOX 5.2

Rebecca (all names have been changed) remembers middle school as three years of constant harassment about her weight. She tried to avoid all physical activity because it ultimately led to increased commenting and gesturing from her classmates. Consequently, she failed physical education classes and increasingly passed up opportunities to socialize with her peers. She left middle school with a sense of self-loathing but also feeling angry and alone.

TOPIC BOX 5.3

Lori (all names have been changed) was approaching ninth grade with great excitement. She would be leaving a small performing arts school for a large high school along with her four best friends. The first week of school, however, things seemed different with her friends. She discovered that Dena had a slumber party the weekend before school started, and Lori had not been invited. When she asked Ashley about the party, Ashley said, "Oh, Dena thought she told you about it. It was just a mistake." Lori tried to accept the explanation, but she began to discover she was left out of increasingly more activities. The girls also seemed cool to her at school and when she left messages to call her, no one returned her calls. Finally, the girls quit pretending and told her they no longer wanted to be friends with her. Lori was crushed because she had been friends with these four since first grade and really had no one else to call a friend.

will acknowledge that exclusion from the group is the most hurtful social obstruction, and it occurs every day in every classroom in every school in America. The image of this bully forces us to broaden our perception of a bully and the behaviors that constitute bullying.

Rachel Simmons (2002 and 2004) and Rosalind Wiseman (2002) have focused their careers around the bullying behavior of girls. The term "relational aggression" is used to describe a type of bullying primarily used by girls to victimize others— a covert use of relationships as weapons to inflict emotional pain. Although girls have often been referred to as the "gentler" sex, research into the bullying behavior of girls refutes that premise; girls' use of aggression is more covert than that of boys. Mary Pipher, a clinical psychologist, brought popular attention to this form of harassment through her best-selling book, *Reviving Ophelia* (1994). The movie *Mean Girls* (2004) and Lifetime TV's movie *Odd Girl Out* (2005) were based in part on the work of Wiseman and Simmons, respectively, and brought "girl bullying" into mainstream consciousness. Although relational aggression is not a new social problem, it has become more widely recognized in large part because of efforts of Simmons and Wiseman.

Cyberbullying, a type of social bullying that involves using various forms of technology to dominate victims, has further complicated an already complex issue (Smith et al., 2008). Many students who would not harass their peers in the hallways of a school would not hesitate to do so on social networking sites. The balance of power is leveled by the use of technology and makes cyberbullying very seductive to the teen who is angry and out for revenge. The veil of anonymity in cyberspace gives courage to the writer of a demeaning text message, and pressing "send" is much less threatening than a face-to-face war of words in the school cafeteria. Often, students who are timid in a physical setting can become fearless in cyberspace.

The fear of getting caught and being punished is diminished as well. Educators know that Internet use is not sufficiently supervised, and cyberspace is a breeding ground for bullying behaviors (American Association of University Women Educational Foundation, 2001; Craig and Pepler, 1997). Cyberspace is certainly less monitored and less structured than any area in a school, thereby increasing the sense of "no one will know it is me." Cyberbullying is akin to the disguising of one's voice on the telephone in years past to communicate a malicious message to someone without revealing your identify. The invention of caller ID may have curtailed the use of one type of technology, but the explosion of new technology in the last twenty years

---

TOPIC BOX 5.4

Angie (all names have been changed) reluctantly allowed her boyfriend, Rob, to take suggestive and revealing pictures of her, which he promised were for his eyes only. After their break-up, Angie's friend showed her a Facebook page that proclaimed "Bad Angie Revealed" with all of the photos posted. Angie reported receiving harassing comments and e-mails. She was sure everyone in school was laughing at her. She finally had to tell her parents about the website.

has provided multiple ways to mistreat others. Sheri Bauman, associate professor at the University of Arizona says:

> The nature of technology magnifies the potential for harm [with cyberbullying]. The size of the audience who could potentially witness the humiliation of a target is enormous. The bullying takes place without restrictions of time and place, so the target has difficulty finding a safe haven (Paterson, 2011, p. 1).

RESPONDING TO BULLYING BEHAVIORS

Scientific interest in bullying behaviors started with Dan Olweus, a psychology professor at the University of Bergen in Norway, who began his seminal work into bullying in the schools in the 1970s. His initial findings clearly demonstrated that bully/victim problems were quite prevalent in school settings, and the 1982 suicides of three Norwegian boys ages ten to fourteen focused the attention of an entire country on bullying in school. Olweus subsequently provided the first systematic intervention study using his own creation, the Olweus Bullying Prevention Programme, to promote the concept that a school environment could be structured to reduce aggression in children. In 1993, Olweus, now considered the world's leading authority on bullying behaviors, wrote *Bullying at School: What We Know and What We Can Do.* His groundbreaking research and intervention programs have played a significant role in increasing awareness around the world that bullying is a growing social problem, one that must be taken seriously by researchers, educators, lawmakers, parents, students, and society in general.

Although Olweus's book *Aggression in the Schools: Bullies and Whipping Boys* was published in the United States in 1978, the United States came late to the party in recognizing bullying in schools as a significant issue and began researching bullying in earnest in the late 1990s. In 2001, Nansel, Overpeck, Pilla, Ruan, Simons-Morton, and Scheidt reported the first nationally representative survey of students in grades six through ten in the United States and estimated that roughly 30% of students were involved in moderate or frequent bullying as the target, the bully, or both. In 2003, the National Association of School Psychologists labeled bullying the most common form of violence in our society (Cohn and Canter, 2003).

Experiences in childhood, with parenting, or from working with schools have made most of us are aware that bullying can have considerable impact on targets and has understandably propelled bullying into a topic of national concern. The link

---

TOPIC BOX 5.5

Nine-year-old Justin (all names have been changed) believes that boys who cry easily are frequent targets of bullying as he was in first and second grades. He says he no longer cries at school if someone says something mean; however, he also admits that now that he is not being bullied, he does not have much compassion for younger boys who cry at school and has called them "sissies."

between prolonged bullying and illness, substance use, and other mental health issues is well documented (Kaltiala-Heino et al., 1999; Nansel et al., 2001; Lyznicki et al., 2004; Gini and Pozzoli, 2009; Duke et al., 2010). Studies have linked being bullied to low self-esteem, feelings of loneliness, higher rates of anxiety, depression and suicide, aggression, and lower academic achievement (Olweus, 1993; Craig, 1998; Kochenderfer and Ladd, 1996; Bond et al., 2001; Baumeister et al., 2002; Ttofi and Farrington, 2011).

McKenna et al. (2011) extended the connection to bullying to include family violence as well as physical and mental health in a report to the Centers for Disease Control and Prevention. The multivariate analysis of data from a Massachusetts Health Survey conducted in 2009 suggested associations between bullying and family violence, for example, witnessing violence or being physically hurt by a family member. The bully-victims, students who both were bullied and bullied others, were more likely to report violent family interactions than bullies or victims. The effects of being bullied do not always end with childhood but can have lifelong consequences. The late Eric Rofes, an assistant professor of education at Humboldt State University, recalled years of being bullied:

> I know a lot about bullies. I know they have a specific social function: they define the limits of acceptable conduct, appearance, and activities for children. They are masters of the art of humiliation and technicians of the science of terrorism. They wreaked havoc on my entire childhood. To this day, their handprints, like a slap on the face, remain stark and defined on my soul (Rofes, 1994, p. 37).

Educators are often reminded that the goal of education is to raise academic performance; however, the educational environment must also nurture a student's emotional well-being to produce healthy, well-adjusted, and productive adults. The link between academic performance and emotional well-being makes sense, because most people have had the experience of not being able to concentrate or be productive under severe stress. This also applies to children: if they are not socially, emotionally, and physically healthy they may not be able to complete their school work, pay attention to their teachers, or perform well on standardized tests. Lacey and Cornell (2011) presented results of an ongoing research study in Virginia that found

TOPIC BOX 5.6

Eight-year-old Seth (all names have been changed) was attacked by a swarm of bees on the school playground and stung numerous times just weeks after he enrolled as a new student. Thereafter, when he was on the playground and saw a bee, he would run to the teacher crying. The other students laughed at his fear of bees and called him a baby. One student was particularly insensitive and would sneak up behind Seth and make buzzing noises to elicit a reaction. Seth began making excuses to stay in during recess to avoid the constant teasing. He would tell his parents that the teachers and the principal hated him.

that in schools where ninth grade students reported higher rates of bullying, passing rates on standardized tests for Algebra I, Earth Science, and World History were 3–6% lower than in comparison schools.

Although a growing body of research exists regarding the prevalence and impact of bullying on targets, research also suggests there is no single cause of why one child bullies another. A child or youth may be at risk of developing bullying behaviors because of individual, family, peer, school, or community factors (Olweus et al., 1999; Limber, 2002).

INTERVENTION RESEARCH

Reducing school bullying has led to the design and implementation of many school-based programs. The Olweus Bully Prevention Programme is considered by many to be the gold standard for intervention programs. The Olweus program was implemented in Norway in 1983, and evaluations in 1991 showed a dramatic decrease in bullying behaviors (Smith et al., 2004). A review by Baldry and Farrington (2007) of sixteen evaluations in eleven countries using the Olweus program determined that eight (50%) produced desirable results, two (12.5%) produced mixed results, four (25%) produced small or negligible effects, and two (12.5%) produced undesirable results.

Numerous intervention programs are based at least in part on the work of Dan Olweus (1997). The basic tenets of the Olweus program are a school-wide approach that offers universal interventions, a value system based on caring, respect, and personal responsibility, a positive approach to discipline with clear behavioral expectations and consequences, skills development, and increased adult supervision and parental involvement. Crucial to his program are early interventions that target specific risk factors and teach positive behavior and critical-thinking skills at the classroom level and intensive individual interventions that provide bullies and victims with individual support through meetings with students and parents, counseling, and sustained child and family supports. Mental health plays a vital role in Olweus's program. Stan Davis (2007) summarizes Olweus's approach to bullying prevention with a visual of a two-tiered house where the foundation of the program relies on a safe

TOPIC BOX 5.7

Jennifer (all names have been changed), Jason, and Jeff thought middle school was a particularly painful time in their lives. In seventh grade, all three students were labeled gay by a peer who taunted Jennifer for her athletic prowess on the soccer field and her seeming lack of interest in boys. The same peer targeted Jason and Jeff and would whisper "fag" at every opportunity. All three students were unaware that the others were being bullied. At the end of the school year, Jennifer asked her mother if she really was gay and just did not know it yet. Jason talked with a counselor about the bullying but asked the counselor not to talk with his teachers, whereas Jeff never told anyone about the teasing for fear that other students would tease him as well.

and affirming school climate accomplished by having consequences for aggression that are inevitable, predictable, and escalating, providing positive feedback to students and a positive feeling tone, and staff spending time with students, especially students at risk. The top tier of Davis's visual requires school staff to help aggressive youth change their behavior, support targets of bullying behavior, and empower bystanders to become involved in stopping the bullying of their peers (Davis, 2007).

Like the Olweus model, the *Steps to Respect* program emphasizes the importance of mobilizing the entire school to reduce bullying, with specific components at the school, class, and individual levels (Hirschstein and Frey, 2006). Research indicating that friendship protects children from being bullied (Hodges et al., 1999; Thompson and O'Neill, 2001) led to the program's dual focus on bullying and friendship. While students learn a variety of relationship skills, including strategies for making and keeping friends and steps for joining a group activity, the *Steps to Respect* also teaches children skills for coping with bullying, including recognizing bullying, using assertive behaviors to respond to bullying, and reporting bullying to adults. Because the school community is made up of bullies, victims, and bystanders, *Steps to Respect* program stresses that all members of a school community must take responsibility for decreasing bullying. Children are taught empathy for the victim of bullying and specific responses for children to use when they witness bullying. Staff training, which increases adult awareness of and ability to respond effectively to bullying, is a central component of the program. This comprehensive prevention strategy has received scientific support in previous studies (Olweus, 1991; Smith and Sharp, 1994) and served as the basis for the development of the *Steps to Respect* program.

Bully-Proofing Your School is a comprehensive school climate program designed to create safe and caring school communities by shifting the power away from bullies and into the hands of the caring majority of students and is available in elementary (Garrity et al., 2000) and middle school (Bonds and Stoker, 2000) versions. Changing the school's climate by using a team approach that involves administrators, teachers, support staff, students, and the community is the heart of the program. It is

---

TOPIC BOX 5.8

At the beginning of a new school year, Ms. Smith (all names have been changed) was informed that Will, a seventh grader, was frequently laughed at and isolated by his classmates. Determined to protect Will from being bullied in her class, Ms. Smith talked with Will and assured him that he should come to her whenever he was being bullied and she would intervene. During the first four months, she monitored Will very closely, intervening when Will complained and supporting him emotionally. However, she began to notice that Will's complaints were becoming more frequent. As she focused her attention more on the other students, she felt that Will was interpreting any student's laughter as "laughing at him." If all the seats at a table were taken when he approached, he saw it as, "they won't let me sit with them." Ms. Smith felt frustrated and wondered whether she was helping Will.

a comprehensive program with five equally important components. To assist adults in determining when to intervene with specific behaviors, Garrity et al. (2000) developed a chart of bullying behaviors based on severity, (mild, moderate, and severe) across six categories (physical aggression, social alienation, verbal aggression, intimidation, racial and ethnic harassment, and sexual harassment). An invaluable training tool, Garrity's visual, which operationally defines specific bullying behaviors, helps teachers and administrators understand the range of behaviors that constitute bullying. With research into illegal and violent behavior suggesting that less serious behaviors are often a prerequisite for more serious behaviors (Elliott, 1994), the chart is particularly useful in the context of developing appropriate responses to early-stage bullying behaviors.

As a school counselor in Maine in 1996, Stan Davis began to question how the school responded to bullying behaviors. In his book, *Schools Where Everyone Belongs: Practical Strategies for Reducing Bullying*, Davis acknowledges that as he shaped his current approach to bullying he was influenced most by the work of Dan Olweus (Davis, 2007). He lists five components in creating a school environment where bullying is minimal:

- Establishing clear rules and procedures about bullying
- Training adults in the school to respond sensitively and consistently to bullying
- Training adults to increase support for socially responsible student behavior
- Improving adult supervision, particularly in less-structured areas such as the playground and cafeteria
- Improving parental awareness and involvement in working on the problem (Davis, 2007)

PREVENTION AND INTERVENTION RESEARCH RESULTS

Universal programs administered to the entire school population with the goal of increasing awareness about bullying and reducing bully behaviors are the basis of most school-based antibullying efforts (Swearer et al., 2010). Some researchers have demonstrated significant and positive outcomes for bully prevention programs (Olweus, 1993; Olweus et al., 1999; Epstein et al., 2002; Frey et al., 2009). However, the lack of positive results from other research efforts (Limber et al., 2004; Smith et al., 2004; Bauer et al., 2007; Vreeman and Carroll, 2007; Merrell and Isava, 2008) suggests that the jury is still out on the overall effectiveness of bullying prevention programs.

To produce a stronger conclusion than can be provided by any individual study, researchers combine several studies that address a shared research hypotheses through meta-analysis. Glass (1976) defines meta-analysis as the statistical analysis of a large collection of analysis results for the purpose of integrating the findings. A meta-analysis by Smith, Schneider, Smith, and Ananiadou (2004) of fourteen studies on the effectiveness of whole-school antibully programs discovered that 86% of the published studies showed that the antibully programs had no benefit or made the problem even worse. Only 14% of the published studies showed that the program produced a minor reduction in bullying. Similarly, a meta-analysis of sixteen studies published from 1980 to 2004, which included data from over fifteen thousand students in Europe, Canada, and the United States, showed positive effects for only one-third of the study variables (Merrell et al., 2008).

Although knowledge, attitudes, and perceptions about bullying were positively affected, the analysis did not find that bullying behaviors were reduced. Additionally, Ttofi et al. (2008) evaluated thirty studies, thirteen of which were based on the Olweus Bully Prevention Programme. The researchers identified twenty intervention elements contained in antibullying programs and determined the frequency of occurrences for each program reviewed. Not surprisingly, the researchers found that the larger the number of intervention elements included in the program, the greater the likelihood that bullying was reduced. Ttofi and Farrington (2011) concluded that the most effective programs were inspired by the work of Dan Olweus and worked best "in Norway, specifically, and in Europe more generally and they were less effective in the USA and Canada" (p. 42).

The limited empirical support regarding the effectiveness of antibullying programs leads to the following question: How did the implementation of the program compare to the original program design? Program fidelity is defined as "the degree of fit between the developer-defined elements of a prevention program, and its actual implementation in a given organization or community setting" (Center for Substance Abuse Prevention, 2001).

> Evidence-based programs are developed and tested over time using theory to build the program components. It is implementation of these program components that is expected to influence program outcomes. Fidelity is the faithful implementation of the program components. Deviations from, or dilution of the program components, could have unintended consequences on program outcomes (Mihalic, 2001, p. 1).

The careful attention to the delivery of a program as it was designed can produce more positive results than programs that are less faithful to the fundamental design (Mihalic, 2001).

Additionally, anonymous self-report is used to measure outcomes in almost all evaluation of school-based interventions. Swearer et al. (2010) suggest more research is needed to determine whether "self-report measures alone are sufficiently sensitive to detect changes in bullying over time" (p. 41). Although a few researchers have used observational measures (Craig and Pepler, 1997; Low et al., 2010) and Frey et al. (2009) used observation in conjunction with self-report, the time and staff costs of observation may prevent the wholesale use of observation.

INDIVIDUAL INTERVENTIONS

The school-wide approach is based on the assumption that bullying is a systemic problem; therefore, interventions are directed at the entire school population. However, students who bully and students who are bullied may need more than the universal curricular activities designed to instill pro-social skills in all children. The oft-heard refrain of "stop the bullying behavior" cannot be interpreted to mean that is the *only* focus for a school. In addition to teaching that bullying is wrong, teaching the skills children need to behave differently has to be a concomitant focus. Interventions should address underlying causes. Bullies and victims may need additional skills development or reinforcement on how to apply the skills they have (Feinberg, 2003).

Countering positive attitudes towards violence is a key for intervention with students who bully. Changing a bully's behavior requires multiple, intensive, focused approaches used over time, for example, training to learn new peer interaction skills, anger and impulse control training, developing empathy, and cognitive restructuring to correct thinking errors (Feinberg, 2003; Boyle, 2005). Students who bully may also be exhibiting signs of more serious problems, such as depression, an anxiety disorder, or being victimized at home, and interventions related to the diagnosis should be prescribed. Group treatment and treating for low self-esteem are not recommended for use with children who bully (Boyle, 2005).

Beyond protection and support, targets of bullying behaviors should be taught skills and strategies to help them avoid being victimized. This is not an attempt to "blame the victim" for the behavior but to empower the victim with tools that can be used when no adult is around to intervene. Targets should be taught to generate appropriately assertive responses to frustrations, provocations, and perceived provocations (Boyle, 2005). Social skills training, including friendship skills, initiating conversations, joining in, strategies to avoid provoking a bully, reading social cues, or walking away, will also benefit most victims. Some targets will also require treatment for depression and anxiety (Boyle, 2005). Izzy Kalman, author of *Bullies to Buddies*, wants educators to teach children not to be victims by equipping students with the skills to handle the challenges of bullying, not protecting them (Delisio, 2007).

PARTING WORDS

Clearly, bullying in schools has captured the nation's attention and rightfully so. Our concerns have been validated in numerous studies, but we are still struggling with finding prevention and intervention efforts that consistently produce positive outcomes. There is still much to do, but the kids we do it for are reason enough to keep us working.

EXERCISES

1. Smith et al. (2004) suggest that inconsistent research findings "reflect a reasonable rate of return on the investment inherent in low-cost, non-stigmatizing primary prevention programs" (p. 557). Given that bullying behaviors are probably highly resistant to change, what components of a school-wide antibullying program would you suggest a school implement?
2. Measurement of program effectiveness in bully prevention has largely used anonymous surveys of students. What are the problems inherent in using self-report measures?
3. Considering that bullying behaviors range from mild to moderate to severe, what are the implications for discipline measures?

REFERENCES

Ali, R. (2010). Dear colleague letter. Accessed October 26, 2010. http://www2.ed.gov/print/about/offices/list/ocr/letters/colleague-201010.html.

American Association of University Women Educational Foundation. (2001). *Hostile Hallways: Bullying, Teasing, and Sexual harassment in Schools*. AAUW Educational Foundation, Washington, D.C.

Americal Psychological Association. (2004). Resolution on bullying among children and youth. Adopted by the APA Council of Representatives in July 2004. Accessed June 11, 2012. http://www.apa.org/about/policy/bullying.pdf.

Baldry, A. C., and Farrington, D. P. (2007). Effectiveness of programs to prevent school bullying. *Victims and Offenders*, 2, 183–204.

Bauer, N. S., Lozano, P., and Rivara, F. P. (2007). The effectiveness of the Olweus Bullying Prevention Program in public middle schools: A controlled trial. *Journal of Adolescent Health*, 40, 266–274.

Baumeister, R. F., Twenge, J. M., and Nuss, C. (2002). Effects of social exclusion on cognitive processes: Anticipated aloneness reduces intelligent thought. *Journal of Personality and Social Psychology*, 83, 817–827.

Bond, L., Carlin, J. B., Thomas, L., Rubin, K., and Patton, G. (2001). Does bullying cause emotional problems? A prospective study of young teenagers. *British Medical Journal*, 323, 480–484.

Bonds, M., and Stoker, S. (2000). *Bully-Proofing Your School: A Comprehensive Approach for Middle Schools*. Sopris West, Longmont, CO.

Boyle, D. J. (2005) Youth bullying: Incidence, impact, and interventions. *Journal of the New Jersey Psychological Association*, 55(3), 22–24.

Carpenter, D., and Ferguson, C. (2009). *The Everything Parent's Guide to Dealing with Bullies: From Playground Teasing to Cyber-bullying, All You Need to Ensure Your Child's Safety and Happiness*. Avon Books, Avon, MA.

Center for Substance Abuse Prevention. (2001). *Finding the Balance: Program Fidelity and Adaptation in Substance Abuse Prevention: Executive Summary of a State-of-the-Art Review*. Substance Abuse and Mental Health Services Administration, Center for Substance Abuse Prevention, Rockville, MD.

Cohn, A., and Canter, A. (2003). Bullying: Facts for schools and parents. Accessed May 2, 2012. http://www.nasponline.org/resources/factsheets/bullying_fs.aspx.

Craig, W. M. (1998). The relationship among bullying, victimization, depression, anxiety, and aggression in elementary school children. *Personality and Individual Differences*, 24, 123–130.

Craig, W. M., and Pepler, D. J. (1997). Observations of bullying and victimization in the school yard. *Canadian Journal of School Psychology*, 12, 41–59.

Davis, S. (2007). *Schools Where Everyone Belongs: Practical Strategies for Reducing Bullying*. Research Press, Champaign, IL.

Delisio, E. (2007). Ending bullying by teaching kids not to be victims. *Education World*. Accessed May 2, 2012. http://www.educationworld.com/a_issues/chat/chat185.shtml.

Duke, N. N., Pettingell, S. L., McMorris, B. J., and Borowsky, I. W. (2010). Adolescent violence perpetration: Associations with multiple types of adverse childhood experiences. *Pediatrics*, 125(4), e778–e786. Accessed June 11, 2012. http://pediatrics.aappublications.org/content/125/4/e778.full.pdf+html.

Elliott, D. (1994). Serious violent offenders: Onset, developmental course, and termination. *Criminology*, 32, 701–722.

Epstein, L., Plog, A. E., and Porter, W. (2002). Bully-proofing your school: Results of a four-year intervention. *Emotional and Behavior Disorders in Youth*, 2(3), 55–56, 73–78.

Farrington, D. (1993). Understanding and preventing bullying. In *Crime and Justice: A Review of Research* (Vol. 17), edited by M. Tonry. University of Chicago Press, Chicago, IL.

Feinberg, T. (2003). Bullying prevention and intervention. *Principal Leadership Magazine*, 4, 1–4.

Frey, K. S., Hirschstein, M. K., Edstrom, L. V., and Snell, J. L. (2009) Observed reductions in school bullying, nonbullying aggression and destructive bystander behavior: A longitudinal evaluation. *Journal of Educational Psychology*, 101, 466–481.

Garrity, C., Jens, K., Porter, W., Sager, N., and Short-Camilli, C. (2000). *Bully Proofing Your School: A Comprehensive Approach for Elementary Schools*. Sopris West, Longmont, CO.

Gini, G., and Pozzoli, T. (2009). Association between bullying and psychosomatic problems: A meta-analysis. *Pediatrics*, 123, 1059–1065.

Glass, G. V. (1976). Primary, secondary, and meta-analysis of research. *Educational Researcher*, 5, 3–8.

Harris, S., Petrie, G., and Willoughby, W. (2002). Bullying among 9th graders: An exploratory study. *NASSP Bulletin*, 86, 3–14.

Hershberger, S. L., and D'Augelli, A. R. (1995). The impact of victimization on the mental health and suicidality of lesbian, gay, and bisexual youths. *Developmental Psychology*, 31, 65–74.

Hirschstein, M. K., and Frey, K. S. (2006). Promoting behavior and beliefs that reduce bullying: The "Steps to Respect" Program. In *The Handbook of School Violence and School Safety: From Research to Practice*, edited by S. R. Jimerson and M. J. Furlong. Erlbaum, Mahwah, NJ.

Hodges, E. V. E., Boivin, M., Vitaro, F., and Bukowski, W. M. (1999). The power of friendship: Protection against an escalating cycle of peer victimization. *Developmental Psychology*, 35, 94–101.

Hoover, J. H., Oliver, R. L., and Thomson, K. A. (1993). Perceived victimization by school bullies: New research and future direction. *Journal of Humanistic Education and Development*, 32, 76–84.

Janssen, I., Craig, W., Boyce, W., and Pickett, W. (2004). Associations between overweight and obesity with bullying behaviors in school-aged children. *Pediatrics*, 113(5), 1187–1194.

Kaltiala-Heino, R., Rimpela, M., Marttunen, M., Rimpela, A, and Rantanen, P. (1999). Bullying, depression, and suicidal ideation in Finnish adolescents: School survey. *British Medical Journal*, 219; 348–351.

Kochenderfer, B. J., and Ladd, G. W. (1996) Peer victimization: Cause or consequence of school maladjustment? *Child Development*, 67, 1305–1317.

Lacey, A., and Cornell, D. (2011). The impact of bullying climate on schoolwide academic performance. Virginia Youth Violence Project. Accessed June 11, 2012. http://www.virginia.edu/uvatoday/pdf/impact_of_bullying_2011.pdf.

Limber, S. P. (2002). Addressing youth bullying behaviors. In *The Proceedings of the Educational Forum on Adolescent Health on Youth Bullying*. American Medical Association, Chicago, IL.

Limber, S. P., Nation, M., Tracy, A. J., Melton, G. B., and Flerx, V. (2004). Implementation of the Olweus Bullying Prevention Programme in the southeastern United States. *Bullying in Schools: How Successful Can Intervention Be?*, edited by P. K. Smith, D. Pepler, and K. Rigby, 55–79. Cambridge University Press, Cambridge, UK.

Low, S., Frey, K., and Brockman, C. (2010). Gossip on the playground: Changes associated with universal intervention, retaliation beliefs, and supportive friends. *School Psychology Review*, 39, 536–551.

Lyznicki, J. M., McCaffree, M. A., and Robinowitz, C. B. (2004). Childhood bullying: Implications for physicians. *American Family Physician*, 70, 1723–1730.

Marini, Z., Fairbairn, L., and Zuber, R. (2001). Peer harassment in individuals with developmental disabilities. Towards the development of a multidimensional bullying identification mode. *Developmental Disabilities Bulletin*, 29, 170–195.

McKenna, M., Hawk, E., Mullen, J., and Hertz, M. (2011) Bullying among middle and high school students. *Morbidity and Mortality Weekly Report*, 60(15), 465–471.

Merrell, K. W., Gueldner, B. A., Ross, S. W., and Isava, D. M. (2008). How effective are school bullying intervention programs? A meta-analysis of intervention research. *School Psychology Quarterly*, 23(1), 26–42.

Mihalic, S. (2001) The importance of implementation fidelity. *Blueprint News*, 2, 1, 1–2.

Nabuzka, O., and Smith, P. K. (1993) Sociometric status and social behaviour of children with and without learning difficulties. *Journal of Child Psychology and Psychiatry*, 34, 1435–1448.

Nansel, T., Overpeck, M., Pilla, R., Ruan, W., Simons-Morton, B., and Scheidt, P. (2001). Bullying behaviors among US youth: Prevalence and association with psychosocial adjustment. *Journal of the American Medical Association*, 285(16): 2094–2100.

National Association of School Psychologists. (2006). Position statement on school violence. Approved by the National Association of School Psychologists Delegate Assembly.

Olweus, D. (1978). *Aggression in the Schools: Bullies and Whipping Boys*. Hemisphere, Washington, D.C.

Olweus, D. (1991). Bully/victim problems among schoolchildren: Basic facts and effects of a school-based intervention program. In *The Development and Treatment of Childhood Aggression*, edited by D. J. Pepler and K. H. Rubins, 11–48, Psychology Press, New York.

Olweus, D. (1993). *Bullying at School: What We Know and What We Can Do*. Blackwell, Oxford, UK.

Olweus, D. (1995). Bullying or peer abuse at school: Facts and intervention. *Current Directions in Psychological Science*, 4, 196–200.

Olweus, D., Limber, S., and Mihalic, S. (1999) *The Bullying Prevention Program: Blueprints for Violence Prevention*. Center for the Study and Prevention of Violence, Boulder, CO.

Paterson, J. (2011). Bullies with byte. *Counseling Today*. Accessed March 3, 2012. http://ct.counseling.org/2011/06/bullies-with-byte/.

Rigby, K. (2002) *New Perspectives on Bullying*. Jessica Kingsley, London.

Rofes, E. E. (1994). Making our schools safe for sissies. *High School Journal*, 77, 37–40.

Simmons, R. (2002) *Odd Girl Out: The Hidden Culture of Aggression in Girls*. Harcourt, Orlando, FL.

Simmons, R. (2004) *Odd Girl Speaks Out*. Harcourt, Orlando, FL.

Smith, J., Schneider, B., Smith, P., and Ananiadou, K. (2004). The effectiveness of whole-school antibullying programs: A synthesis of evaluation research. *School Psychology Review*, 33(4), 547–560.

Smith, P., Morita, Y., Junger-Tas, J., Olweus, D., Catalano, R., and Slee, P., eds. (1999). *The Nature of School Bullying: A Cross-National Perspective*. Routledge, New York.

Smith, P. K., and Sharp, S. (1994). *School Bullying: Insights and Perspectives*. Routledge, London.

Smith, P. K., Mahdavi, J., Cavralho, M., Fisher, S., Russell, S., and Tippett, N. (2008). Cyberbullying: Its nature and impact in secondary school pupils. *Journal of Child Psychology and Psychiatry*, 49, 376–385.

Swearer, S. M., Eseplage, D. L., Vaillancourt, T., and Hymel, S. (2010). What can be done about school bullying? Linking research to educational practice. *Educational Researcher*, 39, 38–47.

Thompson, M., and Grace, C. (2001) *Best Friends, Worst Enemies: Understanding the Social Lives of Children*. Ballantine Books, New York.

Thompson, M., and O'Neill-Grace, C. (2001). *Best Friends, Worst Enemies: Understanding the Social Lives of Children*. Ballantine Books, New York.

Ttofi, M. M., and Farrington, D. P. (2011). Effectiveness of school-based programs to reduce bullying: A systematic and meta-analytic review. *Journal of Experimental Criminology*, 7, 27–56.

Ttofi, M. M., Farrington, D. P., and Baldry, A. C. (2008). *Effectiveness of Programmes to Reduce School Bullying*. Swedish Council of Crime Prevention, Information, and Publications, Stockholm.

U.S. Secret Service. (2002). An interim report on the prevention of targeted violence in schools. Accessed May 2, 2012. http://www.secretservice.gov/ntac/ssi_final_report.pdf.

Vreeman, R. C., and Carroll, A. E. (2007). A systematic review of school-based interventions to prevent bullying. *Archives of Pediatrics & Adolescent Medicine*, 161, 78–88.

Wiseman, R. (2002). *Queen Bees & Wannabes: Helping Your Daughter Survive Cliques, Gossip, Boyfriends, & Other Realities of Adolescence*. Crown Publishers, New York.

6 Cyber Security

Michael Land

CONTENTS

> A nursery worker admitted raping a toddler in his care and luring more than 20 teenage girls into sexual traps on the Internet using a web of false identities. The 20-year-old pedophile also built up a personal library of hundreds of photographs and videos of teenage girls in sexual poses whom he met through chat rooms and social networking sites.
>
> **John Bingham**
> *The Telegraph*

The role of a school is to facilitate learning. Creating a learning environment requires the organization of faculty, staff, students, infrastructure, materials, and growing concerns of safety. An area of growing concern in school safety is the aspect of cyber safety. The education environment has become infused with technology. Students have access to technology in both instructional and personal environments that should be utilized in a safe and responsible manner. Increasingly, it is left to schools to assure students function, mindful of safety and responsibly, in a cyber-safe environment.

Providing a cyber-safe environment for schools and students is a broad topic. Educational environments have many areas to consider in facilitating a cyber-safe environment. Schools have administrative technology systems to retain everything from payroll information to student grades. They have instructional platforms, used by the teachers in the classrooms, and challengingly, schools have a role in facilitating a safe cyber environment in light of faculty, staff, and students' personally owned technology. The role of cyber safety extends from the school's network to the far-reaching networks of the Internet.

Computers and cyber technologies enhance student performance by engaging, involving, and empowering students. Cyber resources foster the development of skills, helping to ensure mastery of educational content and student use of technology every day. Schools without technology are practically nonexistent.

Today students are exposed to technology throughout their life. Both primary and secondary schools provide students basic technology literacy skills to reinforce higher-order thinking skills such as problem solving, synthesis, and analysis of information. Schools and educators incorporate a considerable amount of technology to make sure instructional resources are an enhancing experience for students. Many teachers embrace and master technology to support teaching methods and curricular goals, and they expect students to utilize that technology in a productive manner.

A large measure of student proficiency assumes that the student will use cyber resources in a secure, safe, and ethical manner. Cyber safety is part of the life skills students must possess. However, in most cases, the students are going to have had little formal direction in how to be safe and ethically sound in the cyber environment outside of the classroom. Increasingly it has become the school's role to ensure a safe cyber environment both in and out of the classroom (Gibbs, 2010).

Very few educators teach students about online safety, security, and ethics as part of the learning process. A recent study shows that only 15% of educators taught students about hate speech, and only 18% provided any instruction regarding how to deal with negative posts, videos, or other Internet content. Just 26% taught kids how to handle incidents of cyberbullying. One-third of teachers covered risks with social networking, and another third taught students about safely sharing information on the Internet. However, only 6% taught students about the safe use of geolocation services despite the rise of its use in Internet-enabled mobile devices (National Cyber Security Alliance, 2011).

The diffusion of Web 2.0 technologies and smartphones provide another interesting dimension to cyber safety. Students, as well as faculty and staff, have the ability to access, communicate, and process data via a robust network platform that they can hold in their hands or store in their pockets. Web 2.0 has made the Internet an instantaneous, participatory, interactive community differing from the initial incarnation of the Internet as an environment where you could gather (or post) information and communicate via asynchronous systems. Smartphones, utilizing mobile computer operating systems, provide the user the ability to interact via Web 2.0 applications synchronously. Web 2.0, combined with smartphones, has provided advanced computing and communication tools to its users and an additional challenge to maintain a cyber-safe environment in schools.

In addressing the assurance of a cyber-safe environment, the first section of this chapter will be viewed from three cyber environments: the Internet, social networking, and smartphones each contain their own purpose, attributes, significance, value, and subsequent risks. The second section of this chapter focuses on creating a cyber-safe environment by implementing measures to cultivate secure, safe, and ethical cyber practices.

It is not the intent of this chapter to provide technical instructions for developing a cyber-safe environment in schools. Aspects such as firewalls, malicious code, and access controls need to be addressed by network administrators who can configure those protection systems properly. However, it is proposed that this chapter enlighten educators with a knowledge base that will allow them to understand the various factors that construct the school's cyber environment, so they can articulate an approach reducing a school's cyber risks. Minimizing cyber risks in schools can be best addressed by making educators aware of their role in the process. Because of schools' and educators' place in the lives of children, it is imperative that cyber safety be a portion of the educational process.

THE INTERNET

The Internet is a world-wide public computer network. The Internet was originally founded by the Department of Defense in the 1970s using transmission control protocol/Internet protocol to connect computers and networks. Since the inception of Hyper Text Markup Language (HTML) in the early 1990s, the Internet evolved from text-based communications platforms to graphically interfaced web pages. HTML fueled the development of the Internet into websites that would be publicly accessible, be hosted via Internet-connected network servers, and allow for other media files rather than mere text to be accessed.

The use of the Internet in schools grew rapidly from the inception of HTML. By 1999, nearly 100% of public schools in the United States had access to the Internet compared with 35% four years prior. In 2005, 94% of public school instructional rooms had Internet access compared with 3% in 1994 (Wells and Lewis, 2006). The use of the Internet has since moved from being an instructional platform to a participatory platform used by most students.

The Internet has become a paradox to educators. The Internet is a great resource, full of endless amounts of information and resources. Use of e-mail, chatting, and browsing (for educational research and personal use) via the Internet has diffused throughout the lives of students. Student Internet access exists not only in the school. Research shows that 95% of teens have Internet access and 82% have broadband Internet access at home (Rainie, 2011). Although the Internet provides many positive attributes for students, some sites may contain data to which children should not be exposed. It is easy to find sites that contain pornographic materials, drug information, and any deviate activity imaginable. If that is not enough, pedophiles and criminals often utilize the Internet to find their victims. The paradox is that while the Internet is an environment that is beneficial to modern education systems, it also contains many risks and hazards to students.

E-MAIL

E-mail in the educational environment is frequently used by students, staff, and faculty for sending and receiving electronic messages. E-mail allows the student to keep in touch with teachers, family, friends, and peers. It facilitates getting help with homework, establishing mentoring relationships, receiving online newsletters, and general communication in an asynchronous format.

Potential risk with e-mail is an inherent quality of the Internet. It utilizes a public network architecture where you can communicate with other e-mail users indiscriminately. There is no required validation that users are who they say. Furthermore, e-mail users can communicate with others through unsolicited messages. These unsolicited messages, or spam, can variably include sexually explicit material, products for sale, or moneymaking schemes or serve as a host for a malicious program.

Another facet of e-mail involves a substantial movement by school systems to use cloud-based e-mail systems. Cloud-based services allow students, faculty, and staff to access applications on a network platform that is leased or rented to schools similar to a utility or tenement arrangement. The cloud extends the system so that students, faculty, and staff have constant access to not only their e-mail but also personal files and applications software anywhere they have a computer and Internet access. For example, an assignment using an application does not have to be installed on the remote computer, nor does it have to be connected to a local area network to function. The application is run from the cloud, and the data generated are stored on the cloud, which can be accessed anytime by anyone with Internet service and proper access credentials.

Both Microsoft and Google are providing schools with free e-mail for their students, as well as online communications, applications, and storage. Many schools are using the cloud-based networks because they have a low cost and require only limited personnel while providing a service to the school. There are risks associated with the movement to cloud-based e-mail systems. Although it is easy to monitor servers when they are run by internal data centers and are completely under the control of the school's information technology department, it is more difficult when you have little control over servers that are located somewhere in the cloud. Therefore, it is important to measure and analyze not only the performance but also the safety and security of hosted information. It is very important that students understand this and know how to respond to risks.

INTERNET RESEARCH

Internet browsing provides students, faculty, and staff the means to explore information on world-wide computer networks, usually by using a browser such as Microsoft Internet Explorer, Google Chrome, or Firefox. The browser allows access to rich educational and cultural resources (text, sounds, pictures, and video). This also gives users an improved ability to understand and evaluate information and stay informed by accessing websites.

Risks associated with Internet research relate to sites with inaccurate or misleading information. There is also access to sexually explicit images and other sites promoting hatred, bigotry, violence, drugs, cults, and other things not appropriate for students. The Internet in general has no restrictions on marketing products such as alcohol and tobacco to children. Some Internet sites deceptively collect personal information from kids in order to sell products to them or their parents via requests for personal information for contests and surveys. The Internet is a relatively wide-open interface to share data without any form of censorship.

Another rampant problem regarding Internet research involves plagiarism by students. A national survey published in *Education Week* found that 54% of students admitted to plagiarizing from the Internet; 74% of students admitted that at least once during the past school year they had engaged in "serious" cheating, and 47% of students believe their teachers sometimes choose to ignore students who are cheating (Plagiarism Dot Org, n.d.). This is an issue that follows the primary and secondary student into college. According to the *New York Times*, students raised during the Internet age have developed an extremely lax attitude towards stealing others' work (Gabreil, 2010).

ONLINE CHATTING

Online chatting is a popular communications tool used by many students. Online chatting is reading messages from others as they are typing them, usually in a theme or social network-specific interface. The inclusion of chatting in social networking sites has helped to maintain its appeal as a widely used communications medium. Chatting is popular because it allows users to communicate with people from around the world by synchronous typing of text into a chat interface. Students can connect to others via websites or social networking portals.

Chatting is an especially risky environment because it provides an interface where people can communicate, in real time, with as much anonymity as they desire. Because of its nature, chatting has become the most likely activity online through which children will encounter people who may want to harm them (Wolak, 2008). Predators target chat rooms to engage students in conversation and then lure them into meeting. Social networking sites add a new dimension to chatting because of the student's online profile. Online profiles make searching different demographic groups simple and easy. Predators will befriend children via social networking sites, usually posing as another child or slightly older teen, and gain trust by behaving as an understanding and trusted friend. Once trust is gained in the chat room, the predator will move the conversation to a private area or in person.

Another negative aspect of chatting is that it is a place where bullies can abuse potential victims. Bullies often start out as friends and then get students to give their e-mail address, where the bullying can continue. Or the bullying may occur via instant messaging or in the chat room itself. It may occur on social networking sites attached to the chat interface, where bullies may post untrue and damaging information (MacDonald, 2008). Chatting is sometimes used to post links to pornography. Students may click on links and be taken to an inappropriate site. It is easy to understand that while online, chat rooms have many risks in contrast to their positive attributes.

SOCIAL MEDIA

The rise of Web 2.0, and most notably online social media, has had a profound effect on students. Web 2.0 combined with broadband Internet access has changed the way students communicate, process, and store data. Although the diffusion of Web 2.0 has impacted users across all demographic groups, none has been as greatly affected as students. Web 2.0 technologies include social networking sites, blogs, wikis, video-sharing sites, hosted services, web applications, and tags. For users, these tools, which are typically free or low cost, represent a transition from institutionally provided to freely available technology.

Web 2.0 technologies are possible because of the adaption of programming tools, such as asynchronous JavaScript (AJAX). AJAX is a group of interrelated web development tools used to program interactive web applications. This programming aspect, combined with diffusion of residential broadband Internet access, created an environment that evolved and migrated users to a very interactive form of the Internet that facilitated interactive social networking. This AJAX-infused adaptation of the Internet is simply referred to as Web 2.0.

Web 2.0 is the current rendition of the Internet that provided a "social" approach to generating and distributing Internet content, characterized by open communication, decentralization of authority, and freedom to share and reuse information (Acar, 2008; Subrahmanyam et al., 2008; Madge et al., 2009). The socialization through Web 2.0 interaction is exemplified by photo and video sharing through such web applications as Photobucket, Flicker, and YouTube, and the use of wikis and blogs in such platforms as Wikipedia, TripAdvisor, and UrbanSpoon, as well social networking sites such as Facebook and Twitter.

Online networking has proliferated through the Web 2.0 environment because of an interactive design attribute that has been further propagated by increased residential Internet access and bandwidth (Acar, 2008; Subrahmanyam et al., 2008; Madge et al., 2009). The Internet today has developed into a network of participation that typifies online social networking. There are a variety of online social networks in existence, the most popular being Facebook and Twitter.

Social networking sites combine common characteristics that allow profile creation, friend listing, and public viewing of friend lists. Online social networking websites also allow users to create their unique web presence referred to as their profile. Through the profile, the social network users live an online identity while exploring friendships and relationships with other individuals who also have profiles on that website. Online social networking does not function entirely in real time, like conventional chat rooms and instant messaging, so the interactions that take place are not always instantaneous, even though most have chat room functionality.

Most social networking profiles are developed from responses to questions that request a user to disclose a variety of personal information (Mazer et al., 2007; Mitrano, 2008; Steinfeld et al., 2008). Personal information includes user names or other identifiers such as sexual preference, schools, geographical location, and the extent of the relationship students are currently in or seeking with others. The profile also allows for self-expression through personal photographs and videos. Students can also make available their list of friends and member groups, as well as create an area where individuals can post remarks. The individual's social networking profiles have distinct web addresses that can be bookmarked or linked, allowing others to use and share that data with third parties.

The most popular online social network for students is Facebook, an online directory that connects people through social networks. The website www.facebook.com (Facebook) was initially designed and developed in 2004 by Mark Zuckerburg, a Harvard University sophomore, and was inspired by a widely known paper version of a college face book. The directory consisted of individuals' photographs and names and was distributed at the start of the academic year by university administrations with the intention of helping students get to know each other. Zuckerburg's initial intention was to create an online website to help Harvard co-eds get to know one another for the purpose of finding roommates (Shier, 2005). It is no surprise that the Facebook website has grown in popularity among students because it was initially designed exclusively for students.

From its creation in 2004, Facebook has grown from hundreds of users to over five hundred million active users (*Digital Buzz*, 2011). Facebook was originally designed for college students but is now open to anyone thirteen years of age or older. In 2006, Facebook lifted its educational organization requisites, where users had to have an e-mail address with an education suffix (i.e., an e-mail address ending in .edu). With that development, there was a mass movement to Facebook as the social networking portal of choice for most social network users (Mazer et al., 2007, 2009).

The Web 2.0 aspect of Facebook provides a tool for friends to keep in touch and for individuals to have a presence on the Internet without needing to build a website. Facebook makes it easy to upload pictures and videos, making its use so simple that nearly anyone can publish a multimedia profile (Mitrano, 2008). Facebook has made it easy to find friends using e-mail addresses, to search by name, or to pull up listings based on a variety of demographic variables. With a public profile on Facebook, a student can be found by the other five hundred million users.

Each Facebook profile has a "wall," where friends can post comments. Because the wall is viewable by all of the student's friends, wall postings are basically a public conversation. By default, students can write personal messages on friends' walls or send a person a private message that will show up in their private inbox similar to an e-mail message. Facebook offers tools to develop and maintain relationships that are of particular importance in emerging adulthood. Recently the use of messaging via online social networking has surpassed e-mails as the primary means of communication between students (Solis, 2009).

Facebook allows each user to set privacy settings. For example, students can have their privacy levels set so that not all other users will be able to view their profile; that is, a student's profile may only be viewed by someone the student has added as a friend. A student can adjust the privacy settings to allow other users or peers to view portions or all of the profile. Users can also create a limited profile, which allows them to hide certain parts of the profile from a list of users that an individual selects.

Another feature of Facebook that students like is the ability to add gaming applications to a user page. Facebook applications are programs developed specifically for Facebook profiles. The most popular of these programs include interactive games such as Farm Town, Mafia Wars, and thousands of other interactive multiuser applications. Because most game applications save scores or assets, friends can compete against each other or against millions of other Facebook users.

Many social networking sites require that users are at least thirteen years old. This includes popular sites like Facebook and YouTube, all of whom ask users to confirm that they meet this age requirement when setting up an account. In addition, other websites that contain adult-oriented material such as alcohol-related advertising or sexually explicit material may require the user to be at least eighteen or twenty-one years of age.

Requiring users to be of age does not seem to be a deterrent in social networking access. A recent study (Lenhart et al., 2011a) shows many online teens (44%) admit to lying about their age so they could access a website. Boys and girls are equally likely to say they were older to gain access to a website or service. The youngest group, ages twelve and thirteen, are more likely than seventeen-year-olds to say they have lied about their age (49% versus 30%). Teens who maintain public profiles on social network sites are far more likely than those who have private profiles to report lying about their age (62% versus 45%) (Lenhart et al., 2011a).

Another problem with online social networks relates to the occasional cruel behavior between students. Teenagers who use social networking sites say their peers are mostly kind to one another on such sites (69%). Still, 88% of these teens say they have witnessed people being mean or cruel to another person on the sites, and 15% report that they have been the target of mean or cruel behavior on social network sites. Among the 15% of social media-using teens who have experienced cruelty or mean behavior on social network sites, there are no statistically significant variances by age, gender, race, socioeconomic status, or any other demographic characteristic measured (Lenhart et al., 2011b).

Providing a cyber-safe environment in schools is ever challenging not only because of the dangers inherent in the Internet and social networking but the added dimension of students bringing their personal technology into the educational environment, which can access a host of technological resources. Students are increasingly utilizing their personal technology, most notably smartphones, to access social networking sites.

Smartphones

It has become commonplace for students to have a smartphone by the time they reach high school. The use of smartphones exceeds the use of traditional computers in accessing the Internet (Albanesius, 2011; Weintraub, 2011). A smartphone is a cellular phone that combines the functions of a networked computing system and a mobile phone. The smartphone hosts a wealth of technology and applications (apps) that are as robust as a PC in many situations. In addition to the standard audio and text capabilities, smartphones typically serve as video/still cameras, audio recorders, media players, and mobile computers. They incorporate apps and browsers that can do a number of things ranging from video conferencing, global positioning system navigation, and social networking via wi-fi and broadband Internet protocol network access.

Research shows that teenagers are much more likely to own a smartphone than adults and are more obsessed with the technology. Forty-eight percent of teenagers own a smartphone compared with just 27% of adults. Of those who own smartphones,

60% of teenagers describe themselves as "highly addicted" to their phones (Gilbert, 2011).

A growing issue with cyber life and smartphone usage relates to geolocation tools such as Facebook Places. These services have become very popular apps, especially in the social networking environment. The apps allow users to "check-in" to locations (neighborhood businesses) via their mobile phone. Their location is then sent to their friends and in many cases includes a map showing their exact location.

Geolocating is used primarily as a marketing tool for businesses, giving reduced cost or free merchandise for app users who visit their store. However, there are some obvious safety considerations. Every time someone checks into a location publicly, they are telling the world exactly where they are at that point in time. Location sharing could encourage stalking, as well a host of other hazardous issues because the user is broadcasting their physical location via the Internet or social media site.

Texting also needs to be considered when looking at smartphones and cyber safety. Technically, texting is not an Internet-based technology. However, it has become widely diffused and has become synonymous with modern youth in the proliferation of cell phone technology. In a 2004 survey of teens, 18% of twelve-year-olds owned a cell phone. In 2009, 58% of twelve-year-olds owned a cell phone, whereas 83% of teens aged seventeen owned a cell phone, up from 64% in 2004 (Goldberg, 2010).

Text messaging has become the preferred method of communication for American teenagers, with one in three teens sending more than one hundred texts a day (Goldberg, 2010). Texting, however, like other cyber technologies with many positive benefits for students, can also create problems. Fifteen percent of cell phone-owning teens say they have received sexual images or nearly nude images of someone they know via text messaging. Older teens are much more likely to send and receive these images; 8% of seventeen-year-olds with cell phones have sent a sexually provocative image by text, and 30% have received a nude or nearly nude image on their phone (sexts). Teens who pay their own phone bills are more likely to send "sexts": 17% of teens who pay for all of the costs associated with their cell phones send sexually suggestive images via text (Ngo, 2009).

Wolford (2011) linked sexting to serious psychological problems in a sample of twenty-three thousand students in the Boston area, finding that "sexting can include overtones of bullying and coercion, and teens who are involved were more likely to report being psychologically distressed, depressed or even suicidal" (http://www.webpronews.com/sexting-linked-to-depression-suicide-2011-11). In fact, twice as many teens who reported sexting had depressive symptoms compared with teens who said they did not sext. Thirteen percent of sexting teens reported a suicide attempt in the last year compared with only 3% of nonsexting students. Not only is sexting a problem, but it has a relationship as a variable in other factors of safety concerns for students.

This section has attempted to provide an understanding of cyber use by students with both positive attributes and consequent risk. Providing a technology-safe environment is a very broad topic and includes many aspects of student life. It is also obvious that technology will continue to play a large part in the educational environment and in the lives of students. Therefore, educators should understand the potential role schools can have in facilitating a safe cyber environment.

PROVIDING A SAFE CYBER ENVIRONMENT

The second part of this chapter addresses the role of educators in providing a cyber-safe environment for students. Providing cyber safety is contingent upon educators understanding the environment, risks, and vulnerabilities of technology. Because of the breadth of the issues and complexity of the environment, there is no more valid approach to achieving a safe cyber environment in schools than to teach it to students throughout their education. This approach will use the educator to do what they do best in promoting secure, safe, and ethically sound practices by students. Cyber safety is a life skill that students must possess.

It is not the intent of this section to address cyber safety from the view of a network administrator but to make the educator aware of what is needed to make schools' cyber environments safe. A school must have both policies and procedures to provide for authentication, firewalls, and virus protection on its computer systems. Authentication will provide secure login and allow individuals to access only the data and applications they require. The firewall also serves as an access control device that will limit user access to unwanted programs and data. Virus protection programs will protect users from viruses and malware on the network. Each of these three areas must be addressed by computing professionals in that environment. The best application for the educator is to teach students the risks associated with the cyber environment and subsequently how to make themselves safe, secure, and ethical cyber citizens.

The task of cyber safety is daunting because there are so many aspects of a student's cyber life that are out of reach of the educator. Students have technology, and they use it very frequently outside the realm of mentorship. They use technology to communicate peer-to-peer and often without knowledge, concern, or ethical understanding. Therefore, a curriculum infused with cyber security, safety, and ethics is important to provide students with the ability to make good cyber decisions. Teaching students to be good cyber citizens will better ensure they do not become cyber crime victims in the future. Because technology is diffused at all learning levels, cyber safety should be included throughout the curriculum beginning in kindergarten and extending until the student completes high school.

Basic principles should be stressed throughout their education. An example is to visit only age-appropriate sites. Students under the age of thirteen should not have a Facebook account. Stress to students to search the Internet safely. Even though a school can have a firewall and apply filtering software, inappropriate materials can still come through.

One basic principle of cyber safety is that students should always be taught to avoid strangers in the cyber world. Students must understand that people are not always who they say they are in cyberspace. If someone the student does not know talks to them, they should not respond, and they should make school staff aware. Another basic principle students should understand is that simply because they can access something on the Internet and copy it or download it, that does not make it theirs. Be it audio, video, or someone else's words, it may be acceptable to use them, but they still belong to the owner and that person deserves to be compensated even if it is merely with a citation and bibliography entry. Students must understand this concept of cyber citizenship. At any level of technological interface, students should be taught to be good cyber citizens. An underlying premise is that if students would not do something in real life, they should not do it online.

CYBER SAFETY IN ELEMENTARY SCHOOL

As technology is introduced to students, so should cyber safety education. In kindergarten, students often learn that technology offers the ability to visit new places, learn new things, and collaborate with others worldwide. Even at this young age, students need to learn rules about being safe online, such as not revealing private information. They need to know the value of information and why it should be protected. Students need to know, even at the youngest age, that computers and technology are not "safe," and there are people who will take advantage of them in those environments.

Students even at this age are enthusiastic about their technology (Saçkes, 2011). Educators should connect with that enthusiasm and relate it to being cyber safe. Encourage discussions regarding the positive and negative aspects of cyber life and cyber citizenship. In a world in which everyone is connected and anything created can be copied, pasted, and sent to millions of people, it is important that students bring a sense of ethical responsibility to cyber environments.

Students are never too young to learn how to manage their own privacy and respect the privacy of others. Students should understand the ethics of participating in and building positive online communities, as well as how cyber communities are challenged because of stalking, bullying, and other negative behaviors. Students need to understand that their role as good citizens also extends to the broader communities. It may sound facetious, but it is never too early to promote the respect of creative work, as well as copyright to fair use. Students should understand the ethics of using creative work from others as they are encouraged to practice their own creativity.

CYBER SAFETY IN MIDDLE SCHOOL

As students progress through school, they become more engaged in cyber life. Students continually need to understand the significance of communicating online and proper etiquette. As middle school students, they must learn approaches for managing their information online to keep it secure. Students should know how to guard against identity theft and keep their data safe from malware while protecting themselves from e-mail phishing. This may sound complex, but these are issues middle school students are confronting daily, and they need to know how to deal with these situations.

Adults may think of students' online and technological activities as cyber life but to them it is just life. Technology has always been a part of their existence, and they are enthusiastic about it. Educators should harness this enthusiasm and encourage students to talk about the impact of technology on their lives, their communities, and culture. Students should learn about the positive and negative aspects of cyber life and the concept of cyber citizenship. It is important that students bring a sense of ethical responsibility to the online spaces where they consume, create, and share information. Being cyber safe is a life skill that students must understand and apply.

Students need to continually learn to manage their own privacy and respect the privacy of others. Students in middle school should explore the ethics of participating in and building positive online communities, as well as understanding how communities are upset because of cyber stalking, harassment, bullying, and other damaging behaviors. Educators should explore the impact of students' individual actions, both negative and positive, on their friends and on the broader communities in which they participate. A middle school student needs to know that what they do in the cyber world can impact their physical life. It is imperative that they understand that their candor with technology can have effects far past the comfort of their private physical space. Students must know that the way they present themselves online can affect their relationships, sense of self, and reputations. Also, the way a student treats others in the cyber world can have repercussions in the physical world.

Students at the middle school level should be encouraged to introduce positive behavior in the cyber world by creating and publishing their own writing, music, videos, and artwork. They also should understand issues related to copyright and public domain cyber resources. This includes knowledge of cyber resources and how they can be used. The ability to access and use data is a life skill they must process. Subject matter may involve activities from citation of a newspaper article for school research to distribution of copyrighted music via peer networks. Students need to know what is acceptable behavior and what is not (Conradson and Hernández-Ramos, 2004).

CYBER SAFETY IN HIGH SCHOOL

In high school, students know and appreciate the value of connecting with others online. They recognize inappropriate contact and know how to respond to it. They learn why certain online relationships are risky and to avoid them. Students develop processes for managing their personal data in cyber environments.

High school students will be fully exposed to the concept of cyber citizenship and consider how they can harness the power of cyber life for good. Students continually learn to manage their own privacy and respect the privacy of others. High school students understand how information they post online can affect getting into college or other future opportunities, as well as how it might impact others.

High school students must understand the power of anonymity in the cyber world. They must consider how anonymity and public posting can intensify bullying, hate speech, and abusive relationships online and how to deal with it if it happens to them. High school students learn to think critically about how they present themselves online as they do in the physical life. Students must consider what their profiles, posts, and avatars convey to others about them and reflect on whether this image is who they want to portray. At the high school level, students should be proficient in using citation and bibliographies for Internet-based research. High school students understand that duplicating copyrighted music or video is illegal. The student must understand the value of copyrights and licensing of data as they are encouraged to consider their own opportunities for creating and using new media.

Making students into good cyber citizens is a goal for all educational levels. Computers and cyber technologies enhance student performance by engaging, involving, and empowering students, and addressing cyber safety as a part of the curriculum is essential. Moreover, it is the responsibility of primary and secondary schools to provide their students basic cyber security, safety, and ethical skills as a portion of the educational process. Schools must educate students regarding an acceptable use policy and make clear the consequences for violating it. Schools must also instill a sense of cyber responsibility. The role of a school in facilitating learning is going to be even more challenging in the future. The best time to begin establishing a culture of cyber safety in school is the first day you introduce your students to technology.

CASE STUDY

Lincoln School District in your state has recently grieved the loss of another student to suicide, the third person in the area to commit suicide in under a year. Statewide, there are on average about fifteen to twenty teen suicides each year. A majority of these cases involved students in middle school or high school who suffered cyberbullying. Some were also targeted because they identified as lesbian, gay, bisexual, or transgendered. Members of the community are calling on the district to address the issue. District administrators are split on the issue. Some are saying the school should address the issue, and others are saying they would open a Pandora's box of problems by doing so.

- Is it the role of the district to address this issue?
- Does the issue of "cyberbullying" differ from "bullying"?
- What factors must be considered in the addressing the issue?

EXERCISES

1. Expand on the statement: The Internet is a paradox for education.
2. When should students be exposed to cyber-safe instruction in schools? And why?
3. What does it mean to be "a good cyber citizen"?
4. Why should educators take a proactive role in reinforcing students as "good cyber citizens"?

REFERENCES

Acar, A. (2008). Antecedents and consequences of online social networking behavior: The case of Facebook. *Journal of Website Promotion*, 3(1/2), 62–83.

Albanesius, C. (2011). Smartphone shipments surpass PCs for first time: What's next? PCMag. com. Accessed March 4, 2012. http://www.pcmag.com/article2/0,2817,2379665,00.asp.

Bingham, J. (2011). Nursery paedophile raped toddler and snared teenagers online. Telegraph. Accessed March 4, 2012. http://www.telegraph.co.uk/news/uknews/crime/8561998/Nursery-paedophile-raped-toddler-and-snared-teenagers-online.html.

Conradson, S., and Hernández-Ramos, P. (2004). Computers, the Internet, and cheating among secondary school students: Some implications for educators. *Practical Assessment, Research & Evaluation*, 9(9), 9.

Digital Buzz. (2011). Facebook statistics: Stats and facts for 2011. Accessed July 15, 2011. http://www.digitalbuzzblog.com/facebook-statistics-stats-facts-2011.

Gilbert, D. (2011). Half of all teenagers own smartphones most addicted. Accessed March 4, 2012. http://www.trustedreviews.com/news/half-of-all-teenager-own-smartphones-most-addicted.

Gabreil, T. (2010) Plagiarism lines blur for students in digital age. *New York Times*. Accessed March 4, 2012. http://www.nytimes.com/2010/08/02/education/02cheat.html?_r=1&hp.

Gibbs, J. (2010). *Student Speech on the Internet: The Role of First Amendment Protections*. LFB Scholarly Publishing, El Paso, TX.

Goldberg, S. (2010). Many teens send 100-plus texts a day, survey says. Accessed June 9, 2012. http://articles.cnn.com/2010-04-20/tech/teens.text.messaging_1_text-messaging-cell-phones-teens?_s=PM:TECH. CNN.

Lenhart, A., Madden, M., Smith, A., Purcell, K., Zickuhr, K., and Rainie, L. (2011a). Social media and digital citizenship: What teens experience and how they behave on social network sites. Accessed March 4, 2012. http://www.pewinternet.org/Reports/2011/Teens-and-social-media/Part-2/Section-3.aspx?src=prc-section.

Lenhart, A., Madden, M., Smith, A., Purcell, K., Zickuhr, K., and Rainie, L. (2011b). Kindness and cruelty on social network sites: How American teens navigate the new world of "digital citizenship." Accessed March 4, 2012. http://pewresearch.org/pubs/2128/social-media-teens-bullying-internet-privacy-email-cyberbullying-facebook-myspace-twitter.

MacDonald, M. (2008). Taking on the cyberbullies: Hidden behind online names and aliases, they taunt, even lay down death threats. Accessed March 4, 2012. http://www.cyberbullying.ca/pdf/Taking_on_the_cyberbullies.pdf.

Madge, C., Meek, J., Wellens, J., and Hooley, T. (2009). Facebook, social integration and informal learning at university: It is more for socializing and talking to friends about work than for actually doing work. *Learning, Media and Technology*, 34(2), 141–155.

Mazer, J., Murphy, R., and Simonds, C. (2007). I'll see you on "Facebook": The effects of computer-mediated teacher self-disclosure on student motivation, affective learning, and classroom climate communication education. *Communication Education*, 56(1), 1–17.

Mazer, J., Murphy, R., and Simonds, C. (2009). The effects of teacher self-disclosure via Facebook on teacher credibility. *Learning, Media, & Technology*, 34(2), 175–183.

Mitrano, T. (2008). Facebook 2.0. *Educause Review*, 43(2), 72.

National Cyber Security Alliance. (2011). Press release: 2011 State of cyberethics, cyber-safety and cybersecurity curriculum in the U.S. survey. Accessed March 4, 2012. http://www.microsoft.com/Presspass/press/2011/may11/05-04MSK12DigitalPR.mspx?rss_fdn=Press%20Releases.

Ngo, D. (2009). Study: 15 percent of teens have gotten "sext" messages. Accessed March 4, 2012. http://news.cnet.com/8301-1023_3-10415784-93.html.

Plagiarism Dot Org. (n.d.). Facts about plagiarism. Accessed July 15, 2011. http://www.plagiarism.org/plag_facts.html.

Rainie, L. (2011). The new education ecology. Sloan Consortium Orlando. Accessed June 9, 2012. http://www.slideshare.net/PewInternet/the-new-education-ecology.

Saçkes, M., Trundle, K., and Bell, R. (2011). Young children's computer skills development from kindergarten to third grade. *Computers & Education*, 57(2), 1698–1704.

Shier, M. T. (2005). The way technology changes how we do what we do. *New Directions for Student Services*, 12(1), 77–87.

Solis, B. (2009). Social networks now more popular than email; Facebook surpasses MySpace. Accessed July 15, 2011. http://www.briansolis.com/2009/03/social-networks-now-more-popular-than.

Steinfield, C., Ellison, N., and Lampe, C. (2008). Social capital, self-esteem, and use of online social network sites: A longitudinal analysis. *Journal of Applied Developmental Psychology*, 29(6), 434–445.

Subrahmanyam, K., Reich, S., Waechter, N., and Espinoza, G. (2008). Online and offline social networks: Use of social networking sites by emerging adults. *Journal of Applied Developmental Psychology*, 29(6), 420–433.

Weintraub, S. (2011). Industry first: Smartphones pass PCs in sales. *CNN Money*. Accessed March 4, 2012. http://tech.fortune.cnn.com/2011/02/07/idc-smartphone-shipment-numbers-passed-pc-in-q4-2010/.

Wells, J., and Lewis, L. (2006). Internet access in U.S. public schools and classrooms: 1994–2005 (NCES 2007-020). National Center for Education Statistics, U.S. Department of Education, Washington, D.C.

Wolak, J., Finkelhor, D., Mitchell, K., and Ybarra, M. (2008). Online "predators" and their victims: Myths, realities and implications for prevention and treatment. *American Psychologist*, 63, 111–128.

Wolford, J. (2011). Sexting linked to depression, suicide: You heard right, nudie pics are really bad news according to study. *WebProNews*. Accessed March 4, 2012. http://www.webpronews.com/sexting-linked-to-depression-suicide-2011-11.

7 Information Security

Michael Land

CONTENTS

The Alaska Department of Education and Early Development alerted students and parents this week that an external hard drive containing more than 89,000 students' personal information had been stolen from the department's headquarters in Juneau. In addition to test scores, the Department of Education and Early Development says the stolen hard drive contains a slew of other information, including those test-taking students' names, dates of birth, student identification numbers, school and district information, gender, race/ethnicity, disability status and grade level.

Jason Lamb
Channel 2 KTUU, Alaska

The role of a school is to facilitate learning. In creating a learning environment, schools must collect a large amount of information about each student, which can include demographic, educational, and/or health data. Educational information commonly collected may contain administrative reports documenting student progress and information regarding past or current use of school services, such as special education, social work services, or other supplementary support. For example, schools that participate in a federally assisted school nutrition program collect personal information to determine student eligibility for free and reduced-price school nutrition programs. The fact is that schools must retain a great deal of student information in the educational process.

Information about students is a vital resource for teachers and school staff in planning educational programs and services, such as designing individual education programs, scheduling students into appropriate classes, and completing reports for local, state, and federal agencies. In emergency situations, information must be readily available to school officials to make a valid response to a life safety event. Schools must maintain information integrity, accessibility, and confidentially.

A student's education information is a compilation of records, files, documents, and other materials maintained by an educational agency that contain information directly related to a student. Educational information may be kept in a variety of formats including handwritten, printed, digital files, and video or audio recordings. Schools have an obligation to students and parents to maintain information integrity, accessibility, and confidentiality. Information must be accurate and available to make timely decisions regarding a student's educational progress. In emergency situations, accurate and accessible school information may be an important factor in student safety. Furthermore, the school could have legal consequences if they fail to safely maintain the information of students (U.S. Department of Education, n.d.).

Public schools are further bound by federal legislation that mandates how student information is to be maintained. For students, information safety is specifically addressed by the Family Education Rights and Privacy Act of 1974, commonly known as FERPA. FERPA is a federal law that protects the privacy of student education information. FERPA assures that students have specific rights regarding information and requires that schools strictly adhere to the guidelines. Therefore, it is imperative that schools have a plan regarding information security.

This chapter has three primary purposes. First, it is to serve as a guide to help schools look at information and assign its value. A risk assessment is a methodical tool to help practitioners look at information as assets to the organization and determine a value of that artifact and then understand the threats and vulnerability as the basis for protection. Second, this chapter will address legislative requirements outlining methods for keeping information secure. Finally, fundamental principles in applying policy and procedures in the pursuit of safekeeping of information will be addressed.

SAFEGUARDING INFORMATION

There is little question that schools must securely maintain educational information that is collected. Protecting information requires maintaining confidentiality and integrity while ensuring accessibility and availability of information via secure, safe, and ethical behavior. Confidentiality involves the prevention of disclosure of information to unauthorized individuals. Information integrity means that data cannot be modified without someone being able to detect its modification. Information availability means it must be accessible when it is needed. The information is made secure by providing physical and procedural measures, ensuring that only those with specific needs have access to the information and by promoting an ethical understanding to all involved as to the sensitivity of information.

All schools should develop plans to safeguard information. Developing a comprehensive information security plan requires going through a process to assess information value, vulnerability, and risk. This process is called a risk assessment or risk analysis. The risk assessment is a continual and dynamic process because risk management is an ongoing iterative process. Risk assessment must be repeated indefinitely. Any time new technology is introduced or changes are made to informational processes, the school's information environment changes. These changes present the possibility of new threats and vulnerabilities impacting the safekeeping of information. Therefore, the risk must be re-evaluated. The school must constantly balance countermeasures and controls to manage risk to student information. Information security involves balancing productivity, cost effectiveness of the countermeasure, and the confidentiality, integrity and availability of the information (National Center for Education Statistics, n.d.).

An information risk assessment is a variation of a more traditional risk assessment, which would focus on indentifying assets and looking at the threats and vulnerabilities of each asset to potential loss. In conducting an information risk assessment, the first step is to identify information assets and estimate their value to the school. Indentifying information assets requires investigation and asking questions such as: What type of information does the school maintain? What is its value to the organization? Where is the information stored? Is it stored digitally or as hard copies?

Educational information includes all the data maintained by schools and by other parties acting for the educator. Information collected by a school about a student includes personal information, such as a school-generated identification number, social security number, picture, or list of personal physical descriptors that would make it easy to identify a student. Other information collected may include family information with the name and address of the student, parent or guardian, emergency contact information, date of birth, number of siblings, and date and place of birth.

School information records also contain educational information such as grades, test scores, specializations, activities, and official letters about a student's status in school and records of individualized education programs. School information concerning schools attended, courses taken, attendance, awards conferred, and degrees earned are kept as well.

The school will also maintain information that may be of a more sensitive nature. This could include special education information specifically related to students with special needs. Disciplinary records of student misconduct, as well as corrective measures taken by the school in dealing with negative actions, may be kept. Medical and health information of students should also be addressed as sensitive information.

The educator must not only identify the information but also place a value on it. An educator should calculate the value of each information asset. This would require a qualitative or quantitative analysis of educational information to determine what that the information is worth to the school. Questions such as "How we can meet the primary mission of this educational enterprise in the absence of this information?" and "What legal ramifications exist if we do not safely maintain this information artifact?" should be considered.

The second step is to conduct a threat assessment to the school's information. A threat assessment of information is a daunting task for educators. It requires knowledge of all of the information maintained by the school and potential threats to that information. Threats can be obvious, such as natural disasters, or more ambiguous, such as a computer virus. Information stored in different media has completely different characteristics and subsequent threats. The threats to the file cabinet in the administrative assistant's office are much different from student educational information stored on the network cloud.

A threat assessment should look at each information asset and then determine its liabilities. Threats can include both accidental and malicious acts originating from inside or outside the school. The threat assessment dictates that each situation of concern be viewed and assessed individually. Application of the assessment is guided by the facts of the specific threat and carried out through an analysis of its characteristics. Providing a safe information environment is very different and is based on the medium in which the information is stored.

Threats to digital information are viruses, malware, exploited vulnerabilities, remote access, mobile devices, social networks, and cloud computing. Viruses and malware can render information useless. Most active computer users have fallen prey to viruses and malware that damage and destroy information. Viruses are malicious codes that cause an infected computer to spread the virus to other associated computers via the network or e-mail contact lists. Malware may be destructive or disruptive in nature but will not have the inherent ability to spread itself. Although you may get a malware infection by visiting a website that hosts the malware, it is not actually a virus unless it utilizes your contact list or network directory services to propagate itself. A malware- or virus-infected machine or network can make a school's information useless (National Center for Education Statistics, n.d.).

A vulnerability exploit is another threat to a school's digital information. A vulnerability exploit is where a "hole" is found in applications or systems software that facilitates unauthorized access to information. It will allow an unauthorized user to access information and data that they do not have legitimate rights to access. Vulnerability exploits are often utilized by viruses and malware to penetrate networked computer systems.

Mobile devices are another threat for information safety. Storing and removing sensitive information from a school on a USB drive and then losing it poses a threat. Another good example is the laptop computer. School-owned laptops often house sensitive information. The loss of that information to theft is a major cause of the loss of information integrity. When computer information is outside the school's physical network, its threats change. Having a school-owned laptop with school information on it outside the logical and physical security of the school poses additional threats from both loss and exploitation and must be considered in a threat assessment.

Online social networking poses additional threats to information safety. Social networking sites, such as Facebook and Twitter, can pose serious threats to a school's information both directly and indirectly. Social networking sites are breeding grounds for viruses and malware plus other attacks to school information systems. Social networking sites can also allow school faculty, staff, and students the ability to post sensitive information regarding other students.

Schools using cloud-based resources must be aware of threats caused by utilizing that resource. The cloud is an Internet-based network that allows students, faculty, and staff to access applications on a network platform that is leased or rented to schools similar to a utility or tenement arrangement. The cloud extends the system so students, faculty, and staff have constant access to not only their e-mail but also personal files and applications software anywhere they have a computer and Internet access. For example, digitized information does not have to be installed on the remote computer, nor does it have to be connected to a local area network to function. The application is run from the cloud, and the data generated are stored on the cloud, which can be accessed anytime by anyone with Internet service and proper access credentials.

Educators may also unwittingly pose a major threat to educational information. This could range from a desperate and disgruntled employee who was recently terminated because of funding or performance to the careless employee who unknowingly releases sensitive information. There is no way to completely eliminate the threat of legitimate insiders, but through good safety policies and procedures, information loss could be greatly minimized. Careless and untrained educators could unknowingly release sensitive, if not damaging, information about a school and its students. Policies, procedures, training, and technical measures can play a major role in reducing an organization's threat from its own educators (McCallister et al., 2010).

The threat assessment of information is a significant task. It requires the knowledge and skills to address threats to the information maintained by schools. Regardless if a threat is from natural disasters or a computer virus, it must be addressed. Once the threat is addressed, the school should evaluate the vulnerability of information to that loss.

The third step is to conduct a vulnerability assessment of school information. A vulnerability assessment should calculate the probability that a threat to information assets could occur. The vulnerability assessment could be based on quantitative or qualitative evaluation of threats to information. The resulting data should provide an idea of the likelihood that a loss of confidentiality, integrity, or access may occur so that measures of information safety could be implemented (National Center for Education Statistics, n.d.).

Often the creation of a chart or matrix will help professionals in determining threats and vulnerabilities to information (Table 7.1). Either qualitatively or quantitatively, educators can value information and the threat vulnerability (Elky, 2006).

For example, for each informational asset, assign it a value from low to high. Then, for each threat vulnerability to the information, assign it a value from low to high. The intercept of these characteristics should provide the data to make a decision regarding how the informational asset should be protected from that threat. For example, the threat of electrical surges to render information stored on a computer irretrievable can be evaluated. In some parts of the country, this could be a very high threat vulnerability, and the information could be very highly valued. In these situations, it would be worth spending twenty dollars for a surge protector. But does the value justify buying a five-hundred-dollar surge protection system? From the data in this scenario, we do not know whether we can justify spending more money because the values will have to be balanced in terms of a comprehensive information safety

TABLE 7.1
Risk Assessment Matrix

High Threat Vulnerability			High Risk
Medium Threat Vulnerability		Medium Risk	
Low Threat Vulnerability	Low Risk		
	Low Information Value	Medium Information Value	High Information Value

program and school budgets. The point is that with the risk assessment, we can make value judgments based on more than assumptions.

The data from an accurate risk assessment of information value, threats, and vulnerabilities will provide the educator the ability to identify, select, and implement appropriate security measures. This does not require that the educator be an expert in safeguarding information, but it does assume that the educator has knowledge of information and knows when to provide a proportional response, even if that requires acquiring expert assistance in completing the task. The educator should understand productivity, cost effectiveness, and value of the informational asset and its threats and vulnerabilities and subsequently how to go about protecting it (Elky, 2006).

Implementing a plan to keep information secure in schools is a balancing act for educators. Educators must evaluate the effectiveness of safety measures without discernible loss of productivity. Providing information safety is contingent upon educators understanding the environment, risks, and vulnerabilities of technology. Because of the breadth of the issues and complexity of the environment, there is no more valid approach to achieving an information-secure environment in schools than to promote secure, safe, and ethically sound informational practices by students, faculty, and staff. A school must have both policies and procedures to provide for authentication, firewalls, and virus protection on its computer systems.

Authentication will provide secure login and allow individuals to access only the data and applications they require. The firewall also serves as an access control device that will limit the user access to unwanted programs and data. Virus protection programs will protect users from viruses and malware on the network. Each of these three areas must be addressed by computing professionals in that environment. The best role for the educator is to be aware of the value of information and its associated risks to develop policies and procedures to protect it. For many educational entities, this means working with a service provider or consultant because they usually have access to greater levels of expertise (Elky, 2006).

Information safety in schools is very necessary because of the liabilities of improperly handling information regardless of legislation requiring schools to do so. Federal regulations require that public schools must make information safe regarding who will interact with it and how the interaction will occur. Schools must understand areas of compliance and how to comply with the standards and regulatory requirements. There has been no educational information safeguarding measure as comprehensive as is provided by the Family Education Rights and Privacy Act of 1974.

FERPA

The Family Education Rights and Privacy Act of 1974, commonly known as FERPA, is a federal law that protects the privacy of student education information. Students have specific, protected rights regarding the release of information, and FERPA requires that institutions adhere strictly to its requirements. Therefore, it is imperative that the faculty and staff have a working knowledge of FERPA requirements before accessing school information.

FERPA places educational information into two broad categories: directory information and nondirectory information. Each category of educational record is afforded different safety protections. It is important for faculty and staff to know the type of educational record that is being considered for disclosure (National Association of Colleges and Employers, 2008).

Directory information is part of the student's educational record. Directory information is information that is generally not considered harmful or an invasion of privacy if released. Under FERPA, the school may disclose this type of information without the written consent of the student. However, the student (or guardian until a student is eighteen years of age or attends a postsecondary institution) can restrict the release of directory information by submitting a formal request to the school. Directory information includes (National Association of Colleges and Employers, 2008):

- Name
- Address
- Phone number and e-mail address
- Dates of attendance
- Degrees awarded
- Enrollment status
- Major field of study

Schools should always make students and parents aware that such information is considered by the school to be directory information and, as such, may be disclosed to a third party. Schools should overstate to students and parents that they can prevent the release of directory information (McCallister et al., 2010).

Nondirectory information is any educational record not considered directory information. Nondirectory information must not be released to anyone, without consent of the student or of a parent or guardian until a student is eighteen years of age or attends a postsecondary institution. FERPA insists that faculty and staff can access nondirectory information only if they have a legitimate academic need to do so. Nondirectory information includes (Federal Register, 2011):

- Social security numbers
- Student identification number
- Race, ethnicity, and/or nationality
- Gender
- Transcripts
- Grades

Students' distinct identifiers, such as race, gender, ethnicity, grades, and transcripts, are nondirectory information and are protected educational records under FERPA. Students have a right to privacy regarding information held by the school. Schools must ensure that students have privacy of this information under FERPA. FERPA also gives students and parents (if the students are under eighteen) the right to access educational information maintained by the school. It also provides the right to limit educational information being disclosed, the right to amend educational information, and the right to file complaints against the school for disclosing educational information in violation of FERPA (Federal Register, 2011).

Students and parents have a right to know about the purpose, content, and location of information maintained as a part of their educational records. They also have a right to expect that information in students' educational records will be kept confidential unless they give permission to the school to disclose such information. Therefore, it is important to understand how educational information is defined under FERPA (National Association of Colleges and Employers, 2008).

Prior written consent is always required before institutions can legitimately disclose nondirectory student information. In many cases, schools tailor a consent form to meet their unique educational needs. However, prior written consent must include the following elements (Federal Register, 2011):

- The information to be disclosed
- The purpose of the disclosure
- The party or class of parties to whom the disclosure is to be made
- The date
- The signature of the student (or parent or guardian) whose record is to be disclosed
- The signature of the custodian of the educational record

Prior written consent is not required when disclosure is made directly to the student or to other school officials within the same institution where there is a legitimate educational interest. A legitimate educational interest may include enrollment or transfer matters, financial aid issues, or information requested by regional accrediting organizations (National Association of Colleges and Employers, 2008).

Institutions do not need prior written consent to disclose nondirectory information where the health and safety of the student is at issue, when complying with a judicial order or subpoena, or where, as a result of a crime of violence, a disciplinary hearing was conducted by the school, a final decision was recorded, and the alleged victim seeks disclosure. In order for institutions to be able to disseminate nondirectory information in these instances, FERPA requires that institutions annually publish the policies and procedures that the institutions will follow in order to meet FERPA requirements (McCallister et al., 2010).

The Family Education and Privacy Act was enacted by Congress to protect the privacy of student educational information. This privacy right is a right vested in the student. Institutions may disclose directory information in the student's educational record without the student's consent. Institutions must have written permission from the student in order to release any information from a student's educational record (McCallister et al., 2010).

It is good policy for the institution to notify the student about such disclosure and to seek the written permission of the student to allow disclosure of any educational information, including directory information. Institutions should give the student ample opportunity to submit a written request that the school refrain from disclosing directory information about them.

OTHER REGULATORY FACTORS

Student information is protected by other agencies and groups in addition to FERPA. The Departments of Agriculture, Health and Human Services, and Justice defend information privacy and protect the information of students in schools. State and local entities may require safeguards for handling student information in a secure manner. Professional standards of ethical practice of school doctors and nurses, psychologists, and other professionals may also establish privacy restrictions (McCallister et al., 2010).

For example, information about drug and alcohol prevention and treatment services for students is covered by confidentiality restrictions administered by the U.S. Department of Health and Human Services. Some states have regulations regarding students' rights to seek treatment for certain health and mental health conditions, including sexually transmitted diseases, HIV testing and treatment, pregnancy, and mental health counseling, as well as protecting information pertaining to HIV confidentiality, medical information, child abuse, privileged communications, and state-specific information retention and destruction regulations. (Privacy Rights Clearinghouse, 2010).

MAKING INFORMATION SECURE

A student's education information is a compilation of records, files, documents, and other materials that contain information directly related to a student and maintained by schools. Information about students is a vital resource for teachers and school staff in planning education programs and services. Schools are legally and ethically required to maintain information integrity, accessibility, and confidentially. Educational information may be kept in a variety of formats including handwritten, printed, digital files, and video or audio recordings. Schools have an obligation to students and parents to maintain information integrity, accessibility, and confidentiality. Information must be accurate and available to make timely decisions regarding a student's educational well-being. There could be a legal consequence if a school fails to safely maintain information of students (Privacy Rights Clearinghouse, 2010).

Schools have an ethical responsibility to keep personal information about its students secure. Public schools are further bound by federal legislation that provides mandates regarding student information. Therefore, it is imperative that schools have a plan regarding information security.

The final part of this chapter addresses the role of educators in providing an information-secure environment. Providing information safety is contingent upon educators understanding the environment, risks, and vulnerabilities of technology. This chapter also assumes that most school information is digital. Because of the breadth of the issues and complexity of the environment, there is no more valid approach to achieving a safe information environment in schools than to use a multimode approach. This approach will promote secure, safe, and ethically sound practices. Schools can make information secure by providing requirements in relation to security, safety, and awareness of anyone who deals with information. The most common approach is to use layers to protect information. There are specific tools that can be utilized at each of these layers.

Instituting security will require specific technical information that most school officials will not possess. Because of this, schools will need an expert to install and configure these applications and appliances. Security layers would include things such as firewalls and routers. A firewall will stop unwanted applications and data from entering a school's network. Unwanted applications and data could be anything from known viruses to accessing social networking sites. The firewall can block all network traffic that is unwanted. If you do not want Facebook to be accessed from school computing resources, then it can be added to a list of blocked web sites.

The routers must be protected in a network. Routers are like a train station through which all information passes. They direct information where it needs to go. A router is like a firewall because you can stop information from being dispersed into the network; however, its primary function is to direct network traffic both in and out of the local network. From this perspective, a firewall or a router can both stop unwanted applications and data, but only the routers and switches show data where it needs to go on the network. The router is a very critical link for a network.

Network controls are a third primary layer to address information security. Network controls will include authentication and file-sharing controls. Authentication will provide a user name and password authentication system. This ensures that only authorized users access the network and its information. The file sharing controls set an access level for each authenticated user to access applications and information on the network. From this perspective, you only give information users the rights necessary to perform their tasks. In respect to information, you can give user rights to read, write, and modify discriminately for each information item stored on the network. The keys for network controls are to have a good password system and only give users access to what they need.

Software layers of safety will include virus and malware detection/prevention systems. Malware and viruses can be among the biggest problems for digitally stored information. The best way to prevent losses from these threats is to maintain adequate antivirus and malware protection software and to keep all updates installed on school computers. Antivirus and malware software must be updated frequently. Often a new virus cannot be stopped until the antivirus software has a definition of the virus to stop or quarantine it. Oftentimes, an update to a computer's operating system and applications software is required because the manufacturer has found a vulnerability in the system that must be addressed through updates and patches.

The final layer of information safety involves creating informed information custodians. Information users must understand the value of information to both the school and students. Through training and continual reinforcement, information must be respected and cared for to assure its integrity, accessibility, and confidentiality. Everyone using school information must adhere to an acceptable use policy with consequences for violating it. Administrators must instill a sense of information responsibility. The role of a school in facilitating learning is going to be even more challenging in the future. The best time to begin establishing a culture of information safety is now.

It is not the intent of this chapter to address information security from the view of a network administrator but to make the educator aware of what is needed to create and maintain an information-secure environment in the school. Schools should understand a fundamental process such as an information risk assessment to view and assess a value to threats and vulnerabilities of information. A school must be aware of all federal, state, and local laws regarding the use and maintenance of student information. Finally, schools must have both policies and procedures to institute security systems for authentication, firewalls, and virus protection of computer systems. Only through these measures can schools adequately address information security.

CASE STUDY

Sensitive personal information of 305 Rockwell High students, including names and social security numbers, were accidentally e-mailed to 26 parents Monday. The school director's office released the information of the students, who were all seniors, via an e-mail attachment sent to 26 recipients. Principal Johnson informed the students and parents whose information was disclosed. "We take the security of student information very seriously, and we deeply regret this error," Principal Johnson said. "We are reviewing the incident and are taking steps to prevent similar incidents from occurring in the future. This should not have happened, and we apologize for it." On further investigation, the principal determined that the information was sent to a group of parents of student athletes as an attachment when the intended file attachment was accidentally replaced by the student data file by the coach. It was an honest mistake.

- If you are Principal Johnson, how do you address this issue?
- What factors must be considered in the addressing the issue?
- What can Principal Johnson do to make sure this does not happen again?

EXERCISES

1. What data will an information risk assessment provide?
2. What is the layered approach to information safety? Identify the components or layers.
3. Without federal and state statutes requiring compliance, how do you think information security would be addressed in schools?
4. FERPA places student information into two categories. What are those categories, what are the characteristics of each category's data, and how are they treated differently in terms of information security?

REFERENCES

Elky, S. (2006). An introduction to information system risk management. Accessed November 15, 2011. http://www.sans.org/reading_room/whitepapers/auditing/introduction-information-system-risk-management_1204.

Federal Register. (2001). 34 CFR Part 99, Part V, Family Education Rights and Privacy, Final Rule. Office of Family Policy Compliance, Family Education Rights and Privacy Act. Washington, D.C.

Lamb, J. (2011) Thousands of Alaska students' personal information accidentally released. Accessed November 15, 2011. http://articles.ktuu.com/2011-03-04/hard-drive_28654519.

McCallister, E., Grance, T., and Scarfone, K. (2010). Guide to protecting the confidentiality of personally identifiable information (PII), NIST Special Publication 800-122. Accessed November 15, 2011. http://csrc.nist.gov/publications/nistpubs/800-122/sp800-122.pdf.

National Association of Colleges and Employers. (2008). FERPA primer: The basics and beyond. Accessed November 15, 2011. http://www.naceweb.org/public/ferpa0808.htm.

National Center for Education Statistics. (n.d.). Data security checklist. Accessed November 15, 2011. http://nces.ed.gov/programs/ptac/pdf/ptac-data-security-checklist.pdf.

Privacy Rights Clearinghouse. (2010). Fact sheet 29: Privacy in education: Guide for parents and adult-age students. Accessed November 15, 2011. https://www.privacyrights.org/fs/fs29-education.htm#11.

U.S. Department of Education. (n.d.) Successful, safe, and healthy students. Accessed November 15, 2011. http://www2.ed.gov/policy/elsec/leg/blueprint/successful-safe-healthy.pdf.

8 Event Security

Ronald Dotson

CONTENTS

School events are a necessary piece of the educational experience. They require a combination of understanding barriers, access control, perimeter security, and resource integration. Partnerships with outside organizations are required in order to have the proper resources. Even small events can require this level of planning and coordination. You must first have a strategy that integrates all of your assets and minimizes vulnerabilities. The Department of Homeland Security defines protection strategy in its national plan for infrastructure protection as deterring threats, mitigating vulnerabilities, and minimizing consequences (U.S. Department of Homeland Security, 2006).

School events are mainly about physical security. Philpott (2010) has presented a model for physical security. This model can be used as a guide for planning school events. The steps are to deter, detect, delay, respond, or investigate and then produce countermeasures.

Deterrence is achieved through a host of initiatives. High visibility of uniformed officers, signage, bag screenings, or visible cameras can help deter criminal activity. In normal security of industrial facilities, the strategy is to detect, delay, and then respond. However, as Philpott notes, a school event has numerous invited persons. It is not a matter of detection but rather deterrence that comes first (Philpott, 2010).

Detection of security breaches is usually visual, but cameras can be used to cover more ground and provide centralized coordination. Screenings also serve as detection strategies. Delay is achieved from physically securing equipment and assets, fencing, locked doors, and the systematic use of barriers. Response at a school event may not be from law enforcement alone. Emergency medical service personnel, security officers, and trained school employees can respond and coordinate incidents until more skilled or ranking personnel arrive (Philpott, 2010). Events should always culminate with the production of countermeasures to any incurred problems, and feedback should be given and received from the various levels involved in the security of the event.

ENTRY

Event security may not have a goal of creating a single entry point. When crowds build and waiting occurs, the opportunities for conflict and criminal actions increase. The ability to respond and control these security hazards is hampered by a crowd. Crowd control requires a vast amount of resources, and the outcome may not be favorable for the reputation of the school or community. Planning the entry points for people and traffic is required. The goal for physical entry is to minimize lines and wait periods while not having so many entry points that resources are stretched to cover them. Traffic control is not a recognized strong point for educators. Partnering with local law enforcement for events that are large enough to impact public transportation is a must. Preventing a tragic loss of a student, parent, community member, or any person is a goal of event planning. Traffic control must utilize local streets and traffic control devices to create a steady and even flow that ideally does not require the physical directing of traffic by law enforcement. However, many streets and roadways would benefit from and require physical traffic control for school events. Educators should involve local law enforcement for this aspect of event planning. School personnel would need traffic control training and should be outfitted with proper attire and equipment. It is an endeavor that a school would not want to undertake unless completely necessary. Experience really matters with traffic control.

Many schools utilize Reserve Officers' Training Corps (ROTC) cadets or other student organizations to aid and help control traffic flow and parking once on school grounds. The goal is to facilitate movement for drivers who are unfamiliar with the grounds, arrange parking for easy flow after the event, and limit any exposure of foot traffic to vehicular traffic. Close supervision and a system for communication are important. Students must be trained on watching for and avoiding traffic hazards, outfitted with proper personal protective equipment, and trained on how to use the equipment. Parking for football games and late evening basketball games will require high visibility clothing and flashlights, preferably with directional wands. Communication can be achieved with two-way radios. If enough radios are not available, having the students work in close-proximity teams with only one coordinator with radio can be accomplished effectively. However, a school employee should be mobile and monitoring radio communications with all teams. It may even be necessary to coordinate breaks and supply water when temperature and humidity are high. All students and employees alike should know how to summon aid in the event of an injury or incident. Labeling or identifying the school grounds and establishing first aid points are critical—not just for students and employees, but for others as well.

Foot traffic must be controlled in front of entry points close to the event. Having lines for those with passes or previously purchased tickets can help alleviate crowding. Lines for purchasing can also be divided alphabetically or be open with enough access to support expected attendants. There are strategies for anticipating attendance numbers. Past history is a good gauge, but early tickets sales and calls for information can also give insight. Additional paths of entry are also needed for special populations. Those with wheelchairs or who are handicapped may need special parking as well as entry assistance.

Students or employees can also be used near the entry points for providing directional assistance. It is important that the use of tactical barriers be utilized in conjunction with access points.

BARRIER STRATEGY

Barriers are anything that can force or invite movement or that influences movement or prevents access to something. There are three levels of barriers; strategic, tactical, and protective barriers. Strategic barriers are the first level of barrier. This type of barrier influences human behavior by getting in the way and pointing to an easier route. It can be things like shrubbery, pottery, sidewalks, columns, or garbage cans. They may not be threatening or viewed as a barrier at all. They funnel traffic to a center. Tactical barriers are those that are designed or meant to prevent access without significant effort to surmount. Decorative walls, hedges, rope barricades, or speed bumps are good examples. They slow or prevent movement in unwanted directions. Protective barriers are those that provide a final level of protection. They may be classified as tactical in some situations, but because of close proximity of that which is to be protected, they are utilized in procedures or methods to provide protection at the final point of entry. A good example is a park bench that is bolted down and reinforced in front of main entry doors to prevent an unintended vehicle from crashing into or charging into the access point. Some barriers are both tactical and protective.

The overall philosophy is to establish outer perimeters and an inner perimeter. It is possible to need multiple inner perimeters. Outer perimeters and strategic barriers are used in conjunction to guide foot traffic and vehicular traffic to central areas. The students or others used to assist in parking then take over. The strategic barriers also keep traffic from inadvertently going into areas where access is not desired.

Once vehicular traffic is parked, strategic barriers can also guide foot traffic to safer walking areas. The idea is to keep pedestrians in better lighted areas with as much physical separation from vehicles as possible. At pedestrian crossings or at corners or areas where a driver's view may be impaired because of building corners, signage, vegetation, or other encumbrances, crossing guards or personnel should be stationed. The preferred situation is to provide physical separation. Use sidewalks, rope barricades, fencing, and ditch lines to protect pedestrians. It would be ideal to have tunnels or crossovers for foot traffic to utilize.

Schools should consider the type and scale of events when they design new buildings and grounds. The earlier in the design and construction phases that a hazard or condition is anticipated and considered, the cheaper and less disruptive the addition will be. Pedestrian tunnels, bridges, or fenced walkways can be planned for early in the construction phase.

Tactical and protective barriers increase in use as the perimeter collapses toward the center of the event. Tactical barriers should increase and be utilized by personnel to guide foot traffic toward and at entry points. Tactical barriers may also need to be utilized for controlling access and egress for emergency vehicles and personnel that may be needed at the center of the event but also at designated first aid stations anywhere on the grounds. Emergency vehicles with lights and siren may still be significantly impeded by crowds and traffic, especially when space is limited.

INTEGRATING COMMUNICATION AND PERSONNEL RESOURCES

Each person should have a central point or supervisor to report to. Organizing personnel into teams will increase effectiveness and use fewer resources. Law enforcement personnel will not have the personnel to be the primary agents. Instead they should be partnered with supervisors or at points that coordinate larger areas or more teams. They can then respond to reported incidents that require additional authority.

Observing the event from above is also a good strategy. An option is to place observation points on roofs or in higher story windows that can also serve as primary points for communication. Individuals positioned in such locations can observe and report or direct response efficiently because they have an overview of the event. It is best to have observation points at a height designed to eliminate a fall hazard. Observation at height may require personal fall protection systems. However, scissor lifts on stable ground protected from vehicular traffic are a good strategy for observation. Scissor lifts are not aerial boom lifts and have acceptable fall protection with the built in guardrail around the platform. Storms and winds may make their use not advisable, but otherwise they are tools that a school will likely already own or lease for maintenance activities.

Publishing the structure and mapping of the grounds with locations of first aid stations, central communication stations, and key assets or positions is very helpful. Each person utilized in the security plan should be equipped with this information. As the event progresses and parking duties diminish, those personnel may be folded into other duties and pulled toward the event or inner perimeters. Then parking lots and outer areas can be patrolled or observed with general patrols and fewer personnel. This is not to say that outer areas are to be neglected or forgotten. Incidents such as vandalism and mass vehicle break-ins will negatively affect attendants.

TRAINING

It may be necessary to hire additional security personnel from a private company, but parent volunteers can fill this role if the numbers are sufficient. Students, employees, and volunteers should receive a minimum amount of training for filling such roles. This training is to enhance the plan, limit injury to them, and limit liability to the school.

From a safety perspective, all employees, students, and volunteers should undergo a minimum level of training. Training needs can be identified for safety from a basic job hazard analysis (JHA). Recognition of uncovered hazards on the JHA, how to summon aid, how to abate the hazards, and how to use any equipment necessary to abate the hazard is the duty of the school. From a security aspect, students and volunteers should only be instructed to be a good witness and report all incidents up the chain and to authorities. Breaking up a fight or stopping criminal property damage is not an assigned duty for them. Instead, they should make note of the description of subjects involved, times, and any personal identifiers such as names, license plates, and escape routes and summon the proper authority. They should be directed that personal harm could come to them by taking action.

School employees are slightly different than students and volunteers. School employees who have had training in breaking up physical confrontations may find it necessary to perform such activity, especially when another's personal well-being is at jeopardy. However, school employees that have not had such training should avoid engagement unless the situation is severe. They must be instructed that their duty is to summon proper aid and personal harm may occur if they intervene. Obviously, the training level of personnel resources must fit their assigned positions.

Many local sheriffs' departments may assign a deputy or have auxiliary deputies stationed at the event. However, few law enforcement agencies have the resources to cover their jurisdictions and a school event. More practically they can staff the event at cost to the school. Uniformed officers establish a strong presence and can be a deterrent. At large events, it is necessary to have personnel with the recognized authority to make arrests and advanced training and knowledge on security, emergency response, and use of force.

RISK ASSESSMENT MODEL

The American Society for Industrial Security (ASIS) was founded in 1955 and offers training and certifications and serves as a professional organization for security professionals. It would be a good practice to incorporate their training and guides for some school employees interested in serving on security and event-planning committees. ASIS publishes a model of risk assessment that is practical and demonstrates that event security plans are living documents in much the same manner as emergency response.

The ASIS model for risk assessment is well established in policing and security circles. It is visualized by the following steps (American Society for Industrial Security, 2003):

- Identify assets
- Specify risk events
- Determine the frequency of risk events
- Determine the impact of events
- Identify options to mitigate the events
- Examine feasibility of the options
- Cost/benefit analysis
- Make a decision
- Reassess

Some of the steps in this model make risk assessment and security plans living documents. Some of this model is based on anticipated events or risks that are unique to organizational and personal experience. You must track events and outcomes so that security plans and responses improve. The recommendation is that you follow or mimic a general 911 or regional dispatch center system. All known or reported incidents should be documented on an incident control log and assigned an incident control number. All incidents should have a minimal assessment completed that helps for future planning. Many incidents will require a full investigation.

A good start is to set goals or priorities for the event security. This is achieved by assigning a value to the physical assets that need protection. Security assets may be concentrated more on the facilities or event functions with higher values (Philpott, 2010).

With strategic planning of any type, the identification of available assets and of needed assets is critical. If a plan cannot properly mitigate or respond to those incidents that will be frequent or those that are infrequent but have critical or immense consequential outcomes, then the needed assets must be bridged. Determining how to obtain the needed assets is part of event planning. The ASIS steps are foundational, but identifying needed assets during the planning phases and during reassessment is vital for continued improvement. Continued improvement should be a goal for any safety- or security-related plan.

It may also be necessary to prioritize the acquisition of needed resources. This requires an actual comparison of risk. The Department of Homeland Security sets the formula for risk as equaling the criticality rating multiplied by the frequency multiplied by the vulnerability rating. In this consideration, it is important to consider all three in determining a rating for risk. A numerical score could be assigned for each, depending on possible outcomes. It would then be possible to rate them more formally. The main point is that all three criteria should be evaluated when existing resources limit the acquisition of needed resources.

CASE STUDY

This year's football team is highly ranked in the state. They are also playing two powerhouses that are very well-known for their sports programs. The game attendance is expected to grow. The superintendent is concerned about incidents outside of the football stadium and on school property. During a boys' regional basketball tournament game last spring, a large number of cars were broken into. Both communities from the schools involved witnessed a lot of criticism of the school district for allowing this to happen. Although the local police responded and handled the reports, many from the rival community displayed disgust with the school district and the community at large. Your new commanding officer for the ROTC program recommends that the cadets be utilized to assist in security and parking details.

- How would you address the situation this year?
- What key elements must be included in your strategy?

EXERCISES

1. In relation to the case study, draft a memo in support of the commanding officer's suggestion or a memo listing the reasons that you do not support his suggestion.
2. Design a lesson plan for teaching ROTC cadets how to stay safe while performing parking lot duty. Specifically, you should consider how they will

be supervised, how levels of rank and supervision coincide, how they will communicate, how they will summon aid, and what equipment they will need to be issued.

3. Because of the incident last spring at the high school, the superintendent is mandating that you have a security plan for a fall festival planned at your elementary school. If you did not have the employee resources for this type of duty, list three other possible sources for the manpower to accomplish this task.

REFERENCES

American Society for Industrial Security (2003). *General Security Risk Assessment Guideline.* ASIS International, Alexandria, VA.

Philpott, D. (2010). *School Security.* Government Training Inc., Longboat Key, FL.

U.S. Department of Homeland Security. (2006). *National Infrastructure Protection Plan.* Accessed March 4, 2012. http://www.dhs.gov/nipp.

9 Scalable Approach to School Security

Gary D. Folckemer

CONTENTS

There are a number of similarities between providing security for a K–12 school environment and a university environment. Both environments have the shared purpose of providing educational experiences for our citizens and typically have property that is open to the public. In this chapter, I will review a scalable approach to school security that can apply to the K–12 school environment and is based on my experience as a university police officer primarily tasked with emergency-management responsibility.

CREATING A PHYSICAL SECURITY SYSTEM

Public school systems could be considered part of our nation's critical infrastructure. Critical infrastructure encompasses the assets, systems, and networks, whether physical or virtual, so vital to the United States that their incapacitation or destruction would have a debilitating effect on security, national economic security, public health or safety, or any combination thereof (U.S. Department of Homeland Security, 2010). The school environment is a bastion of freedom, where free-thinking, learning, expression, and action based on one's values, beliefs, and feelings are encouraged and celebrated. This is where we influence society and build the future. It is in this environment that precautions must be taken to safeguard the individuals who live, learn, and work there, and there are physical and informational assets of the institution that must be protected from loss and subversion. Numerous threats can manifest themselves, and through an array of physical security system components, measures can be taken, in whole or in part, to best meet the needs of protecting people, property, and the environment. These measures can be viewed from the perspective of various environments.

FROM HOSPITAL EMERGENCY DEPARTMENTS

A key risk factor for violence, and a key for its prevention, may be long waiting times and frustrated individuals. A well-managed hospital emergency department can reduce that tension and the resulting risk of acting out. The inability to meet patient or family expectations contributes to violence—patients want information, access to a wide range of services, good discharge planning, compassion from staff, involvement in decision making, and high-quality care, and they want it all quickly.

Although a lot of the focus is on patients who are drunk, on drugs, mentally ill, or gang members, there are other potential sources of violence as well. Domestic violence situations involving staff members can spill onto the hospital campus. Disgruntled workers or former employees can cause trouble. The security department has to be prepared for all of these scenarios.

Among the common options for improving security are metal detectors, security cameras, and dedicated security patrols. Hospitals can also set up separate and secure areas for psychiatric patients and prisoners from jails or prisons. Some hospitals choose to restrict access not just to the emergency department but to all parts of the hospital using an electronic badge system. At Community Medical Center in Fresno, metal detectors, cameras, limited access, and a manned security booth substantially decreased the number of weapons confiscated in the patient care area. However, the study did not find a reduction in reported assaults by patients, indicating the continued need for staff training in the management of violent patients and potentially violent situations (Greene, 2008).

FROM THE CHEMICAL HEALTH AND SAFETY INDUSTRY

It is important to assess facilities' attractiveness as a potential target. This addresses the nature of any special hazards present at the facility and considers the potential outcomes of an attack. Are there obvious security weaknesses? Is information on the facility readily available to the public? Answering these questions will help determine whether security vulnerability analysis is necessary for the facility.

Every facility can benefit from a review of its infrastructure. This includes all utilities, entrances/exits, process and production equipment, telephone and data lines, water supply, backup power systems, process controls, hazardous material storage tanks and pits, fire alarm systems, and sprinkler systems. Making a list of all infrastructure details is helpful when evaluating security vulnerability, as well as such mundane tasks as budget analysis, routine and preventive maintenance, and personnel responsibilities.

Security systems can include employee/student identification processes, intrusion detection, cameras, alarms, and visitor/contractor identification. Schools are obviously more vulnerable than industrial facilities in terms of open access, but the challenges of securing academic facilities are best met by clearly understanding the most vulnerable physical areas and providing appropriate security. The most dangerous target of any terrorist activity at a laboratory facility is likely to be hazardous materials that could be used to initiate a fire, explosion, or toxic release. This could include obvious targets such as bulk flammable liquids in tanks or drums, flammable gases in cylinders or tanks, or smaller quantities of flammable, reactive, or explosive chemicals.

Any evaluation of response systems should begin with a review of personnel responsibilities and training. The individuals who are responsible for internal response must be properly trained at one or more levels, including basic hazard awareness, the use of appropriate equipment, and the locations where response equipment and supplies are stored. Other key facets of response systems include drills, operational status of alarms and automatic monitoring equipment, arrangements with outside agencies and contractors, and routine and emergency communication within the facility.

Key employees need to be assigned responsibilities for monitoring evacuation of personnel, and provisions should also be in place for sheltering in place when evacuation is impossible or inappropriate. All facilities should be aware of the worst-case scenarios associated with terrorist or vandal activity. It is important to acknowledge that in many cases, the worst possible situations can be caused by internal problems such as disgruntled employees, employee disagreements, or strikes. It is employees and students who often are the most aware of vulnerabilities within the facility and have the accessibility to such areas, and it is not enough simply to plan for disruption from outside sources. Usually, a worst-case scenario involves a fire, explosion, or toxic release. Having a clear understanding of the implications of such a disaster is the first step in prevention.

Each facility is different, and any attempt to evaluate security vulnerability is necessarily site-specific. Small chemical businesses and academic facilities should prepare by understanding their vulnerabilities and recognizing the need to protect their chemical, biological, radiological, and nuclear resources (Phifer, 2007).

FROM OUR SCHOOL SYSTEMS

Several characteristics make some schools more vulnerable to violent student behavior. School size, location, physical condition, diversity, and policies all play a role in the amount, type, and severity of violence. Steps can be taken to prevent violence in regard to the school building itself. Physical environment has shown to affect student actions, attitude, and motivation. Schools that are clean, free of graffiti, and in good condition are significantly less likely to have instances of students acting out. There are several ways to improve the environmental factors within the school. Adjusting traffic flow in hallways can decrease the potential for adverse student encounters. Dividing the entrance and exit of the cafeteria, staggering the beginning and ending of the lunch period, keeping time spent in line limited, and increasing staff supervision during periods of high traffic flow can all help to reduce negative interactions between students.

Violent acts tend to occur in isolated areas, such as the end of a hallway or a hidden corner on the playground. Faculty should be very aware of these areas, and students should be prohibited from them. Limiting access to, or completely closing, the school building and campus during non-school hours also can increase safety. If steps can be taken to improve the overall climate of the school and prevent small disruptions, schools may be able to reduce the overall occurrence of the more violent acts.

Various prevention and intervention programs are suggested, such as creating a school-based student intervention team, renovating the school environment, changing teaching strategies, providing social skills training, and providing training and awareness for cultural sensitivity (Eisenbraun, 2007). Therefore, while physical security systems may be ineffective in schools, perhaps because of the immaturity and impulsivity of the resident population, there is still an expectation on the part of the parents and society that physical security systems be implemented.

After giving consideration to issues brought to us from various environments, our physical security system should begin with a consultation meeting and partnership with the people in charge of the area in question. We will meet with the area stakeholders in their working environment to evaluate their current facilities and procedures, and we will offer suggestions for their consideration. Our suggestions would include the following measures to guard against the risks and vulnerabilities described previously.

LOCKS ON DOORS AND WINDOWS

It seems simple and obvious, but it is important that the work environment be capable of being secured. There should be locks on all doors and windows that the occupant can readily engage should a dangerous incident take place. In addition to the physical safety of the occupants when the area is in operation, it is important to be able to secure the area when it is not open or when it is not staffed. This concept extends to the use of fences around outside areas of activity, operation, storage, and protection.

TELEPHONES IN ALL AREAS

All structures in the work environment should have working, land-line telephones. This ensures that occupants will have a means of communication independent of cellular telephones. Hard-line phones maintain enough electricity within the phone line to keep the phone operational during times of a power outage. If someone is threatening you or you otherwise become concerned for your safety, you can remove the handset from the cradle, place it on the desktop, dial 911, and leave the line open. Standard protocol for the emergency dispatcher will be to try to make contact with the person on the other end of that call. If there is no response by the caller, the dispatcher should be able to read the number and the location that the call is coming from and dispatch a police officer to assess the situation. In an office or classroom, the phone can be concealed from the view of a potential aggressor. Leaving the line open allows the dispatcher to hear what is happening at the other end of the call. Police can then be updated about the situation while en route.

PEER OVER WATCH AND DURESS PROCEDURES

It is important to discuss how faculty and staff members should look out for one another. Awareness is a key. Individuals should inform each other about people or situations that may pose a threat or risk of harm. This can include letting others know of domestic violence issues that could intrude upon the workplace. This should also include individuals who may be disgruntled, aggressive, unstable, and the like. Potential situations should be discussed so that individuals in the work environment know what things to look for and what actions to take to protect themselves, others, and summon appropriate help if an incident occurs.

SECURITY TRAINING

All individuals can be considered security officers to some degree. It is worthwhile having training sessions or providing guidelines to help individuals know what they can do to provide for their own security. A portion of such a basic guideline follows:

- Strengthen physical security—Do not let thieves ruin your school. Schools make attractive targets for burglars, so you need to take extra precautions to protect the premises.
- Secure the perimeter—Create a secure physical perimeter to keep out thieves using locks, self-closing doors, window locks, lockable internal doors between rooms, security curtains or window shutters, and alarms.
- Monitor visitors—Do not let visitors inside the secure perimeter without an escort. Vet contractors and support personnel. Restrict access to sensitive areas, such as server rooms or human resources records. Staff should be encouraged to query unescorted strangers in secure areas (*Get Safe Online*, 2011).

ACCESS CONTROL

At a minimum, a visitor sign-in log can be kept. This would require that all individuals record their identity, contact information, and reason for being in the area. This can serve as a record of visitation and can be a deterrent for illegal activity. A sign-in log can be augmented with visitor badges and escort protocols. At a more secure end of the spectrum, access control can include proximity card badges with readers and electronic locks on secured areas. Metal detectors can be used at entry points provided competent staff members are engaged to operate the devices.

ALARM SYSTEM

An intrusion alarm can be considered for the school environment that includes some or all of the following options. Door and window contacts can both monitor the closure of doors and windows after hours, when the area should be secured, and during operational hours, where a chime can sound to indicate when a portal is opened. Motion sensors or glass-break sensors can detect someone entering the premises after hours. Panic buttons can be included in both fixed locations or in portable formats. The buttons can be pressed to send a duress signal to a central monitoring station, such as a reception desk, an alarm-monitoring agency, and/or the police department.

CAMERA SYSTEM

Camera systems can be installed with device number, type, and placement being limited only by the available budget. At a minimum, a camera on each entrance and each exit to the school should be used. Additional cameras should be considered for areas of special concern, such as cash-handling areas or chemical-storage locations. The camera feeds should be routed to a monitoring station for real-time viewing, and they should be recorded for later review if an incident occurs.

SECURITY PERSONNEL

Consideration can be given to security personnel. This can be in the form of security officers, proprietary or contracted, plain-clothed or uniformed, and unarmed or armed, hired to protect the work environment. It could be an established protocol of calling on security professionals to be at the school and available during a time of sensitive activity or intervention where trouble might reasonably be anticipated.

PHYSICAL SECURITY SYSTEMS À LA CARTE

As stated earlier, the forgoing has been a set of suggested physical security system components. The purpose for developing this menu of options is to have an overall system that is scalable depending on the individual school environment, as in a building, a department, or an office. It is meant to give consideration to the particular needs of the area and the budget available. The risks meant to be addressed revolve around the protection of individuals from physical harm resulting from assault and the protection of assets from theft, corruption, or subversion for unlawful purposes. As I have tried to describe here, providing for the physical security of a school can be a difficult process because of the open and free environment that exists. As such, a flexible set of devices, services, and options seems to be the best approach. There simply is no one size that fits all in this sort of endeavor.

SITUATIONAL CRIME PREVENTION

The key elements in situational crime prevention are as follows. There must be a motivated offender, there must be a suitable reward or goal, and there must be an absence of appropriate controls. Situational crime prevention posits that all three elements may be assessed to determine the crime vulnerability of a location or a situation.

Possible controls or mitigating factors for a motivated offender can include denying access to sensitive areas. One can publicize warnings of punishment for illegal behavior, and one can prosecute apprehended offenders. Possible controls or mitigating factors for a suitable reward or goal can include decreased availability of assets that might be stolen from a potential victimization site. One can render valuable assets less attractive to thieves or alter the behavior of potential victims so that they might be less likely to be victimized. One can also make available assets impossible for thieves or other offenders to benefit from. Possible controls or mitigating factors for the absence of appropriate controls can include assigned security officers to protect assets and locations. One can install or upgrade a security system, and one can educate nonsecurity employees and others to participate willingly in loss-prevention strategies (McCrie, 2007; Dunlap, 2011).

CRIME PREVENTION THROUGH ENVIRONMENTAL DESIGN

The concept of crime prevention through environmental design (CPTED) plays an important and recurring role in security operations. Some important CPTED guidelines are presented by the U.S. Department of Justice, the Office of Community Oriented Policing Services, and the Prince William County, Virginia, police department.

Crime prevention through environmental design is an approach to problem solving that considers environmental conditions and the opportunities they offer for crime or other unintended and undesirable behaviors. CPTED attempts to reduce or eliminate those opportunities by using elements of the environment to 1) control access; 2) provide opportunities to see and be seen; and 3) define ownership and encourage the maintenance of territory.

It is necessary to control access by creating both real and perceptual barriers to entry and movement. The environment must offer cues about who belongs in a place, when they are supposed to be there, where they are allowed to be while they are there, what they should be doing, and how long they should stay. Users/guardians can also serve as access control if they pay attention to people and activities and report unwanted behaviors to the appropriate authorities.

Taking advantage of design to provide opportunities to see and be seen is another good strategy. This includes opportunities to see from adjacent properties or the site perimeter onto the site and possibly to see parking areas and buildings; opportunities to see from one part of the school to another; and opportunities to see parking, walkways, and other areas of the site from various locations inside the building. These design elements need to be supported by potential observers (they actually need to look for and then report unusual behavior) and by policies and procedures, for example, related to landscape maintenance.

Design can be used to define ownership and encourage maintenance of territories. As mentioned previously, the design should provide cues about who belongs in a place and what they are allowed to do. Administrative support in the form of rules and regulations about use and maintenance can be critical to the success of various design applications (Zahm, 2007).

CPTED design strategies have evolved over time. Although many actual techniques have been in use for hundreds of years, it has only been in the last few decades that urban-planning experts such as Jane Jacobs and Oscar Newman have explored the relationship between the built environment and criminal behavior.

Each of the following CPTED strategies offers guidelines that property owners, design professionals, and developers or remodelers may apply to reduce the fear and incidence of crime and improve the quality of life. There are four overlapping CPTED strategies. They include natural surveillance, natural access control, territorial reinforcement, and maintenance (Casteel and Peek-Asa, 2000).

NATURAL SURVEILLANCE

The placement of physical features, activities, and people in a way that maximizes visibility is one concept directed toward keeping intruders easily observable and therefore less likely to commit criminal acts. Features that maximize the visibility of people, parking areas, and building entrances are unobstructed doors and windows, pedestrian-friendly sidewalks and streets, open entry areas, and appropriate night-time lighting.

NATURAL ACCESS CONTROL

A security objective can be to deny access to crime targets and create a perception of risk for offenders. People are physically guided through a space by the strategic design of streets, sidewalks, building entrances, landscaping, and neighborhood gateways. Design elements are very useful tools to clearly indicate public routes and discourage access to private areas.

TERRITORIAL REINFORCEMENT

Physical design can also create or extend a sphere of influence. Users are encouraged to develop a sense of territorial control, whereas potential offenders, perceiving this control, are discouraged. This concept includes features that define property lines and distinguish between private and public spaces using landscape plantings, pavement designs, gateway treatments, appropriate signage and "open" fences. A six-foot-high straight bar tubular steel fence can be used to provide a barrier that establishes territoriality. This type of fencing will also deter graffiti, would be attractive, and would not obstruct common observation. If a large area is fenced, chain-link fencing could be used, but it is less attractive, and the maintenance cost for repair and removal of graffiti would be higher. A wooden fence is typically not recommended because of the high cost of repairs and graffiti removal. Signage should be placed at the entrance and along the fence line. The signs should state no loitering or trespassing and that all violators will be prosecuted and by whom. All signage and fencing should be well maintained and repaired promptly.

MAINTENANCE

Lastly, care and maintenance enable continued use of a space for its intended purpose. Deterioration and blight indicate less concern and control by the intended users of a site and indicate a greater tolerance of disorder. Proper maintenance prevents reduced visibility caused by plant overgrowth and obstructed or inoperative lighting while serving as an additional expression of territoriality and ownership. Inappropriate maintenance, such as over-pruning of shrubs, can prevent landscape elements from achieving the desired CPTED effects. Communication of design intent to maintenance staff is especially important for CPTED to be effective.

PARKS, TRAILS, AND OPEN SPACES

Parks, trails, and open spaces provide a number of design challenges for personal safety, because they are typically large and used by a variety of people. Direct observation is not always possible or even desired in natural settings. Often, there is a conflict between safety principles and preserving the naturalness of the resource. Designing for safety in these areas should be focused on pathways, parking areas, and other areas of concentrated activity (Prince William County Police Department, 2009).

SECURITY GUARDS

A final major component of a security operations strategy involves the use of a guard force. Proprietary, contracted, or blended security services may be considered. A blended approach can be considered for areas on a school campus that desire a static security presence.

McCrie (2007) informs us that for most of the twentieth century, security workers were permanent proprietary employees with the same expectations and corporate relationships as any other employee. Beginning early in the 1900s, and most notably since the early 1950s, security guard positions have increasingly been provided by contract services. Although proprietary and contract personnel both have their own distinct advantages and disadvantages, it is important that all factors be weighed in making this decision.

The main reasons cited for contracting out security employees are as follows. Contracting security services involves less direct cost. Employers generally believe that it costs less to contract out for services than it does to employ a full-time staff. There is administrative ease. The contract firm is responsible for recruiting and vetting security employees. The security service should be highly proficient at this task. The security services firm handles routine details for contract security employees, similar to those for proprietary workers. It provides for criminal records screening. In some states and geographic areas, the law dictates that security personnel be screened through criminal justice databases for the presence of convictions that would bar them from working in the field. Many service providers are thus able to assure employers that security workers have no evidence of significant criminal records in the jurisdictions where such information is checked.

The recruiting and vetting is transferred. The process of recruiting and vetting new security personnel is the responsibility of the service provider. The training is transferred. The security vendor is expected to have the commitment and expertise to train security personnel to the level required for the school's needs. Additionally, the contractor may provide specific additional training to meet the needs of the assignment. The supervision is transferred; security officers usually are supervised by the service provider. Many large contracts include a full-time site manager who handles routine administration. There is specialized liability insurance. Security service providers normally should possess comprehensive liability insurance as a safeguard against potential lawsuits. There is specialized protective experience. The security services provider should serve as a general resource, as needed, in security matters. Security services firms may share their practices and resources, acting as informal consultants on procedures and policies on a limited basis.

There is flexibility in personnel scheduling. When the contractor requires additional personnel for special purposes, such as a conference, annual meeting, or an unexpected event, the services contractor can add employees on short notice. Similarly, if an employee fails to meet the needs of the school, the employee often may be replaced rapidly. There is less likelihood of collusion and fraternization. Contract security personnel are hired and managed by a separate organization from the clients they serve. This managerial separation makes collusion less likely than if security workers were proprietary staff members. Emergency staffing is available. If the school requires additional security personnel for brief assignments, security services may provide additional officers as needed. This could be less burdensome than hiring additional personnel on a regular basis.

The factors cited by organizations that have elected to establish and maintain a proprietary security service are as follows. It facilitates personnel retention. Security directors generally prefer to have low employee turnover. This is for many reasons, including the time and cost of recruiting, training, and guiding workers. Proprietary employees are more likely to remain on the job longer. There is a perception of greater quality in employees. Many employers believe that proprietary workers, in general, are superior workers because they are attracted to the normally better compensation and career opportunities within a proprietary organization. There is greater site knowledge. An aspect of greater worker retention is that such employees are likely to know the people, procedures, and principles better than those with a shorter tenure. Logically, this produces more reliable service.

There are more flexible controls. In proprietary programs, personnel may be transferred from one location to another as a condition of employment. Contract employees also may be shifted with ease from one site to another. However, some security directors believe that this process is easier for employees who are permanent workers. There is greater loyalty to the employer. Many directors of proprietary programs believe that staff workers are more loyal than contract workers. This view cannot be quantified though the argument is appealing on its surface. There is reliability of service. A contract security firm may prove to be disappointing after an initial period of meeting school standards.

There can be cost savings. Normally, managers expect to achieve significant savings from contracting out the security guard and patrol services relative to in-house equivalents. This may be true of direct cost, such as hourly wages paid in the two models. However, greater quality of work among proprietary security officers could result in greater cost savings through preventing events at a greater level.

Combining proprietary and contract staffs can be an effective and efficient answer. Many security directors conclude that proprietary and contract services have complementary benefits. Therefore, they include both types of services in their operations strategy.

There are core expectations for security officers. Security officers have numerous obligations to their employers, each of which is of critical importance. The following is a description of their primary obligations. Officers should deter undesirable activity. The primary goal of security personnel is to deter or prevent harm to people, property, and the environment. Officers should delay undesirable activity. In the event that offenders commit a crime, security personnel may delay their successful flight, leading to apprehension. Officers should detect undesirable activity. When an incident occurs or when procedures are not followed properly, security officers can ascertain the violation quickly, mitigating or reducing the chances of loss.

Officers should respond to undesirable activity. Security personnel are trained to respond to detected incidents or calls for service while on duty. They are expected to take action at such times to protect people, property, the environment, and to make the public feel safer. Officers should report activity. In the event of an incident, a report from an independent observer, like a security officer, provides important information for management. The report serves as a possible factor in changing internal procedures, the basis for an insurance claim, or as possible evidence in an arrest or lawsuit. These core competencies are necessary for proper implementation of situational crime prevention.

BLENDED SECURITY MODEL

In many cases, a blended security approach would be appropriate. The primary on-site security function could be carried out by unarmed contracted security personnel, with an on-site supervisor from the contracted company on duty if required by the area stakeholders. A school system employee, in responsible charge of the work area, would be assigned the function of oversight of the contracted security program.

A critical concern in this proposed model is the importance of quick and reliable communication to and a close working relationship with the school system security department. The school system security department would be summoned to evaluate and assist with safety, security, and emergency management risks, threats, or incidents. Some primary considerations used in determining this blended model are discussed later in this chapter.

A shift is taking place within many Fortune 500 companies and smaller firms alike. Companies are placing a heightened focus on their core competencies and outsourcing support functions not associated with their primary business. This move to outsource is especially prevalent in the area of personnel-related support functions, which include security services. Using a contracted provider for security purposes protects employees and assets, saves money, and allows the school to devote its attention and resources to its core competencies (Zalud, 2007). Part of the financial picture is spreading risk and liability. This is a significant reason to outsource security services. Organizations with proprietary programs are liable for any incident resulting from the action or inaction of their officers, as well as deficiencies in the training, operation, and management of their program (Zalud, 2006). A contracted security company is responsible for keeping up with regulatory standards. Some may assume that professional training is a key tool in the ongoing process of reducing the risks faced by an organization. This is certainly the case, but it is equally important to recognize the ways in which the converse can be true. Inadequate training is, in and of itself, an additional risk factor (Villines and Ritchey, 2010).

With massive budget reductions, overhead trimming, downsizing, right-sizing, and less-euphemistically disguised mass layoffs, everyone is feeling pressure to demonstrate their value and keep costs low. Security has rarely enjoyed a carte-blanche access to funds, and although we have been living with budget constraints for years, this time is slightly different. It is not just "more for less" that end-users demand, it is "more for less and better" (McDargh, 2009). Carefully evaluating the value of security services provided and the costs associated with managing those services, taking into account liability issues, regulatory compliance issues, and training issues can be an immense and complicated undertaking. A basic cost analysis is necessary to determine exactly how we can achieve more for less and better.

CASE STUDY

South High School is in an urban environment. Gang activity appears to be on the rise, based on the presence of graffiti and increased acts of violence on school property. The school was built in 1950, and few repairs have been made since original construction. Principal Hayes is under pressure by the board of education to improve what is occurring at the school.

- What should Principal Hayes's primary objectives be?
- What tools are available to help her accomplish her objectives?

EXERCISES

1. What elements are available to establish a physical security system? Which of these would be most appropriate in a school with which you are familiar? Why?
2. What is "situational crime prevention"?
3. What are the four components of CPTED? Which component would assist in improving security at a school with which you are familiar? Why?
4. When evaluating contract security officers, proprietary security officers, and a blended model, which option would you chose as a component of a school security management system? Why?

REFERENCES

Casteel, C., and Peek-Asa, C. (2000). Effectiveness of crime prevention through environmental design (CPTED) in reducing roberies. *American Journal of Preventive Medicine*, 18(4S), 99–115.

Dunlap, E. S. (2011). *Lecture Notes from SSE 827: Issues in Security Management.* Eastern Kentucky University, Richmond, VA.

Eisenbraun, K. D. (2007). Violence in schools: Prevalence, prediction, and prevention. *Aggression and Violent Behavior*, 12, 459–469.

Get Safe Online. (2011). Strengthen physical security: Don't let thieves ruin your business. Accessed June 28, 2011. http://www.getsafeonline.org/nqcontent.cfm?a_id=1098.

Greene, J. (2008). Violence in the ED: No quick fixes for pervasive threat. *Annals of Emergency Medicine*, 52(1), 17–19.

McCrie, R. D. (2007). *Security Operations Management.* Elsevier, Amsterdam, the Netherlands.

McDargh, J. N. (2009). Making more for less better. *Security Technology Executive*, 19(6), 30–31.

Phifer, R. W. (2007). Security vulnerability analysis for laboratories and small chemical facilities. *Journal of Chemical Health & Safety*, 12–14.

Prince William County Police Department. (2009). *CPTED Strategies: A Guide to Safe Environments.* Special Operations Bureau, Crime Prevention Unit, Prince William County Police Department, Manassas, VA.

U.S. Department of Homeland Security. (2010). Critical infrastructure. Accessed June 30, 2011. http://www.dhs.gov/files/programs/gc_1189168948944.shtm.

Villines, J. C., and Ritchey, D. (2010). Training: Asset or risk? *Security*, 47(10), 44–48.

Zahm, D. (2007). *Using Crime Prevention through Environmental Design in Problem Solving.* U.S. Department of Justice and Office of Community Oriented Policing Services, Washington, D.C.

Zalud, B. (2006). Officers: In-house or outsource? Security, 43(11), 58–60. Accessed June 9, 2012. http://search.proquest.com/docview/197815606?accountid=10628.

Zalud, B. (2007). Security officers as a business strategy. *Security*, 44(11), 79–83.

10 School Violence and At-Risk Students

Kelly Gorbett

CONTENTS

On April 20, 1999, two high school seniors entered their high school located in the small town of Littleton, Colorado, with guns, knives, and bombs with the intention of killing hundreds of their classmates and teachers. The end result was the deaths of twelve students, one teacher, and the two attackers with several others wounded. Widespread attention to the massacre led to panic among Americans and caused school personnel, administrators, researchers, and policy makers to examine and improve prevention and safety efforts. The tragedy at Columbine High School resulted in the most significant change to school violence prevention efforts and strategies in the United States.

School violence, however, has not been limited to just one single event. In fact, several schools have experienced crises to this devastating proportion, such as the Chowchilla school bus kidnapping in California in 1976 and school shootings in Paducah, Kentucky; Springfield, Oregon; and Santee, California, to name a few. Fortunately, the incidence of serious, violent attacks at this level on school campuses remains extremely low. However, given the degree of unpredictability and uncertainty of the occurrence of these events, schools must be prepared. As recently as during the writing of this chapter, continued reports of student conspiracies to commit violent attacks similar to the attack on Columbine High School in 1999 exist (e.g., a seventeen-year-old student from Tampa, Florida, was discovered by police in August 2011 carrying out plans to bomb and kill hundreds of students and staff on the first day of school at his former high school). Furthermore, although this degree of violence in schools is rare, other instances of violence and disruptive behavior are not, including but not limited to, name-calling, bullying, teasing, and threats (Furlong et al., 1997; Dwyer et al., 1998). Schools must have a plan to address behaviors of at-risk youth and provide interventions at all levels to help increase the likelihood of developing safe, effective school environments.

In this chapter, an overview of the major findings in the literature related to school violence is presented. Beyond the scope of this chapter, several textbooks written by experts in this field address these issues more extensively and are recommended for further review (Loeber and Farrington, 1998; Brock et al., 2002; Shinn et al., 2002; Sprague and Walker, 2005).

SCHOOL SAFETY

Despite the overwhelming fear that violent attacks such as the one at Columbine High School have caused for parents, school faculty, staff, and students, school is still the safest place that children go. According to the U.S. Bureau of Justice Statistics and the National Center for Education Statistics (2000), more children are victimized away from school than at school. Current statistics suggest that fewer than 1% of child deaths occur at school (U.S. Department of Education and U.S. Department of Justice, 1999). Overall rates of school crime have been steadily declining since the early 1990s (Snyder and Sickmund, 1999). Nevertheless, violence and disruptive behavior at varying levels of intensity and frequency still exist, leaving numbers of students feeling unsafe in their schools (Centers for Disease Control and Prevention, 2000). Discipline for these behaviors remains an area of public concern and scrutiny as numerous schools continue to focus on reactive, punitive measures rather than preventive ones.

Schools hold the ultimate responsibility of providing a safe environment for children. As human beings, safety is unarguably one of our most fundamental needs. Most developmental theorists (e.g., Erikson, Maslow, Ainsworth, Bowlby) stress the importance of developing feelings of safety, security, and trust early on so that future milestones and developmental tasks may be achieved. This holds true to children in schools as well. As Stephens (2002) argued,

> The quality of a child's education can be severely affected if a child is not in a safe and welcoming learning environment. Teachers cannot teach and students cannot learn in an environment filled with fear and intimidation (p. 49).

Therefore, student aggression and violence not only affect the individual student's own emotional well-being but also significantly affect everyone in the school environment (Batsche, 1997). These behaviors lead to disrupted learning and compromise the instructional time for all students (Martini-Scully et al., 2000). Schools promote learning when children feel safe and supported. A need remains for schools to identify and intervene with students at risk for violence.

RISK FACTORS

Early development is a critical time in a child's life when they develop important school-readiness skills, learn social rules, and form attachments. Positive models in the home and school setting are vital during this time of development. In fact, the majority of longitudinal research suggests that adolescent youth problem behavior begins in early childhood (Kazdin, 1987; Loeber and Farrington, 1998). For example, Capaldi and Patterson (1996) reported that children exhibiting antisocial behavior problems during their elementary school years were the most likely to have an early onset of police arrest. For this reason and as discussed later in this chapter, experts in the field stress the importance of early intervention efforts that work collaboratively within the home, school, and community as the best means of prevention in the area of at-risk youth.

Clearly, numerous obstacles and limitations exist for researchers to be able to uncover or pinpoint an exact cause of youth violence. Factors that contribute to violence for at-risk youth are usually complex and include a variety of sources. However, researchers have been able to identify several risk factors for at-risk youth. The risk factors were defined by Kingery and Walker (2002) as "family, neighborhood, and society-level behaviors, attitudes, and events that can be measured, that are present in the life of an individual student before or during a violent or aggressive outcome" (p. 74). As these authors noted, risk factors are not necessarily the causes of violence but demonstrate a causal relationship instead. Consensus among the research has been that the longer the exposure and greater number of risks involved, the more likely a child is to be at risk for violence (Patterson et al., 1992; Loeber and Farrington, 1998; Walker and Sprague, 1999). Walker and Sprague (1999) also pointed out that "risk factors operate at different levels and sometimes overlap" (p. 68).

Research has demonstrated through longitudinal studies that young children travel a destructive path from childhood to negative outcomes later in life. Walker and Sprague (1999) have developed a useful illustration that clearly portrays the familiar path of at-risk youth that begins with exposure to several risk factors beginning early in life, which leads to development of maladaptive behavioral manifestations (e.g., lack of school readiness and problem-solving skills), producing negative short-term outcomes, such as truancy and school discipline referrals, and eventually leads to the negative and destructive long-term outcomes, such as violence. In their work, Walker and Sprague stated,

> There are strong and clearly established links between these risk factors, the behavioral manifestations and reactions that result from exposure to them, the short-term negative effects on the developing child that flow from this exposure, and finally, the destructive, long-term outcomes that (a) too often complete this developmental progression and (b) ultimately prove very costly to the individual; to caregivers, friends, and associates; and to the larger society (p. 68).

In 2001, the U.S. Department of Health and Human Services released the surgeon general's report on youth violence, which indicated several risk factors (both personal and social) that predict youth violence. In the primary grades, risk factors included: 1) being male; 2) substance abuse; 3) aggression; 4) low intelligence; 5) antisocial parents; 6) poverty; 7) psychological conditions such as hyperactivity; 8) weak social ties; 9) antisocial behaviors, attitudes, beliefs, and peers; 10) exposure to TV violence; 11) poor school performance; 12) abusive parents;

13) poor parent-child relationships; and 14) broken homes. Secondary grade risk factors were identified as 1) crimes against persons; 2) family conflict; 3) academic failure; 4) physical violence; 5) neighborhood crime; 6) gang membership; 7) risk-taking behaviors; and 8) poor parental monitoring.

Beyond personal and social characteristics, Hawkins et al. (1992) described risk factors expanding across several contexts, including family, school, neighborhood, community, and the larger society. Walker and Shinn (2002) developed a comprehensive list of risk factors that expanded across these differing contexts, including child factors, family factors, school context, and community and cultural factors. These authors pointed out that the more proximal risk factors have the greater influence on the individual (e.g., family-based risks as compared with community factors). Child factors included such risks as insecure attachment, difficult temperament, low intelligence, and premature birth. Family factors included characteristics of the parents, such as teenage mothers, psychiatric disorders, and substance abuse, family environment (e.g., family violence), and parenting style (e.g., rejection of child, neglect, or abuse). School context included bullying, deviant peer group, and school failure, whereas community and cultural factors indicated risks such as lack of support services, socioeconomic disadvantage, and exposure to media portrayals of violence.

At this point, it is important to discuss that there has been a clear link established between school failure and at-risk youth. Udry (2000) reported that students who said they had frequent problems with their schoolwork were more likely to use alcohol, smoke cigarettes, become violent, carry weapons, and attempt suicide. Additionally, Walker and Shinn (2002) discussed the following school risk factors as having a direct relationship with the development of antisocial or destructive behavior: 1) getting in trouble with the teacher; 2) failure to engage and bond with the process of schooling; 3) being socially rejected by teachers and peers; and 4) failing academically, especially in reading. Studies have also demonstrated a clear relationship between dropping out of school and an increase in delinquent activity (Rumberger, 1987; Kirsch et al., 1993; Costenbader and Markson, 1998).

Other risk factors, such as dysfunctional family environments, poverty, and abuse, have an indirect relationship to school failure and clearly fit within the path of at-risk youth leading to the short-term and later long-term outcomes described by Walker and Sprague (1999). These risk factors likely exist from birth or a very young age; therefore, schools have little influence on these factors until the child enters school. However, Walker and Shinn (2002) argued that an important role for schools then is to

> enhance protective factors in academic, social-emotional, and mentoring-support domains in order to buffer and offset the negative effects of risk factors, particularly in the areas of school adjustment and achievement (p. 9).

Protective factors prevent or reduce the likelihood that a child will engage in antisocial, criminal behaviors later in life. Walker and Shinn included a list of protective factors across contexts that are associated with antisocial and criminal behavior.

Child protective factors included social skills, attachment to family, empathy, problem-solving skills, easy temperament, school achievement, and good coping style. Family factors included supportive, caring parents, secure and stable family, and strong family norms and morality, whereas school context included positive school climate, prosocial peer group, and a sense of belonging. Community and cultural factors included access to support services, community networking, and participation in church or other community groups.

Finally, schools must also be willing to examine their current practices and implement change where needed. Experts agree that it is never too late to implement change to support children and youth in our schools (Loeber and Farrington, 1998). Several areas of school practice have been identified by Sprague and Walker (2005, p. 60) that may contribute to antisocial behavior and the potential for violence, including:

- Ineffective instruction that results in academic failure
- Inconsistent and punitive classroom and behavior-management practices
- Lack of opportunity to learn and practice prosocial interpersonal and self-management skills
- Unclear rules and expectations regarding appropriate behavior
- Failure to effectively correct rule violations and reward adherence to them
- Failure to individualize instruction and support to accommodate individual differences (e.g., ethnic and cultural differences, gender, disability)
- Failure to assist students from at-risk backgrounds (e.g., those in poverty, racial/ethnic minority members) to bond with the schooling process
- Disagreement about and inconsistency in implementation among staff members
- Lack of administrator involvement, leadership, and support

EARLY INTERVENTION

Research focused on early intervention efforts has been widespread and has led to a growing awareness of the importance of early intervention on the prevention of later problems. These efforts have fortunately led to a number of policy and program changes, including some federal and state funding for early intervention programs. In fact, there have been changes to federal law that mandate and regulate delivery of services to children eligible for special education to extend to children in these early years. With the enactment of Public Law 99-457, amendments to the Individuals with Disabilities Education Act (IDEA) of 1997, children from birth to twenty-one years of age are now eligible for state and federal funding for special education and related services. This law also mandated a comprehensive child-find system, including descriptions and definitions of the early identification and assessment requirements. As a result, statewide early intervention programs (e.g., First Steps) have been developed to intervene for children ages birth to three years old with developmental delays and disabilities under IDEA part C and early childhood special education (IDEA, Part B-619).

Efforts have also been made to extend preschool programs to include children in poverty. For example, the most widespread and systematic are the federal programs serving children in poverty (e.g., Head Start, Early Head Start, Title I).

State-funded preschool programs linked to area K–12 schools are supported in many states as well.

Although this reflects a positive step in the right direction, further programming is needed to support early intervention efforts given the evidence on how important early intervention can be in the prevention of academic and behavior problems later in school. The positive effects of early intervention have been demonstrated in the areas of cognitive, language, and social-emotional development, as well as approaches to learning, including attention (Greenwood et al., 2011). Belfield (2005) reported that attending preschool was associated with a 12% fall in the rate of special education identification. Zigler et al. (2006) postulated that early childhood education provides not only the school readiness skills for academic learning but also a critical foundation in social and emotional regulation that makes learning possible.

PREVENTION

Recent literature has focused on prevention and early intervention efforts in promoting school safety. In fact, programs have been developed, such as the Safe Schools/ Healthy Students Initiative, funded by the Departments of Health and Human Services, Education, and Justice, to provide programs to school districts at the prevention, early intervention, and treatment levels to address social, behavioral, and mental health issues in cooperation with community partners and law enforcement agencies (U.S. Department of Education, 1999; Thornton et al., 2000).

As described above, risk factors for at-risk youth develop at a very young age, and these factors can eventually lead down that path of destructive long-term outcomes. Therefore, the need arises to divert students from this path very early in their lives. Because of the significant link between school failure and at-risk youth, schools must promote programs to improve early literacy skills, as well as behaviors that promote school success. Walker and Shinn (2002) stated that, "the two greatest risks for school failure are (a) the display of a very challenging behavior pattern (i.e., antisocial behavior, aggression, opposition-defiance, bullying, etc.) and (b) early school failure, especially in learning to read" (p. 11). Therefore, we must target early intervention and prevention efforts in these areas.

Research has demonstrated that children who have problems learning to read are at a greater risk for experiencing behavior problems in their early elementary years (McIntosh et al., 2006). Therefore, prevention efforts are aimed at the significant body of research that promotes the development of early literacy skills and disputes the belief that reading instruction begins when children start school. In fact, precursors for reading skills are taught in the early school years and begin developing prior to formal schooling. Typically, reading research had focused on discrete skills that are prerequisite to reading and have left out the important role that early childhood literacy plays in facilitating reading and writing acquisition (Gunn et al., 1995). Currently, emergent literacy is a popular topic that is being discussed and researched in the field of reading and education. Emergent literacy emanated from cognitive psychology and psycholinguistics and suggests that, "reading and writing develop concurrently and interrelatedly in young children fostered by experiences that permit and promote meaningful interaction with oral and written language" (Sulzby and Teale, 1991, p. 729). In addition, Teale and Sulzby (1986) added that within an emergent literacy framework, children's early attempts at reading and writing, although unconventional, are viewed as legitimate beginnings of literacy.

Beyond promoting and integrating early intervention for literacy and academic success, schools must also continue to work on current practices that address problem behaviors. Traditionally, schools have primarily adopted punishment practices, such as referrals and suspension. Unfortunately, these practices fall short of positive long-term effects and serve merely as a means of removing the child from the situation for a short period of time to provide respite (Sprague and Walker, 2005). Little evidence exists to support the long-term effects of punishment practices in reducing antisocial behavior (Skiba et al., 1997; Irvin et al., 2004). Sprague and Walker also argue that these practices are more likely, in the long term, to impair child-adult relationships and attachment to schooling, which only exacerbate the problem rather than help prevent it.

School discipline practices instead need to focus on teaching positive behavior and finding appropriate replacement behaviors for students. Development of behavior solutions should be a school-wide effort partnered with family and community resources. Sprague and Walker (2005) stated that, "punishment alone, without a balance of support and efforts to restore school engagement, weakens academic outcomes and maintains the antisocial trajectory of at-risk students" (p. 61). Therefore, these authors suggested that discipline measures should instead "1) help students accept responsibility, 2) place high value on academic engagement and achievement, 3) teach alternative ways to behave, and 4) focus on restoring a positive environment and civil social relationships in the school" (p. 61).

Research has focused on promoting the development of prevention models and frameworks to specifically address school violence within school settings. Although these programs will look completely different for every school, in order to address the school's individual needs and interests, Furlong et al. (2002) conceptualized school violence prevention frameworks to address at least the following eight domains: security, screening and assessment of aggressive behaviors, relationship building and bonding, individual student skill development, and developing nonviolent campus norms, schooling process and structure, school discipline and positive support, and enhancing school climate.

COMPREHENSIVE THREE-LEVEL APPROACH TO INTERVENTION

In order to promote intervention efforts aimed at prevention in a positive and inclusionary manner, rather than the traditional punitive and exclusionary manner (e.g., suspension, expulsion), Walker et al. (1996) revised the three-tiered model for use in schools in preventing violence and antisocial behavior patterns. Walker and his colleagues believed that in order to produce consistent, effective behavior changes, interventions needed to be comprehensive and target the entire population within and across all school settings, including the total student population as well as the entire school staff, not just individual teachers, and integrated within a well-developed district-level plan. They argued that interventions needed to be directly coordinated with each other and matched appropriately with the complex nature and level of severity of student behavior and academic problems. The three-tiered model includes intervention at the levels of primary prevention, secondary prevention, and tertiary prevention.

Primary prevention strategies are universal and address the whole school, aimed at meeting the needs of 75–85% of the student population, such as school-wide behavior-management systems. The goal of primary prevention strategies is to enhance protective factors and teach prosocial behaviors and skills for school success. According to Walker and Shinn (2002), although primary prevention strategies are underutilized by many schools, these prevention approaches may have "the greatest potential for use by schools in establishing a positive school climate" and "diverting at-risk students from a path leading to later negative or destructive developmental outcomes" (p. 16). Intervention aimed at this level can also help produce important behavioral data to help drive school-wide decisions on needed programming and intervention at the remaining two, more intensive tiers. There data also help identify the at-risk children needing more support at the secondary and tertiary levels.

Secondary prevention strategies provide more intensive academic and/or behavioral support to those students who do not respond to the universal interventions provided from primary prevention. Secondary prevention strategies are designed to meet the needs of 10–15% of students identified as at-risk within the school population. Interventions at this level become more intensive and individualized, as well as more expensive. Examples of secondary prevention strategies include small-group counseling and social-skill instruction and targeted reading interventions.

Finally, tertiary prevention strategies are designed for the most severely at-risk students, comprising the needs of 3–5% of all students. Generally, these students require intensive wraparound services to help the whole child and work with the family (Walker et al., 1996). At this level, intensive services should include resources and partnerships among families, schools, and community resources.

CONCLUSION

In this chapter, an overview of the major findings in the literature related to school violence was presented. Overall, schools are generally the safest place for children; however, to the extent to which violence and disruptive behavior occur, schools must be equipped with solid prevention efforts. There must be a plan to address behaviors of at-risk youth and provide interventions at all levels to help increase the likelihood of developing safe, effective school environments. Schools can no longer be reactive and present punitive approaches to address problem behaviors. Instead, school staff in every position must challenge current practices and be willing to promote change. It is never too late to promote change in the hopes of keeping school a safe, positive environment where children can learn and grow.

CASE STUDY

Micah is an eight-year-old, first-grade student at We Learn Elementary. He was retained in kindergarten because of a lack of school readiness, difficulty following instructions, and trouble listening and following daily school routine. His kindergarten teacher described him as "friendly, but quiet." She also stated that he seemed "immature."

This school year, Micah is beginning to fall further behind academically, especially in reading. He was recommended for a small-group reading program to help him with his reading deficits. His current first-grade teacher, Ms. Jones, asked the principal for a student assistance team (SAT) meeting because she was concerned about Micah academically, as well as his lack of consistent attendance.

At the SAT meeting, Ms. Jones shared with the school staff that she has had a difficult time getting in touch with Micah's parents to share her concerns. She also wanted to share with the parents information about a free reading-tutor program available after school. Ms. Jones also shared with the SAT work samples, including drawings that Micah had completed in art class. She told the SAT, "he really seems interested in drawing, and he is so talented." Results of recent screenings were shared with the SAT, and Micah had passed his hearing and articulation screening but failed his vision and language screening.

- If you were a member of this SAT, what other information would you want to obtain about Micah? Be sure to consider both home and school environments.
- From the information that has been provided in the case study, what would you identify as possible risk factors for Micah? What might be considered protective factors?
- Outline a plan for helping Micah be more successful in school, keeping in mind the three tiers of prevention. What areas need further exploration? In what areas does Micah need further help and support? What programs or resources would you recommend?

EXERCISES

1. Reflect back on the list of nine school practices that have been identified by Sprague and Walker (2005) that may contribute to antisocial behavior and the potential for violence. Thinking about the current discipline practices in your school, do any of the nine school practices identified fit as areas of concern and potential change at your school? Discuss ideas for promoting change in each of the areas of concern at your school.
2. What resources or programs are available in your school district and community that help promote early intervention and prevention for at-risk students? What other resources or programs do you think would be helpful in your school district or in your community?
3. Thinking about the three tiers of prevention—primary, secondary, and tertiary—how does your school already fit within this model of prevention? What programs or interventions do you have available at your school for students at each tier? Can you think of any areas of need?

REFERENCES

Batsche, G. M. (1997). Bullying. In *Children's Needs II: Development, Problems, and Alternatives*, edited by G. G. Bear, K. M. Minke, and A. Thomas, 171–179. National Association of School Psychologists, Bethesda, MD.

Belfield, C. R. (2005). The cost savings to special education from preschooling in Pennsylvania. Pennsylvania Build Initiative, Pennsylvania Department of Education, Harrisburg, PA. Accessed March 4, 2012. http://www.portal.state.pa.us/portal/server.pt?open=18&objID=381892&mode=2.

Brock, S. E., Lazarus, P. J., and Jimerson, S. R., eds. (2002). *Best Practices in School Crisis Prevention and Intervention*. NASP Publications, Bethesda, MD.

Capaldi, D. M., and Patterson, G. R. (1996). Can violent offenders be distinguished from frequent offenders?: Prediction from childhood to adolescence. *Journal of Research in Crime and Delinquency*, 33, 206–231.

Caplan, G. (1964). *Principles of Preventive Psychiatry*. Basic Books, New York.

Caponigro, J. R. (2000). *The Crisis Counselor*. Contemporary Books, Chicago, IL.

Centers for Disease Control and Prevention. (2000). 1999 youth risk behavior surveillance system. Accessed May 4, 2012. www.cdc.gov/mmwr/preview/mmwrhtml/ss4905a1.htm.

Constenbader, V., and Markson, S. (1998). School suspension: A study with secondary school students. *Journal of School Psychology*, 36, 59–82.

Dwyer, K., Osher, D., and Wagner, C. (1998). *Early Warning, Timely Response: A Guide to Safe Schools*. U.S. Department of Education, Washington, D.C.

Furlong, M., Morrison, G., Chung, A., Bates, M., and Morrison, R. (1997). School violence. In *Children's Needs II: Development, Problems, and Alternatives*, edited by G. G. Bear, K. M. Minke, and A. Thomas, 245–256. National Association of School Psychologists, Bethesda, MD.

Furlong, M. J., Pavelski, R., and Saxton, J. (2002). The prevention of school violence. In *Best Practices in School Crisis Prevention and Intervention*, edited by S. E. Brock, P. J. Lazarus, and S. R. Jimerson. National Association of School Psychologists, Bethesda, MD.

Greenwood, C. R., Bradfield, T., Kaminski, R., Linas, M., Carta, J. J., and Nylander, D. (2011). The response to intervention (RTI) approach in early childhood. *Focus on Exceptional Children*, 43(9), 1–22.

Gunn, B. K., Simmons, D. C., and Kameenui, E. J. (1995). *Emergent Literacy: Synthesis of Research*. National Center to Improve the Tools of Educators, U.S. Office of Special Education Programs, Washington, D.C.

Hawkins, J. D., Catlano, R. F., and Miller, J. Y. (1992). Risk and protective factors for alcohol and other drug problems in adolescence and early adulthood: Implications for substance abuse prevention. *Psychological Bulletin*, 112(1), 64–105.

Individuals with Disabilities Education Act Amendments. (1997). 20 U.S.C. §1400 et seq.

Irvin, L. K., Tobin, T. J., Sprague, J. R., and Vincent, C. G. (2004). Validity of office discipline referrals measures as indices of school-wide behavioral status and effects of school-wide behavioral interventions. *Journal of Positive Behavior Interventions*, 6(3), 131–147.

Katz, A. R. (1987). Checklist: 10 steps to complete crisis planning. *Public Relations Journal*, 43, 436–447.

Kazdin, A. (1987). Treatment of antisocial behavior in children: Current status and future directions. *Psychological Bulletin*, 102, 187–203.

Kingery, P. M., and Walker, H. M. (2002). What we know about school safety. In *Interventions for Academic and Behavior Problems II: Preventive and Remedial Approaches*, edited by M. Shinn, H. M. Walker, and G. Stoner, 71–88. NASP Publications, Bethesda, MD.

Kirsch, I., Jungeblut, A., Jenkins, L., and Kolstad, A. (1993). Adult literacy in America: A first look at the results of the National Adult Literacy Survey. Report prepared by Educational

Testing Service with the National Center for Education Statistics, U.S. Department of Education, Washington, D.C.

Loeber, R., and Farrington, D. P., eds. (1998). *Serous and Violent Juvenile Offenders: Risk Factors and Successful Interventions.* Sage, Thousand Oaks, CA.

Martini-Scully, D., Bray, M., and Kehle, T. (2000). A packaged intervention to reduce disruptive behaviors in general education students. *Psychology in the Schools*, 37, 149–156.

McIntosh, K., Horner, R. H., Chard, D., Boland, J. B., and Good, R. H. (2006). The use of reading and behavior screening measures to predict nonresponse to school-wide behavior support: A longitudinal analysis. *School Psychology Review*, 35(2), 275–291.

Patterson, G. R., Reid, J. B., and Dishion, T. J. (1992). *Antisocial Boys.* Castalia Press, Eugene, OR.

Rumberger, R. (1987). High school dropouts: A review of issues and development. *Review of Educational Research*, 57, 101–121.

Shinn, M. A., Walker, H. M., and Stoner, G., eds. (2002). *Interventions for Academic and Behavior Problems II: Preventive and Remedial Approaches.* National Association of School Psychologists, Bethesda, MD.

Skiba, R.J, Peterson, R. L., and Williams, T. (1997). Office referrals and suspension: Disciplinary intervention in middle schools. *Education and Treatment of Children*, 20, 295–315.

Snyder, H. N., and Sickmund, M. (1999). Juvenile offenders and victims: 1999 national report. Office of Juvenile Justice and Delinquency Prevention, U.S. Department of Justice, Washington, D.C.

Sprague, J. R., and Walker, H. M. (2005). *Safe and Healthy Schools: Practical Prevention Strategies.* Guilford Press, New York.

Stephens, R. (2002). Promoting school safety. In *Best Practices in School Crisis Prevention and Intervention*, edited by S. E. Brock, P. J. Lazarus, and S. R. Jimerson, 47–65. NASP Publications, Bethesda, MD.

Sulzby, E., and Teale, W. H. (1991). Emergent literacy. In *Handbook of Reading Research*, edited by R. Barr, M. L., Kamil, P. B. Mosenthal, and P. D. Pearson, Vol. 2, 727–757. Longman, New York.

Teale, W., and Sulzby, E. (1986). *Emergent Literacy: Writing & Reading.* Ablex, Norwood, NJ.

Thornton, T, N., Craft., C. , Dahlberg. L. L., Lynch. B. S., and Baer, K. (2000). *Best Practices of Youth Violence Prevention: A Sourcebook for Community Action.* Centers for Disease Control and Prevention, National Center for Injury Prevention and Control, Atlanta, GA.

Udry, J. (2000). The National Longitudinal Study of Adolescent Health: Research report. University of North Carolina, Chapel Hill, NC. Accessed March 4, 2012. www.cpc.unc.edu/projects/addhealth.

U.S. Bureau of Justice Statistics and the National Center for Education Statistics. (2000). *Indicators of School Crime and Safety.* (2000). U.S. Departments of Education and Justice, Washington, D.C.

U.S. Department of Education and U.S. Department of Justice. (1999). 1999 annual report on school safety. Washington, D.C.

U.S. Department of Education, Health and Human Services. (1999). Safe Schools/ Healthy Student Initiative. Accessed May 4, 2012. http://www.sshs.samhsa.gov/initiative/default.aspx.

U.S. Surgeon General (2001). Youth violence: A report of the surgeon general. U.S. Department of Health and Human Services, Substance Abuse and Mental Health Administration, Center for Mental Health Services, National Institute of Health, National Institute of Mental Health. Rockville, MD. Accessed March 4, 2012. http://www.surgeongeneral.gov/library/youthviolence/report.html.

Walker, H. M., Horner, R. H., Sugai, G., Bullis, M., Sprague, J. R., Bricker, D., and Kaufman, M. J. (1996). Integrated approaches to preventing antisocial behavior patterns among school-age children and youth. *Journal of Emotional and Behavioral Disorders*, 4, 193–256.

Walker, H. M., and Shinn, M. R. (2002). Structuring school-based interventions to achieve integrated primary, secondary and tertiary prevention goals for safe and effective schools. In *Interventions for Academic and Behavior Problems II: Preventive and Remedial Approaches*, edited by M. A. Shinn, H. M. Walker, and G. Stoner. National Association of School Psychologists, Bethesda, MD.

Walker, H. M., and Sprague, J. R. (1999). The path to school failure, delinquency, and violence: Causal factors and potential solutions. *Intervention in School and Clinic*, 35(2), 67–73.

Zigler, E., Gilliam, W. S., and Jones, S. M. (2006). *A Vision for Universal Preschool Education*. Cambridge University Press, New York.

11 Mental Health in the Schools

Kelly Gorbett

CONTENTS

The surgeon general's report on mental health estimates that 11% of U.S. children ages nine to seventeen have a diagnosable mental or addictive disorder resulting in significant functional impairment (U.S. Surgeon General, 1999). These statistics, coupled with the potential for disruptive or troubled youth, make a strong case for staffing of mental health support personnel in the schools and offering advanced training for school staff. Additionally, schools typically benefit from partnering with community agencies and programs, such as community mental health support for children, wraparound services, and mentoring programs.

Staffing of trained mental health professionals within schools is beneficial in providing extra support for teachers and staff in meeting the needs of students. School psychologists, school counselors, and school social workers are among the most trained in providing mental health support for students in the school setting. Family resource specialists also play an important role in meeting the mental health needs of students by helping link families with community support and providing extra training and resources for families. Many community mental health centers and programs are also now developing positions within the schools, known as school mental health therapists, to help meet the mental health needs of students.

In addition to the importance of staffing trained mental health professionals within schools, teachers and staff also need to have training to be able to respond to the mental health needs of their students, from basic classroom management and discipline strategies to working with troubled youth and responding to disruptive behavior, as well as suicide risk identification and intervention. At any time, a student may confide in a teacher that he or she may want to harm him or herself or a student may get in a violent verbal and physical confrontation with another student and threaten violence in front of a paraprofessional or teaching assistant. Teachers and school personnel must be trained to know the specific procedure and/or protocol for responding to these types of situations.

To meet the ever-growing needs of students and to foster a stronger learning environment, school personnel are recognizing the need to address behavior and mental health issues at the forefront. Recent literature suggests that the best way for schools to accomplish such a sizeable task is to develop a school-wide plan. The school-wide plan should be integrated within a well-developed district plan and would benefit from structuring interventions within the three-tier framework (as discussed in Chapter 10: School Violence and At-Risk Students) developed by Walker and his colleagues (1996). The three-tier framework, including primary, secondary, and tertiary prevention, provides a comprehensive framework for targeting interventions among the entire school population within and across all school settings (Walker et al., 1996).

Throughout the rest of this chapter, an overview of topics is discussed that should be considered and integrated within a comprehensive school-wide support plan for meeting the mental health needs of students. Beyond the scope of this chapter, several textbooks written by experts in this field address these issues more extensively and are recommended for further review (e.g., Brock et al., 2002, 2009; Shinn et al., 2002).

WHOLE SCHOOL PROGRAMS AND CLASSROOM LEVEL SUPPORT

At the primary prevention level of a school-wide plan, interventions should target the whole school population. Primary prevention programs should incorporate training of social skills and life skills into a classroom curriculum with topics such as good decision making and conflict resolution. The goal of this type of intervention is to promote a positive school climate with emotionally healthy students. Among the most strongly supported prevention programs in the research, *Second Step* (Moore and Beland, 1992) is also one of the most widely used programs. *Second Step* consists of a violence prevention curriculum that is structured and systematic for preschool through eighth grade children. The program is developed to teach students empathy, impulse control, problem solving, and anger management.

School personnel should incorporate whole-school activities and programs to promote social skills, good citizenship, and good behavior and to help students develop pride in their school. An emphasis should be placed on teaching skills to children and helping them learn to be accountable for their own choices and actions. Weekly classroom guidance activities can provide further support and teaching in these areas. Additionally, many schools have adopted peer mediation programs to help students feel ownership and responsibility in working through conflict and identifying reasonable solutions to problems among peers. Antibullying programs have also been given considerable attention in the research, and many schools are working on increasing antibullying efforts.

TARGETED INTERVENTIONS

The goal of primary prevention is to meet the needs of the majority of students; however, the need will exist for more targeted interventions among some students with greater risk factors and/or mental health needs. At the secondary prevention level, Walker and Shinn (2002) identified examples, such as small-group social-skills lessons, behavioral contracting, and assignment to alternative classroom placements. At the tertiary level, Walker and Shinn (2002) described intervention as collaborative wraparound, interagency approaches.

STUDENT SUPPORT TEAMS

Lewis, Brock, and Lazarus (2002) emphasized the need for schools to "be prepared to identify, refer, and intervene" with students identified as needing interventions at the secondary and tertiary levels as early as possible. Student support teams, also commonly referred to as student assistance teams, help address the individual needs of at-risk students. Student support teams include teachers, special educators, interventionists, school counselors, agency providers, family, and other relevant school personnel. The student support team considers all relevant factors related to the whole child, including but not limited to health information and screenings, attendance, educational records, behavior and social/emotional functioning, academic performance in the classroom, and universal screening performance.

Dwyer (2002) defined the responsibilities of the student support team as "evaluating student academic and behavioral needs, consulting with teachers and families, and generating effective planned interventions" (p. 174). The team approach is useful for gathering all relevant information and creating an intervention plan that can be carried through with consistency and integrity.

THREAT ASSESSMENTS

Threat assessments are a common approach among school- and district-wide plans to systematically assess risk among troubled students. Reddy et al. (2000) assert that threat assessment is the most promising approach for providing school personnel with a systematic risk assessment of targeted violence in schools.

The threat assessment model was developed by the U.S. Secret Service. Reddy et al. (2001) outlined the three guiding principles of the threat assessment approach: 1) there is no specific profile or single type of perpetrator of targeted violence; 2) there is a distinction between making a threat and posing a threat; and 3) targeted violence is not random or spontaneous. Reddy et al. (2001) defined the threat assessment approach as information gathering, using specific questions about the specific case, to "determine whether there is evidence to suggest movement toward violent action" (p. 169). They also provided examples of key questions utilized in a threat assessment, including questions that focus on:

> 1) motivation for the behavior that brought the person being evaluated to official attention, 2) communication about ideas and intentions, 3) unusual interest in targeted violence, 4) evidence of attack-related behaviors and planning, 5) mental conditions,

6) level of cognitive sophistication or organization to formulate and execute an attack plan, 7) recent losses (including loss of status), 8) consistency between communications and behaviors, 9) concern by others about the individual's potential for harm, and 10) factors in the individual's life and/or environment or situation that might increase or decrease the likelihood of attack (p. 169).

Threat assessments are utilized when "students engage in specific behaviors that draw the attention and concern of others, suggesting the potential for violence and need for intervention" (Lewis et al., 2002, p. 256). The goal of a threat assessment is to distinguish between a student making a threat and posing a threat or, in other words, actually engaging in a plan to harm oneself or others (Reddy et al., 2001 Lewis et al., 2002). Any time a direct threat of violence is made, the threat must be evaluated by the threat assessment process.

School- and district-wide plans that incorporate the threat assessment process utilize a team approach, including school psychologists, school counselors, school mental health therapists, administrators, law enforcement, and teachers. Various sources of information, such as reports from witnesses, student interviews, parent interviews, teacher interviews, and review of records should be utilized during the threat assessment process. School- and district-wide plans include specific protocol, usually involving a flow chart, for assisting staff in the threat assessment process. Record keeping is essential during the process. Threat assessment teams also need to be familiar with federal laws on duty to warn that require third parties or authorities to be informed when threats are made to or about a specific, identifiable individual (see *Tarasoff v. Regents of the University of California*, 1976).

SUICIDE RISK ASSESSMENT

Threat assessment procedures are utilized when a threat is made toward another individual. However, when a threat of suicide is made, different procedures are typically established within a school district. Much like the threat assessment process, a district establishes a standard, systematic protocol for school personnel to follow in response to a student suicide threat.

Suicide prevention efforts stress the importance of all school personnel becoming familiar with risk factors and precipitants (Kalafat and Lazarus, 2002). Kalafat and Lazarus also stress the promotion of protective factors, such as contact with caring adults and a sense of connection with school or community, as an important component of suicide prevention. Suicide prevention efforts can be implemented within whole-class curriculum involving social skills and problem solving.

When responding to a threat of suicide from a student, standard protocol should utilize a team approach with participation and resources from trained mental health professionals. In order to keep a student safe, the student should remain with school personnel after a threat of suicide has been made. Interviewing questions are aimed at assessing the student's motivation and intent, level of organization to formulate and execute a plan, prior suicidal attempts or behaviors, suicide-related behaviors (e.g., writing a note, giving away personal items), recent losses, or other factors in the individual's life that might increase or decrease the likelihood of suicide, coping skills, and feelings of hopelessness. Various sources of information are also important in assessing suicide risk, including student, parent, and teacher interviews and a review of records.

Assessing a person for suicide risk involves a "comprehensive analysis of risk factors and warning signs" (Granello, 2010). Granello further states that becoming familiar with the risk factors and warning signs is extremely important to effective suicide risk assessment. According to Granello, there are more than seventy-five identified suicide risk factors in the literature, such as impulsivity, poor problem-solving ability, recent loss, and an ongoing stressor. Examples of warning signs include withdrawal, giving away prized possessions, and developing a plan. Schwartz and Rogers (2004) concluded that the more warning signs and risk factors an individual possesses, the greater risk for suicide.

Because of the complexity and involvement of suicide risk assessment, trained mental health professionals should lead this type of assessment. Contacting parents and connecting with mental health resources are necessary components of suicide risk assessment. Furthermore, suicide risk assessment is ongoing (Granello, 2010). Therefore, follow-up is necessary from trained mental health professionals for students that are a suicide risk.

CRISIS PLANNING

Finally, an important part of any school- and district-wide plan is to focus on crisis planning. A critical first step of crisis planning is to develop a crisis response team. Brock et al. (2002) suggest that building a strong crisis team involves defining roles, obtaining administrative support, and gaining training and staff development. There is an abundance of research on crisis intervention, and consistently throughout the literature, experts in the field of crisis intervention promote the development of a crisis plan and training a crisis team before a crisis happens (Pitcher and Poland, 1992; Brock et al., 2001). Stephens (2002) described a good crisis plan as focusing on "crisis prevention, preparation, management and resolution" (p. 60). He further explained that a crisis plan should include a definition of a crisis, develop an understanding of the different types of crises a crisis team would respond to, and then provide a detailed, step-by-step process for all possible types of crisis situations.

Crisis plans should be proactive and preventative. However, the majority of U.S. schools are still relatively reactive in response to a crisis. During a crisis, school staff members are equally affected by the crisis, making it extremely difficult to organize an effective crisis intervention response. Therefore, research emphasizes the need for schools to put more efforts into planning and prevention of school crises (Pitcher and Poland, 1992; Brock et al., 2001).

Caplan (1964) identified three general classifications of prevention that guide crisis planning efforts: primary, secondary, and tertiary prevention. According to Caplan, primary prevention "involves lowering the rate of new cases of mental disorders in a population over a certain period by counteracting harmful circumstances before they have a chance to produce illness" (p. 26). Therefore, Caplan is describing the process of identifying risk factors of at-risk youth and trying to provide a positive school climate with emotionally healthy students. These activities and interventions occur before a crisis actually happens and are designed to help.

Secondary prevention, as described by Caplan (1964), refers to programs designed to reduce the rate of mental disorders by "shortening the duration of existing cases through early diagnosis and treatment" (p. 89). This is the immediate response offered to those who have entered into the crisis state. Examples include individual and group crisis intervention. Tertiary prevention as described by Caplan refers to programs designed to rehabilitate trauma victims to return their productive capacity as quickly as possible to its highest potential" (p. 113). At this level, therapies are provided by trained mental health professionals, and these students usually require long-term treatment.

School crisis prevention efforts have been strengthened throughout recent years with a new model and curriculum developed through collaborative efforts of work-groups sponsored by the National Association of School Psychologists (NASP). The PREPaRE curriculum was developed for educators and school-based mental health professionals to provide "training on how best to fill the roles and responsibilities generated by their membership on school crisis teams" (Brock et al., 2009, p. viii). The PREPaRE acronym represents interventions in a sequential and hierarchical order, specifically (Brock et al., 2009, p. ix):

- **P**revent and prepare for psychological trauma
- **R**eaffirm physical health and perception of security and safety
- **E**valuate psychological trauma risk
- **P**rovide interventions
- **a**nd
- **R**espond to psychological needs
- **E**xamine the effectiveness of crisis prevention and intervention

The curriculum is based on three assumptions: 1) it is critical to prepare for the crisis-related needs and issues of children; 2) multidisciplinary teams make the best use of professional skills; and 3) schools have their own unique structures and cultures. NASP offers training workshops to teach the PREPaRE curriculum and guidance for schools in developing their own crisis response plan.

CONCLUSION

In this chapter, an overview of topics was discussed that should be considered and integrated within a comprehensive school-wide support plan for meeting the mental health needs of students. To best respond to the mental health needs of children, school personnel must consider important staffing issues as well as training needs of all school employees. A school-wide plan, integrated within a district plan, is essential for helping to address the behavior and mental health needs of students. Interventions should be designed and implemented at a primary, secondary, and tertiary prevention level. School-wide plans should also consider building student support teams, threat assessment teams, and crisis response teams in order to be most effective in their efforts to meet the mental health needs of students.

CASE STUDY

As the new school principal of Pretend Elementary, you have been working with district personnel, administration, school staff, and teachers in reducing the number of behavior referrals and suspensions within your school. You have realized that Pretend Elementary has a high number of at-risk students. However, school efforts have failed at trying to meet the students' overall mental health needs. You are working on a school-wide plan, integrated with the district plan, to increase efforts to meet the mental health needs of the students at Pretend Elementary.

- What steps would you consider important in building your school-wide plan?
- Within your school, who would you consider to be key personnel in meeting the mental health needs of students?
- What resources and interventions would you want to include in your school-wide plan?
- What training would you want to provide to school staff?
- How would you work on building relationships and partnerships with community resources and programs?
- What teams would you create within your plan?

EXERCISES

1. Think about how your school meets the mental health needs of students. Consider and discuss these areas:
   a. Staffing needs—Who works within your school to help meet the mental health needs of students? What are their specific roles and responsibilities?
   b. Training needs—What training has already been provided to school personnel? What further training do you think would be important in helping to meet the mental health needs of students? Does the staff at your school feel comfortable responding to at-risk students?
   c. Intervention needs—What interventions are already being provided within your school to help meet the mental health needs of students? Do they span across all three tiers of primary, secondary, and tertiary prevention? What further ideas do you have to meet the intervention needs of your school?
2. What resources or programs are available in your school district and community that help meet the mental health needs of the students within your school? What other resources or programs do you think would be helpful in your school district or in your community? Does your school have a good partnership with community resources or programs?
3. Discuss your school's plan to meet the mental health needs of students. Does it involve a student support team, crisis response team, and/or threat assessment team? What other components might be important in relation to prevention efforts at your school?

REFERENCES

Brock, S. E., Lazarus, P. J., and Jimerson, S. R., eds. (2002). *Best Practices in School Crisis Prevention and Intervention*. NASP Publications, Bethesda, MD.

Brock, S. E., Nickerson, A. B., Reeves, M. A., Jimerson, S. R., Lieberman, R. A., and Feinberg, T. A. (2009). *School Crisis Prevention and Intervention: The PREPaRE Model*. NASP Publications, Bethesda, MD.

Brock, S. E., Sandoval, J., and Lewis, S. (2001). *Preparing for Crises in the Schools: A Manual for Building School Crisis Response Teams* (2nd ed.). Wiley & Sons, New York.

Caplan, G. (1964). *Principles of Preventive Psychiatry*. Basic Books, New York.

Dwyer, K. P. (2002). Tools for building safe, effective schools. In *Interventions for Academic and Behavior Problems II: Preventive and Remedial Approaches*, edited by M. A. Shinn, H. M. Walker, and G. Stoner, 167–211. NASP Publications, Bethesda, MD.

Granello, D. H. (2010). The process of suicide risk assessment: Twelve core principles. *Journal of Counseling and Development*, 88, 363–371.

Kalafat and Lazarus. (2002). Suicide prevention in schools. In *Best Practices in School Crisis Prevention and Intervention*, edited by S. E. Brock, P. J. Lazarus, and S. R. Jimerson, 211–223. NASP Publications, Bethesda, MD.

Lewis, S., Brock, S. E., and Lazarus, P. J. (2002). Identifying troubled youth. In *Best Practices in School Crisis Prevention and Intervention*, edited by S. E. Brock, P. J. Lazarus, and S. R. Jimerson, 249–271. NASP Publications, Bethesda, MD.

Moore, B., and Beland, K. (1992). *Evaluation of Second Step, Preschool-Kindergarten: A Violence Prevention Curriculum Kit*. Committee for Children, Seattle, WA.

Pitcher, G., and Poland, S. (1992). *Crisis Intervention in the Schools*. Guilford Press, New York.

Reddy, M., Borum, R., Berglund, J., Vossekuil, B., Fein, R., and Modzeleski, W. (2001). Evaluating risk for targeted violence in schools: Comparing risk assessment, threat assessment and other approaches. *Psychology in the Schools*, 38, 157–172.

Schwartz, R. C., and Rogers, J. R. (2004). Suicide assessment and evaluation strategies: A primer for counseling psychologists. *Counseling Psychology Quarterly*, 17, 89–97.

Shinn, M. A., Walker, H. M., and Stoner, G., eds. (2002). *Interventions for Academic and Behavior Problems II: Preventive and Remedial Approaches*. National Association of School Psychologists, Bethesda, MD.

Stephens, R. (2002). Promoting school safety. In *Best Practices in School Crisis Prevention and Intervention*, edited by S. E. Brock, P. J. Lazarus, and S. R. Jimerson, 47–65. NASP Publications, Bethesda, MD.

U.S. Surgeon General. (1999). Mental health: A report of the surgeon general. U.S. Department of Health and Human Services, Substance Abuse and Mental Health Administration, Center for Mental Health Services, National Institute of Health, National Institute of Mental Health, Rockville, MD. Accessed March 4, 2012. http://137.187.25.243/library/mentalhealth/home.html.

Walker, H. M., Horner, R. H., Sugai, G., Bullis, M., Sprague, J. R., Bricker, D., and Kaufman, M. J. (1996). Integrated approaches to preventing antisocial behavior patterns among school-age children and youth. *Journal of Emotional and Behavioral Disorders*, 4, 193–256.

Walker, H. M., and Shinn, M. R. (2002). Structuring school-based interventions to achieve integrated primary, secondary and tertiary prevention goals for safe and effective schools. In *Interventions for Academic and Behavior Problems II: Preventive and Remedial Approaches*, edited by M. A. Shinn, H. M. Walker, and G. Stoner. National Association of School Psychologists, Bethesda, MD.

Section II

School Safety

12 Fire Hazards

Greg Gorbett

CONTENTS

This chapter introduces fire safety concepts from the perspective of fuel, heat, and oxygen. The statistics regarding fires within educational properties is included here as support for the fire safety recommendations made throughout the chapter. Finally, a discussion of general fire safety aspects will be included to assist the school official in making appropriate decisions when faced with the risks of fire.

THE FIRE PROBLEM IN SCHOOLS

There were 6,260 structure fires annually reported to municipal fire departments in educational properties between 2005 and 2009 (Evarts, 2011). The classification of an educational property by the National Fire Protection Association (NFPA) for purposes of the statistics is broken into three subcategories: 1) day-care centers; 2) nursery, elementary, middle, high, and nonadult schools; and 3) college classroom buildings and adult education centers. These fires resulted in an average of eighty-five civilian fire injuries and in excess of $112 million in direct property damage annually. The vast majority of these fires (72%) occurred in nursery, elementary, middle, high, or nonadult schools. Fires in this category also resulted in 80% of the injuries and 85% of the total property loss. It is obvious from these statistics that preK–12 schools are the most susceptible to fire and may benefit the most from fire safety principles.

A review of the statistics from fires within preK–12 educational properties will assist in identifying the greatest threats. Studies revealed that over half of the fires (51%) were intentionally set (incendiary). The next leading causes of fires was cooking equipment (21%), playing with a heat source (18%), and heating equipment (10%). Cooking equipment fires were the second leading cause; however, they resulted in the most civilian injuries (50%). The leading ignition source for these school fires was a lighter (21%). It was found that almost one-third (31%) of all fires originated in a bathroom or lavatory. The most common first fuels ignited were solids, including rubbish, trash, or waste (25%) and magazines, newspapers, or other writing paper (11%). The majority of fires (38%) started between 10 a.m. and 2 p.m. In summary, the leading cause of fires within K–12 schools is incendiary in nature, where a lighter is used to ignite a solid, with the fire originating in a bathroom or lavatory, and the fire usually occurs around lunchtime.

FIRE TRIANGLE—CORNERSTONE OF FIRE PROTECTION

Fire is a rapid oxidation process, which is a chemical reaction resulting in the evolution of heat and light in varying intensities (NFPA 921, 2011). The fire triangle is the most popular portrayal of fire and is considered the cornerstone of fire protection. The fire triangle consists of three equal sides: 1) fuel, 2) heat, and 3) oxygen. These three elements are required for ignition and continued combustion. The fire triangle concept is so comprehensive that it is taught to elementary students for simple memorization, yet it also serves as the principle for many doctoral dissertations.

In its basic application, the fire triangle defines when ignition of a fuel is possible, whether the fire will continue, and when suppression of the fire is possible. Ignition can only occur when these three elements are present and in the correct quantity. A fire continuing to burn is also only possible if these three elements are in the correct quantity and present. Firefighters and fire protection engineers use the depiction of the triangle to explain how a fire can be extinguished by simply removing one of the three legs of the triangle.

The fire triangle must have fuel, heat, and oxygen present and in sufficient quantity or state in order to develop into a chemical chain reaction. This chemical chain reaction serves as the fourth component when the triangle is represented by what is known as the fire tetrahedron. The fire tetrahedron consists of fuel, heat, oxygen, and an uninhibited chemical chain reaction. In other words, for ignition to occur there must be a fuel in the correct state (i.e., gaseous) that has mixed with an adequate amount of oxygen to form a flammable mixture, which when presented with a suitable heat source (i.e., ignition) will ignite and undergo combustion. This chapter begins by dividing out these three sides of the triangle to discuss the hazards associated with the fuel(s), heat, and oxygen commonly found within schools.

INTRODUCTION TO THE FIRE SAFETY CONCEPTS TREE

The fire safety concepts tree is a simple way to evaluate fire safety within structures where the ultimate objective or goal is to save lives from fire (Figure 12.1). It is a simple decision framework that outlines the various avenues that can be taken to accomplish the main objective. The main objective—saving lives—can be attained by either preventing the ignition of the fire or managing the impact of the fire.

In an ideal world, fire protection professionals would always choose the path of preventing ignition, where heat sources and fuel would be separated from each other completely. This is almost impossible in today's society, but it should be attempted nevertheless. Consequently, fire safety must often evaluate the other path of the tree where it is assumed that ignition may occur and structures and systems must be designed and maintained to manage the impact of the fire. Managing the fire is broken into managing those aspects exposed to the fire or managing the fire itself.

Much of the next chapters within this text will focus in on the design and general issues associated with managing the fire itself. Managing a fire can be accomplished in three different ways: active, passive, or a combination of both. Active fire protection includes elements that require some sort of action, be it mechanical, manual, or electrical, for a system to begin its protective operation. These systems most commonly include fire detection and alarms or sprinkler systems. The features of fire protection chapter will focus on sprinkler systems and active systems. Passive fire protection includes elements that do not require an external activation for their protection feature to operate. According to the *NFPA Fire Protection Handbook* (Cote, 2003), these elements can be broken into three categories: rate of fire growth, containment or compartmentation, and emergency egress. These concepts will be covered in greater detail in the features of fire protection and life safety chapters.

It is imperative that there are multiple layers of protection against fire. If the choice is to only rely on one, there is a very good chance that people will die. For instance, the Seton Hall University dormitory fire that killed three and injured more than fifty students is an example of what happens when a school relies on one means of protection.

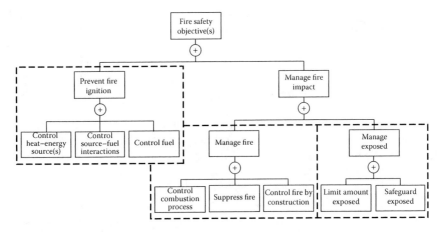

FIGURE 12.1 Fire safety concepts tree.

The dorm was not fitted with sprinklers, nor was it properly compartmentalized, but it had an automatic detection and alarm system that was supposed to ensure the alerting of occupants for moving them outside to safety. However, the fire protection professionals had not planned on many of the students not awakening or just plain ignoring the alarm and not taking action. With the lack of sufficient backup protection in the dorm, the fire quickly spread from the lobby area to the residential areas causing the injuries and deaths. Had the dorms been properly compartmentalized, the spread of the fire would have been slowed considerably, allowing more time for the fire department to arrive and control the fire and for the students to be rescued.

FUELS

The first side of the fire triangle discussed here is fuel. Fuels must be in their gaseous state for ignition and combustion to occur. There are three common states of matter that a fuel can initially exist as: solids, liquids, or gases. Fuels that are initially in the solid or liquid state must be heated first before gases are emitted and ignition can occur. Therefore, consideration of how a solid or liquid changes phase to a gas will be taken into account here. The discussion begins with fuels already in the gaseous state, because ignition is ultimately driven by this state, and it applies to all fuels regardless of their initial state of matter.

A method to evaluate a fuel's potential ignition and combustion hazards is to review the Material Safety Data Sheet (MSDS) for the material. It is recommended that all facilities, including educational properties, have a collection of these sheets to properly evaluate the potential hazard and proper usage for different materials. In this section, many of the general concepts about fuels will be discussed with a focus towards covering the information provided by the MSDS. Finally, each section will have a brief description on the preventative measures that can be implemented to prevent future fires.

GASES

Matter in a state where atoms collide but are not in a fixed arrangement and assume the shape or their container is in the gaseous state (Meyer, 2005). Gases are the simplest state of matter and are therefore the easiest fuel to discuss. However, gases are elemental to the understanding of how combustion occurs in all states of matter and will be discussed first. Fuels can typically only enter the combustion reaction when they are in a gaseous state; liquids and solids must turn into a gas before combustion is possible.

A mixture of fuel gases and oxygen must be in specific percentages in order to support a combustion reaction. Most commonly, combustion reactions occur with oxygen in the air and can occur within a range reported as a volume percentage of gas to air. The percentage of fuel and air where the lowest amount of fuel will ignite and undergo combustion is called the lower flammable limit. Conversely, the highest percentage of fuel to air where combustion can occur is the upper flammable limit. A mixture with insufficient fuel gas is too lean to burn; mixtures with insufficient air are too rich to burn.

The two most common fuel gases that may be encountered in schools are propane and natural gas. These fuels are commonly used for cooking and heating appliances. The lower and upper flammable limits for propane gas are commonly reported as 2.1 and 9.5% fuel in air (Gorbett and Pharr, 2011). In other words, when an atmosphere is filled with 2.1–9.5% of propane gas, combustion can occur. Outside of these limits, propane will not ignite. Natural gas is primarily composed of methane gas, which has commonly reported flammable limits of 5 and 15% fuel in air. These limits exist for local concentrations, as well as serve as the required concentrations in larger volumes (i.e., rooms). For example, if a large-enough volume is filled between these limits because of a gas leak, then the mixture can become explosive, resulting in a combustion explosion. It is important to notice that the two most common fuel gases found within schools require a very low percentage of gases in the area for ignition of a fire or a combustion explosion to initiate.

The characteristic of gases that determines whether they rise or fall in air is termed their vapor density, sometimes referred to as gas specific gravity. Dry air is assigned a vapor density of 1. More-dense gases (higher mass to volume ratio) have a vapor density greater than 1 and tend to displace the air and concentrate at lower levels, sinking in air. A common example of a gas that is heavier than air is propane, which has a vapor density of 1.5. If enough propane has leaked out, it tends to collect in the lower portions of compartments. Conversely, less-dense gases have vapor densities of less than 1 and tend to rise in air. An example of this is methane or natural gas, which has a vapor density of 0.553 (Gorbett and Pharr, 2011).

Gases are typically easier to ignite because they are already in the correct state for combustion, with the only restriction being whether the flammable limits are met for combustion to occur. This is one of the reasons that gases are commonly used as the primary fuel for residential, commercial, and industrial applications.

Because of their ease of ignition and their abundant usage, there are some safety precautions that have been implemented on these fuel gases in the United States. Both of these gases are naturally colorless and odorless, which means that a gas leak could occur, and no one would be aware. However, because of their danger and wide usage, federal regulations have been placed on the transport and delivery of these gases where an odorant is required to be inserted into the gases prior to delivery to the customers. This added odorant to the gas, usually methyl mercaptan, causes it to smell like rotten eggs, allowing it to be detectable if a gas leak occurs. Employees that work in areas where these gases are used should be educated on this danger and instructed to contact the local fire department.

Flammable limits apply to all fuels regardless of their initial physical state. Thus, liquids and solids also have a required minimum amount of fuel that must be present before ignition can occur.

LIQUIDS

Liquids that produce vapors that can undergo combustion with relative ease are considered ignitable liquids. Liquids do not burn; rather the vapors produced by the liquids actually undergo combustion. Because it is the vapors that ignite, flammable limits and many of the same principles that were discussed in gases are imperative when dealing with ignitability and flame spread of liquids. As the liquid increases in temperature, vapors are evolved from the surface of the liquid. Once these vapors are within their flammable limits, ignition can occur. Combustion of these vapors causes heating of the liquid mass, which causes the continued release of vapors that can enter the combustion reaction. Heating can come from an external source, atmosphere, or radiant heat feedback from a flame to the fuel surface.

Liquids volatize (change to vapor) when sufficient energy is present to cause conversion; an example of this is water left uncovered evaporating into the atmosphere over time. The rate of conversion is the main consideration when differentiating the major classes of liquids. Some liquids can vaporize from the atmospheric temperature alone and be hazardous at normal temperatures. For example, there is enough energy transferred to gasoline for it to emit vapors at a relatively low temperature.

One physical concept that is recorded on a liquid fuel's MSDS is specific gravity. Specific gravity describes a comparison of a liquid substance's density with the density of water. Liquids with specific gravity of less than 1 are less dense than water and tend to float on water, whereas liquids with a specific gravity greater than 1 will sink. Oils and most ignitable liquids will have a specific gravity less than 1, meaning that they will float on top of the water. This concept is similar to that discussed regarding the vapor density of gases. The vapor density concept also relates to the vapors emitted from the liquid and is also listed on the MSDS. Those liquid vapors that have a vapor density greater than 1 tend to sink in air. For example, gasoline vapors have a vapor density of approximately 3–4 and tend to concentrate in the lower levels of a compartment. This is one of the reasons why codes require water heaters in garages or utility rooms to be placed 18-inches above the floor, because in the event of a gasoline spill the vapors would concentrate near the floor and could possibly be ignited by the open pilot flame.

The flash point of a liquid is the test for determining the relative hazard of a liquid substance. The flash point (f.p.) of a liquid is the lowest temperature of a liquid at which it gives off vapors at a sufficient rate to support a momentary flame across its surface, as determined by specific laboratory tests (Gorbett and Pharr, 2011). In other words, the flash point of a liquid is the temperature that the liquid must reach to vaporize enough vapors to reach the lower flammable limit. Flash points are determined by specific laboratory test protocols that produce a momentary flash of flame across the surface of the liquid. There are several different test apparatus for flash point tests, but they are typically generalized as two types: open cup or closed cup. The MSDS of a liquid lists the flash point temperature with the designation of open or closed cup. Each individual test method may produce slightly different flash points for the same liquid. The classification system for flash points is based on a 100°F separation. Those liquids that have flash points under 100°F are considered flammable liquids, and those liquids with flash points above 100°F are classified as combustible liquids. The 100°F split was chosen because of the relative hazards associated with maximum normal atmospheric temperatures. For example, gasoline is a commonly used flammable liquid that has a recorded flash point of −45°F, whereas kerosene is a commonly used combustible liquid with a recorded flash point of 115°F.

Ignitable liquids are commonly found within schools including both flammable and combustible liquids. Flammable liquids, such as gasoline (f.p. = −45°F) may be commonly found in utility rooms or facility services buildings, whereas acetone (f.p. = 1°F) may be found in chemistry laboratories or even within cleaning solvents. Combustible liquids, such as kerosene, diesel fuel, or paints (f.p. = 100–200°F) may be commonly found in utility rooms, whereas cooking oils (f.p. = 200°F) are found in kitchens. Just because something is listed as a combustible liquid does not mean it is safe. It is relatively safer than the flammable liquids but can be equally dangerous depending on its use. Some of the uses where combustible liquids can easily ignite are described later in this chapter.

Liquids with high flash points (combustible liquids) are less likely to ignite under typical atmospheric conditions, although when the ambient temperature is elevated, those liquids are prone to act similarly to flammable liquids. Examples include diesel fuel (fuel oil #2) with a flash point of 120°F dispersed on asphalt pavement during hot summer days. External heating can raise the fuel above its flash point; thus sufficient vapors will be present to sustain a flame should the minimum ignition energy be introduced.

Characteristics associated with ignitable liquids are relatively simple to understand when a single element or compound is involved; however, mixtures are more difficult to understand. Flammability characteristics of the most volatile element or compound may persist; this becomes the danger indicator for the entire volume of mixture. This is a common problem with mixing chemicals for laboratories or when cleaning with multiple cleaning solvents.

Finely divided particles of liquid dispersed in air are known as aerosols. Aerosols convert to vapor with greater ease compared with a pool of the liquid because there is no need to heat a significant volume of the liquid to its flash point. An aerosol is composed of very small particles of the liquid and does not lose energy to the surrounding mass, as is found within a pool of the same liquid. Once ignition of a few aerosol droplets occurs, flame propagation can ensue. Aerosols are not always intended to be in the form, such as with aerosol dispensed from pressurized cans and spray nozzles. Often, high-pressure fluid transfer systems experience small ruptures, which provide the opportunity for small particles to be released. Fluid from hydraulic lines and other pressurized liquids have been shown to easily ignite because of this phenomenon.

The preventative steps for liquid fuels are to first identify their locations within the building. Second, ensure that the material safety data sheets are acquired and current for each of these materials. Next, establish policies for the safe storage and use of these liquid fuels. For example, if there are facility issues that require diesel fuel or gasoline, make certain that the fuel remains out of the occupied buildings and is locked away in an appropriate flammable liquids storage cabinet. Limit or prevent student interaction with this location.

SOLIDS

Solids are matter that has a definite shape and volume. When heat is applied to a solid material, the temperature begins to increase at a rate dependent on the material properties and the heat applied. As heat is applied and the temperature begins to increase, the energy may be great enough to cause the molecular bonds to break down into lower molecular weights and release from the solid as a vapor. The breakdown or decomposition of the molecular bonds by thermal energy is better known as thermal decomposition. Solids, depending on their composition, are expected to undergo one or more of the following changes when exposed to a sufficient heat source: melt, dehydrate, char, and/or smolder.

The majority of solids will undergo a chemical decomposition, better known as pyrolysis. Pyrolysis is the chemical decomposition of the solid fuel by the application of a heat source. As the molecular bonds begin to break down, smaller molecules (lower molecular weights) are released in the gaseous form and ignited if released from the solid at a sufficient rate that the flammable limit is reached. Most solids will undergo this process of pyrolysis when exposed to a sufficient heat source; some, however, may melt first and vaporize like a liquid. For example, cellulosic materials (i.e., wood, paper) and thermoset polymers will undergo pyrolysis and leave behind a carbonaceous residue (char) in the presence of a sufficient heat source, but thermoplastics (plastic bottles) will melt when exposed to heat and vaporize similar to liquids.

The surface-area-to-mass ratio is an important concept that must be considered when analyzing combustion of solid fuels. As the surface area increases and the mass is lessened, the fuel particles become more finely divided. When heat is applied to a solid or even a liquid, the heat is conducted into that material. If a large mass exists behind the surface, the material has a greater ability to dissipate the energy away from its surface. If the material has been cut or divided out of the large mass, then its surface-area-to-mass ratio has increased. When the same heat as before is now applied to this lesser mass, the energy cannot dissipate as much because of the lack of mass for the heat to be conducted into. Therefore, the energy is able to increase the temperature at the surface at a much quicker rate, which may result in ignition of the material at a much faster rate.

For example, take a wood log and apply a match. Would you expect the wood to ignite? Most likely not, because the energy imparted by the flame is being dissipated through the large mass of the wood and does not allow the temperature to increase to the point to cause pyrolysis. However, when we cut or sand this wood log, wood shavings may be collected. If we apply that exact same match, do you expect the wood shavings to ignite now? The only thing that has changed is the surface-area-to-mass ratio. More likely than not, the wood shavings will ignite. When the match is exposed to the lesser mass, there is not as much mass to allow for heat to be conducted away and dissipated from this localized location. Because the energy cannot be dissipated, the temperature of the wood increases, enabling the pyrolysis process.

The same principle applies when trying to ignite a material on a bulk surface area versus an edge. For example, when trying to ignite a piece of paper, it is much easier to ignite the paper on its edge rather than in the center. The material properties of the paper do not change from the center of the paper to its edge. The only item that changes is the ability of the energy to easily be dissipated through the material. The center of the paper allows heat to be conducted in every direction away from the heat source, preventing a fast temperature rise. The edge of this paper has only certain directions that the heat can be conducted into the remaining mass, which delays the energy dissipation from its surface.

The higher surface-area-to-mass relationship can be found in almost every wood shop in schools. Housekeeping is an important safety precaution in these areas. It is important for the school to have a strict policy on housekeeping in these locations.

The lessened mass found with loose pieces of paper has been shown to be a problem with schools, especially when these fuels are placed on bulletin boards, hung around classrooms, and along the walls. Decorations within classrooms and along corridor walls have become a major fire safety issue within schools (Burnside, 2008), so much so that some states have fire safety laws in place against covering a certain amount of the wall surface area within classrooms and along hallways. Massachusetts passed a law that forbids schools from having classrooms with more than 20% of their classroom walls covered with any decoration or student work and 10% of the corridor walls (Coan, 2011). This regulation originates from the International Building Code where it is required because of the ease of ignition and fast flame-spread hazard associated with paper (Burnside, 2008). Additionally, the orientation of this fuel creates a faster flame spread and increased heat release rate, further increasing their hazards to life safety. Strict adherence to limiting the amount of decorations and student work along walls and hanging from the ceiling is one method of preventing fires. The other option is to place student work and decorations behind a proper glass covering, such as a shadow box or display case. One caution with this tactic is that the use of Plexiglas for these display coverings is strongly prohibited because of its ease of ignition.

Storage of solid fuels throughout the school is a potential fire hazard. Excessive storage of paper products, trash, decorations, cardboard boxes, plastic totes, wooden pallets, and other similar items may provide a fuel that could overwhelm the sprinkler system in the event of a fire. A sprinkler system is designed based on the expected fuel load for a given area. If an area becomes a catchall for overflow storage, then the fuel load for that given area may be too much for the sprinkler system to handle, and the fire may not be suppressed. For example, storage closets, janitor closets, backstage areas, multipurpose rooms, and extra classrooms usually become cluttered with various items that do not have any other place to be stored. To prevent this issue, institute policies where items will be thrown away when they are stored outside of their designated locations. Also, idle pallets present a major fire hazard. Pallets are the wooden or plastic frames that are used for shipping large items or a collection of large items through freight. Idle pallets are stacks of pallets that are no longer in use and waiting to be removed or thrown away. Those areas where unloading and loading may occur (dock areas) and outside trash areas adjacent to the school building are areas where idle pallets may be stored. These present one of the largest possible fuel loads in a school, and it is guaranteed that the sprinkler system was not designed for this fuel load. For these areas, do not allow idle pallets to accumulate or be stacked for any length of time. Sprinkler systems will be discussed in greater detail in the life safety chapter.

HEAT

Heat is the transfer of energy based on a temperature difference between two objects or different regions of a single object (Gorbett and Pharr, 2011). The flow of energy is always in the direction from a high temperature to a lower temperature. Temperature is simply a measure of the average kinetic energy of the molecules within a body and should not be confused with heat.

Heat transfer is the most important concept to understand when analyzing fire phenomena. It is the driving factor for every aspect of fire development, including ignition, growth, detection, flashover, and suppression. The transfer of energy in the form of heat causes fuels to begin pyrolyzing (vaporizing). When heat transfer is intense enough for long enough duration, a flammable mixture can form that permits ignition. Heat is also the driving factor for flame spread across the initial fuel and/or radiant ignition of secondary fuels. From a suppression standpoint, when the heat can be removed from the fuel's surface, then the production of gaseous fuels decreases, and the fire is extinguished. Some of the basic sources of heat can come from electrical (resistance, arcing, and sparking), chemical reactions, mechanical (compression and friction), lightning, hot surfaces, open flames, and the sun.

There are three different modes or processes by which heat can transfer between bodies including conduction, convection, and radiation. Conduction and convection require an intervening medium (i.e., solid, liquid, or gas), whereas radiation requires no intervening medium.

Conduction heat transfer is the transfer of energy through direct molecular excitation, which is most prevalent through solids because of their molecular density. An example of conduction is the transfer of energy through a metal cooking utensil that has been placed in a fire. Convection heat transfer is the transfer of energy through a fluid (gas or liquid). Examples of convection heat are boiling water and hot air balloons. Radiant heat transfer is the transfer of energy via electromagnetic waves and requires no intervening medium. The best example of radiant heat transfer is the sun.

Common ignition sources that are found within schools include open flames, hot surfaces, chemicals, heaters, and cooking. Open flames are statistically the greatest threat for schools. The statistics show that incendiary fires and playing with a heat source accounted for 69% of all fires. As discussed in the fire safety concepts tree section, removal of the ignition source is the primary mechanism to prevent ignition. Because lighters were shown to be the number one ignition source for fires within educational properties, it is reasonable to ban lighters from school grounds. It would also be reasonable to ban all open flames from school grounds, including matches.

The second most common ignition source within these occupancies is cooking equipment. Cooking fires resulted in the greatest number of civilian injuries. Therefore, it is essential for schools to consider training their kitchen staff on how to effectively extinguish cooking fires. There will be a more comprehensive discussion on fire extinguishers and their safe use in the features of fire protection chapter. Annual fire extinguisher training may be a simple solution to this problem.

Portable heaters, both convective and radiant, are often used as supplemental heat sources during the winter months. They are also common ignition sources typically because of careless use. Heaters should never be placed within three feet of a combustible item. Heaters threaten college dormitories more than K–12 school facilities. However, with the initiative to keep heating costs low because of economic and environmental concerns, it is expected that more of these portable heaters will be used to supplement a building's system. There are two ways of dealing with this problem: 1) to ban these portable heaters altogether or 2) to provide safety guidelines to all regarding the proper use of these heaters.

The preventive steps here are simple: ban the use of that specific ignition source or implement procedures to regulate its use. The ability to ban all ignition sources is not feasible; therefore, it is important for school officials to be careful with the storage and types of fuels that are allowed within the buildings.

OXYGEN

Atmospheric air is composed of 21% oxygen. The oxygen within the air serves as the principle oxidizer for most combustion reactions. The removal of oxygen as a means to extinguish or suppress a fire is impractical for most circumstances and is definitely not a choice when life safety is to be considered. Therefore, the removal of oxygen during a fire or to prevent a fire from occurring in educational properties is impractical. It is a side of the triangle that we must manage but cannot remove.

Medical oxygen is often used for individuals with breathing problems, which introduces elevated oxygen atmospheres. The oxygen within the cylinder is not a fuel. However, it serves to increase the rate of combustion of whatever fuel is burning nearby. These elevated areas of oxygen allow fuels to ignite easier and burn better. Therefore, it is important to increase the control of ignition sources when medical oxygen cylinders are in use. Increased awareness and control of these ignition sources is the primary mechanism of preventing ignition in these circumstances.

In addition, if the school has a shop class or facility services area, it is likely that an acetylene torch may be present. Oxygen cylinders are used in conjunction with the acetylene fuel to permit higher temperatures for cutting and welding purposes. These cylinders should remain in locked areas at all times and should never be used without proper supervision.

PROTECTING AGAINST INCENDIARY FIRES

The majority of fires in educational properties are intentionally caused. Thus, this section of the chapter provides recommendations with how to design against such fires. The term *incendiary* is the correct term for classifying a fire that was "intentionally ignited under circumstances in which the person knows that the fire should not be ignited" (NFPA 921, 2011, p. 140). Dr. John DeHaan noted that there are three principle elements that are required to establish an incendiary fire as falling within the legal condition of *arson* (2011, p. 508).

1. There has been *a burning of property*. This must be shown to the court to be actual destruction.
2. The *burning is incendiary in origin*.
3. The *burning is shown to be started with malice*, that is, with the specific intent of destroying property.

An important aspect in analyzing incendiary fires is to determine what are the most common fuel and ignition sources utilized by the arsonist. Studies show that paper, rubbish, and/or trash are the first fuels ignited by the arsonist typically with a lighter or open flame. Arsonists will attempt to either ignite the largest combustible within the room or combine several combustibles for the first item ignited (DeHaan, 2011).

Generally, arsonists attempt to start fires in locations that are secluded or hidden within the structure (DeHaan, 2011). This is why bathrooms and lavatories are the most susceptible for incendiary fires in schools. Contrary to popular belief, gasoline, kerosene, and other ignitable liquids (flammable or combustible) are seldom listed as the material first ignited (U.S. Department of Commerce, 1978). Following is a set of recommendations for guarding against incendiary fires.

SECURITY

Security is not typically thought of as a building design aspect or as a fire protection element. However, according to a literature review, security was the most often quoted incendiary fire protection/prevention method. In fact, Underdown (1979, p. 119) stated that, "the prevention of fires of doubtful origin is purely one of security." Some of the most common and functional security features include (Factory Mutual System, 1977):

1. Illuminating exterior and entrances
2. Painting buildings a light color
3. Installing a combination burglar alarm and fire alarm with an automatic dialer
4. Trimming or removing shrubbery and signs that obstruct the view of the building from the street
5. Keeping doors locked and bolted
6. Keeping windows locked or protected with wire screens
7. Restricting access to the roof and upper floors
8. Fencing the property
9. Ensuring flammable and combustible liquids are strictly controlled in a properly safeguarded, locked area
10. Maintaining lighting in combustible storage areas
11. Using a monitored closed circuit television

One special security control mentioned several times in the literature is the locking of all sprinkler control valves in the open position. Many times, arsonists will render or attempt to render the fire protection systems ineffective by sabotaging the automatic sprinkler system (NFPA 921, 2011). The locking of all sprinkler valves will help to ensure that the sprinkler system has a water supply and will be effective during a fire.

Special attention must be given to the design of yard storage and trash areas. There is a strong possibility of a yard storage or trash fire leading to an exposure fire (DeHaan, 2011). Yard storage and trash locations will be among the most difficult areas to protect against incendiary fires because of their ready-made concentration of combustibles, easy accessibility, and easy escape (Factory Mutual System, 1977). Specific security controls include fencing the entire yard and locating storage at least 50 feet (15 meters) from the perimeter of the fence, as well as installing lights along the fence or in the locations of the trash bins (Factory Mutual System, 1977).

In 1979, defensible-space theory was studied to determine whether crime, specifically burglaries, could be discouraged for high-rise structures by simple design changes. This theory essentially espouses that by changing the design of a structure (e.g., landscaping, walkways, and fences), crime could be diverted. Further studies showed that the addition of low shrubs and fences was enough to divert many crimes. Percentages of crime decreased dramatically after these design features were implemented. Therefore, the security issues implemented as a part of the overall design of the structure can help to prevent criminal activity. Many of the same design features implemented during these studies were also described previously. Therefore, from these studies, it is reasonable to assume that incendiary fires would also be diverted (Newman, 1993).

COMPARTMENTATION

One of the most important fire protection aspects of a building design is to contain the fire within a limited area, typically the room of origin. The process of containing the fire and the combustion by-products (i.e., smoke, soot, narcotic gases) is known as compartmentation (Janssens, 2003). This is accomplished by providing effective barriers, which include protecting any openings and penetrations. Compartmentation is even more important during an incendiary fire because arsonists will typically utilize a large fuel package or combine several smaller fuel packages for the first item ignited (DeHaan, 2011).

The barriers (including the walls, floors, and ceiling assemblies) are any surface that can delay or prevent fire or smoke moving from one space to another (Fitzgerald, 2004). Similar to the design of buildings against accidental fires, effective barriers are the ideal passive fire protection for controlling and preventing smoke, hot gasses, and fire from spreading into the next compartment (Watts, 2003). However, the effectiveness of these barriers depends on the penetrations and opening protection that are in place (Watts, 2003). Therefore, the best method to analyze the effectiveness of the barriers and the possible performance during an actual fire is to analyze buildings on a case-by-case basis.

However, there are some general fire protection elements that can be implemented into the building design that will aid in the protection against or mitigation of multiroom fires. All barriers should be of fire-resistive composition, including all floors, walls, and ceiling assemblies (Janssens, 2003). Special attention must be given to the protection of the openings and penetrations. Therefore, there are also some opening protection devices that may assist in providing more effective barriers, including: self-closing doors, thermal lock pins, intumescent caulk for penetrations, dampers, and pressurization systems for the prevention of smoke movement.

The majority of fire deaths in incendiary fires are related to the narcotic gases and incapacitating smoke and not from thermal injuries (NFPA 921, 2011). Therefore, smoke control systems should be designed into structures that are prone to incendiary fires. Smoke confinement can best be achieved by, "providing a physical barrier, such as a wall with self-closing doors and dampers that restricts smoke movement, and by using a pressure differential across the physical barrier to prevent smoke from entering the nonfire area" (Jaeger, 2003). This can be accomplished by using the building's HVAC system, which can be designed to provide this pressurization.

DETECTION AND ALARM AND MEANS OF EGRESS

The primary goal of a detection and alarm system is to notify occupants for prompt evacuation. A working fire detection and alarm system is critical for any structure.

However, the automatic detection and alarm systems are even more important during incendiary fires. Arsonists will typically ignite the largest fuel package within the compartment or arrange multiple fuel packages for use as the first fuel ignited, which allows for more destructive fires (DeHaan, 2011). Also, in those cases when an ignitable liquid is used, the fire growth may be much faster, which will limit the margin of safety for safe evacuation of occupants.

Every corridor and compartment in a structure should be equipped with some type of detection device. This detection device should be equipped either with an alarming device that is interconnected to the remaining devices or functions as a signaling device that initiates an alarm panel to trigger a general building alarm. Combustible storage areas, flammable and combustible liquid storage areas, refuse areas, and remote locations that are secluded or hidden from view should be equipped with detection devices. Smoke detection devices must be installed and regularly inspected in bathrooms and lavatories. In those structures that are more prone to incendiary fires, activation of smoke and heat detectors or a building alarm should automatically phone the fire department or a monitoring agency. Therefore, detection systems serve not only as the primary notification system for the building occupants but also the beginning of the manual suppression stage (Fitzgerald, 2004).

Detection and alarm of a fire are important, but maintaining a clear and effective means of egress is just as important. Oftentimes, arsonists will block or obstruct the entrance/exit (NFPA 921, 2011). Therefore, every structure needs multiple means of egress. Depending on the occupancy type and the authority having jurisdiction, codes and standards may already require alternate means of egress. Smoke pressurization systems can be used to provide means of egress that are clear of smoke.

SUPPRESSION

Automatic sprinklers are particularly effective for life safety because, "they warn of the existence of fire and, at the same time, apply water to the burning area" (Hisley, 2003). Standard sprinklers will typically detect a fire much later than a smoke or heat detector. Therefore, a combination of a detection/alarm system and suppression system is a dependable method of protection.

An analysis in 1977 of all Factory Mutual incendiary losses over $100,000 showed that "the primary factors contributing to the extent of the loss were the lack of automatic sprinklers and closed sprinkler control valves" (p. 14). These were valves not closed by the arsonist. This study further states that "valve supervision is one of the most important forms of protection against incendiary losses" (1977, p. 14). Even in those buildings that are equipped with sprinklers, incendiary fires can be devastating if the owners do not regularly maintain and inspect their fire protection systems. Therefore, regular inspections by the building owner, as well as strict code enforcement, should control this problem.

In addition to the automatic sprinkler system, fire extinguishers should be installed throughout the structure. Obviously, these extinguishers should be located according to the codes and standards. However, it may be prudent to equip those locations that have higher risk with multiple extinguishers. Personnel should be trained on how to use the fire extinguishers and more importantly on fire-related safety aspects.

IGNITION CONTROL AND FUEL CONTROL

Ignition control "involves identifying all sources of heat that could cause ignition in combustibles and isolating these sources from the fuel" (McDaniel, 2003). This is a prevention method currently used in the airline industry. Buildings that are susceptible to incendiary fires should forbid smoking. Also, these buildings may want to forbid cigarette lighters and matches from the premises (McDaniel, 2003).

Fuel control means controlling the type, arrangement, and burning characteristics of potential fuels (Jaeger, 2003). A method of fuel control for educational occupancies is to remove trash and rubbish from the bathroom trash cans. One procedure that schools could implement is a routine emptying of the bathroom trash cans and removal of paper towels from those areas.

CODE ENFORCEMENT

Requiring business owners to meet more strict regulatory codes has been put forward as a means to decrease incendiary fires. Properly enforced codes require more inspections of the structures. According to the Department of Justice, these inspections are important to prevention in three ways (U.S. Department of Justice, 1980):

1. Statutory mandates that require automatic fire detection and alarm systems, as well as automatic sprinkler systems, will help reduce the potential damage from an incendiary fire
2. The failure of a particular building to meet standards may indicate that the building has lost its economic viability and therefore is a possible target for an incendiary fire
3. The mandated upgrading of a building, if completed, reduces the profitability of an arson fire because of increased investment in the property by the owner

From these recommendations, the Arson Early Warning System was formed. This system is used to organize and develop information needed to identify specific properties that need to be included in the incendiary fire prevention strategies and programs (Icove, 2002). Essentially, this program was formed in cooperation with the inspection programs. Inspectors would attempt to predict which structures were on the brink of losing their value. These inspectors would then focus more attention and perform more inspections of these properties.

In addition to the code enforcement system, most of the protection aspects mentioned previously are already required by the codes and standards in place for educational properties. Therefore, the need to enforce these codes and standards cannot be overstated. Better code enforcement and more strict regulatory codes will aid in reducing the potential for incendiary fires and mitigate the losses of incendiary fires.

CONCLUSIONS

The principle of fire safety is the simple depiction of the fire triangle. The fire triangle is composed of heat, oxygen, and fuel. To prevent ignition or control a fire after ignition, the school official will need to be conscious of removing one of the three sides from the triangle. If a location has abundant fuel, then the mitigation tactic will be to prevent ignition sources and heat from coming into contact with this fuel. In those locations where heat or ignition sources are expected, then it is important for fuels to be controlled. The majority of fires within educational properties are occurring in the K–12 schools because of someone intentionally igniting trash or paper with a lighter. It is important for the schools to use this information to focus on preventing incendiary fires by removing the fuel and/or heat source from the equation for effective fire safety.

CASE STUDY

Three high school students left the lunchroom to steal items from the woodshop classroom (i.e., tools and appliances). After stealing the items, they worried about leaving fingerprints and decided to collect as much fuel as they could and pile it in the center of the room to intentionally set the room on fire. The students piled up paper, trash, and wood from the classroom. Afraid that this fire would not be big enough, they went into the hallway and started opening up lockers and pulling out papers and books. They strung these fuels from the hallway into the classroom, trying to cause the fire to spread out of the classroom and into the hallway. The fuels were ignited by a plumber's torch found within the woodshop. The fire caused a tremendous amount of damage to the woodshop classroom but was suppressed before causing major damage to the hallway.

- What interventions can be utilized to address the prevention of such an event?
- What items of awareness would be important for faculty and staff to consider in preventing such an incident?

EXERCISES

Identify the first fuels ignited, the ignition source, and oxygen from the case study above. Take a tour throughout the easily accessible public areas of your school.

1. Identify those solid, liquid, and gaseous fuels that exist and are easily accessible.
2. Locate the MSDS for each of these substances and identify the vapor density, specific gravity, flash point, and other fire hazards that may be listed for that fuel.
3. Identify those classrooms that have abundant amounts of fuel just lying around and those classrooms where decorations or artwork are hanging on walls or from the ceiling.
4. Identify sources of heat that could serve as potential ignition sources.

Next, take a tour throughout the not as easily accessible areas of your schools and evaluate the potential for those same four items above. Compare and contrast your findings. Identify what are the biggest problems and locations within your school.

For each location, identify the best mechanism to prevent a fire from occurring in these different locations (i.e., remove the fuel, heat, or oxygen). Consider how to ensure the security of these locations.

REFERENCES

Burnside, J. (2008). Determining fire hazards when educators decorate their classrooms in Clinton, Mississippi. Accessed December 1, 2011. http://www.usfa.fema.gov/pdf/efop/efo42524.pdf.

Coan, S. (2011). Explanation of 527 CMR 10.09 governing school work. The Commonwealth of Massachusetts Executive Office of Public Safety Department of Fire Services. Accessed May 4, 2012. http://www.mass.gov/eopss/docs/dfs/osfm/cmr/schoolwallregexp.pdf.

Cote, A., ed. (2003). *NFPA Fire Protection Handbook*. National Fire Protection Association, Quincy, MA.

DeHaan, J. (2011). *Kirk's Fire Investigation*. Brady/Prentice Hall, Upper Saddle River, NJ.

Evarts, B. (2011). *Structure Fires in Educational Properties*. National Fire Protection Association, Quincy, MA.

Factory Mutual System. (1977). *Arson*. FM Engineering and Research, Boston, MA.

Fitzgerald, R. (2004). *Building Fire Performance Analysis*. John Wiley & Sons, West Sussex, UK.

Gorbett, G., and Pharr, J. (2011). *Fire Dynamics*. Brady/Prentice Hall, Upper Saddle River, NJ.

Hisley, B. (2003). Storage occupancies. In *NFPA Fire Protection Handbook*, edited by A. Cote. National Fire Protection Association, Quincy, MA.

Icove, D. (2002). *Incendiary Fire Analysis and Investigation*. National Fire Academy, United States Fire Administration and Federal Emergency Management Agency, Emmitsburg, MD.

Jaeger, T. (2003). Detention and correctional facilities. In *NFPA Fire Protection Handbook*, edited by A. Cote. National Fire Protection Association, Quincy, MA.

Janssens, M. (2003). Basics of passive fire protection. In *NFPA Fire Protection Handbook*, edited by A. Cote. National Fire Protection Association, Quincy, MA.

McDaniel, D. (2003). Cultural resources. In *NFPA Fire Protection Handbook*, edited by A. Cote. National Fire Protection Association, Quincy, MA.

Meyer, E. (2010). *Chemistry of Hazardous Materials, 5th Edition*. Prentice Hall, Upper Saddle River, NJ.

National Fire Protection Association (NFPA) 921. (2011). *Guide for Fire and Explosion Investigation*. National Fire Protection Association, Quincy, MA.

Newman, O. (1993). Defensible-space modifications at Clason Point Gardens. In *Applications of Environment-Behavior Research: Case Studies and Analysis*, edited by Paul D. Cherulnik. Cambridge University Press, Cambridge, UK.

Underdown, G. (1979). *Practical Fire Precautions* (2nd ed.). Gower Press Handbook, Westmead, UK.

U.S. Department of Commerce. (1978). *Fire in the United States*. United States Fire Administration, National Fire Data Center, Washington, D.C.

U.S. Department of Justice. (1980). *Arson Prevention and Control*. National Institute of Law Enforcement and Criminal Justice. National Institute of Law Enforcement and Criminal Justice, Washington, D.C.

Watts, J. (2003). Fundamentals of fire-safe building design. In *NFPA Fire Protection Handbook*, edited by A. Cote. National Fire Protection Association, Quincy, MA.

13 Fire Protection Systems

William D. Hicks and Greg Gorbett

CONTENTS

Monday, December 1, 1958, was like any other day at Our Lady of Angels School in Chicago, Illinois. The building, built prior to the 1949 Chicago Building Code, lacked sprinklers, fire doors, and other features of fire protection granted by buildings designed after the code. A local alarm, consisting of an on-site bell, and some pressurized fire extinguishers were all that was present to protect the occupants. Around 2:30 p.m., the janitor noticed smoke, and before the fire department could be notified, the fire quickly cut off the egress of the occupants by traveling up an unprotected stairway, trapping them on the second floor. An all out effort by the Chicago Fire Department, one of the best equipped and prepared of the time, could not catch up with the spread of the raging fire. By the dawn of December 2, Chicago's worst fears were realized as eighty-seven children and three nuns had perished, and ninety other children and three more nuns were injured (Cowan, 1996).

In the built environment, several systems and approaches to fire prevention and protection are employed. There are two basic categories of systems that are employed: passive systems and active systems. Many of these systems are invisible to the untrained eye and remain unrecognized for their contribution. Others make their presence very evident once they have "functioned," and either alerts occupants to an emergency or functions to suppress a fire. Also of great importance is fire prevention, which incorporates a combination of education of the occupants, as well as conscientious choices about how we interact with fuel, heat, and oxygen. It is important that any person responsible for the fire and life safety of a school obtain a copy of the *Fire Protection Handbook* from the National Fire Protection Association (NFPA). This is a primary resource for fire and life safety in the built environment and has chapters on almost every topic related to fire prevention and suppression.

PROTECTING THE BUILT ENVIRONMENT

During the design phase of a building, many factors are considered regarding fire prevention. Many of them are prescriptive in nature, meaning they come from applicable fire, life safety, and building codes that have been adopted into law. These are finite and set in stone, unless a variance can be arranged with the local authority having jurisdiction, such as a fire marshal or other building code official.

First, the type of occupancy is identified. Occupancies are based on criteria set up in the applicable code identifying the types of activities, age, and capability of occupants and types of fuels that are likely to be present. For example, an educational occupancy (K–12) would be expected to have children ranging from five to eighteen years old under adult supervision (NFPA 101, 2012). There would be a significant amount of paper and plastic in the building, with some flammable liquids. Next, the building and fire code requirements for the appropriate occupancy are reviewed, and the required features of fire protection are identified (NFPA 1, 2012). These features will include both passive and active systems and components.

Building plans are reviewed and approved prior to construction for compliance with these design requirements. Additionally, plans are reviewed for the installation of all passive and active fire protection assemblies and systems to ensure their appropriateness. This is also where the locations of hydrants and emergency vehicle access will be assessed. During construction, periodic inspections will occur to ensure the systems are installed appropriately. The final inspections occur once the building has been constructed and completed. It is during this time that acceptance testing will occur, and all features of fire protection will be tested and inspected closely to ensure they are functioning properly and installed correctly (Farr, 2008).

Passive features include fire-rated assemblies, fire doors, smoke barriers, and fire stoppage of openings in walls and floors. Fire-rated assemblies are expected to have been tested in accordance with a number of American National Standards Institute testing standards. A fire-rated assembly consists of components that, when installed together, achieve a fire-resistive rating (NFPA 80, 2010). Very specific test protocols are followed, and specially designed apparatus are used during testing.

According to Davoodi (2008), interior walls are required to be fire rated and are the most prevalent passive system throughout a building. Called fire barrier walls or fire partitions, they are not a structurally supportive assembly and normally do not extend to the ceiling. For example, when a certain thickness of dry wall is properly secured to a wall, the seams are taped and mudded, and the wall is painted, a fire resistive rating is achieved. Ratings are given in minutes or hours and may range from thirty minutes to several hours. In addition to having a fire rating, interior wall coverings must meet flame-spread ratings for the occupancy type, which determine what materials can be used to cover or be posted in hallways and stairs. Flame-spread ratings are measurements of how fast a flame will move across the material, and builders must strictly adhere to these ratings.

NFPA 80 identifies fire doors as closures placed at given points in a building to compartmentalize, or contain, fire and smoke in the area of origination to prevent it from spreading. They also serve to protect the path of egress, or escape. Fire doors consist of assemblies that include handles, seals, closer assemblies, door frames, and sometimes windows, that are tested together as one unit. A fire door frame must be paired with its matching door model, because this pairing forms an assembly that has been tested and received a fire rating as a unit. There will be a label on the door and/or the door frame that identifies the rating of the assembly. The frame will consist of fire or smoke seals to prevent passage of the by-products through the assembly. Also of importance is that the fire door be installed into a part of the structure with at least an equal fire-resistance rating.

NFPA 80 requires that a fire door will have a self-closing device, usually magnetic, that holds the door in the open position until the alarm sounds. This closing device will also close the door behind occupants as they pass through it. It is of importance that these doors are not held open by door chocks of any kind, because this will defeat the purpose for having the door. A positive latching door mechanism helps the door seal tight with the frame. Fire doors may also incorporate a window, which will be rated as part of the assembly. It may contain wired glass or a special type of heat-resistant glass but must come installed by the manufacturer. No modifications of the door are allowed, including decorating, painting, or making any holes in the door, frame, or the assembly in which it is installed.

FIRE DETECTION

NFPA 72 (2010), the National Fire Alarm and Signaling Code, addresses the detection of fire through many different technologies. The selection of the appropriate technology is based largely on the types of fuels, the ambient conditions expected to be present, and the activities that will be regularly occurring. For example, smoke detection technology would not be appropriate in areas where tobacco is smoked. It is also important to understand that there are several types of tests performed on all detectors. The parameters for this testing are set out in the manufacturer's recommendations that accompany the product. Acceptance testing is performed after the initial installation of the system to ensure that it functions correctly. Reacceptance testing occurs after a major change or repair occurs to the system itself or after remodeling occurs or significant changes to the layout of occupancy occur.

Detectors are also required to undergo annual functional testing, which only tests the functionality of the detector and not its sensitivity. This field testing is performed using aerosols or smoke for smoke detectors, heat guns (much like a hair dryer) for heat detectors, or matches in cases of flame detectors. It is important to understand that some detectors do not reset themselves, so a clear understanding of the type of detector is critical for a successful test.

The manufacturer or local fire codes may require sensitivity testing of a sampling of detectors over a certain age. This is usually done in a laboratory setting and is a true test of the condition of the detector in terms of the manufacturer's setting. Always follow the manufacturer's testing requirements and only use trained personnel for such testing. Also found in these requirements is the proper spacing of the detector in a building to ensure adequate coverage.

Smoke and heat detection is the most common approach in educational occupancies to detect fires. Smoke detectors utilize either photoelectric or ionization technology, both of which we will discuss. First, there are smoke detectors and smoke alarms. Smoke detectors are connected to a fire alarm system in order to activate a fire alarm notification system. When the detector activates, it sends an initiating signal to the fire alarm panel, which in turn initiates the fire alarm audible and visual notification devices. Smoke alarms, which are most commonly found in residences, contain both detection abilities and an audible alarm.

Photoelectric smoke detectors operate by one of two designs, both of which utilize a light emitting diode cell, called a sender cell, and a receiving cell. The first design incorporates a sending cell and receiving cell, which are in line with each other. During normal conditions, the light-emitting diode (LED) shines on the receiving cell, which creates a small electrical current. When smoke enters the chamber, it obscures the LED, thereby interrupting the current flow, which in turn causes activation of the alarm. The second design includes the same approach, but the cells are not in line and no current is present normally. When smoke enters the chamber, it refracts light onto the receiving cell creating current flow, causing an initiating signal to be sent. Ionization detectors use a small amount of radioactive material to produce an electrical current between two plates, completing a circuit. When smoke enters the detector, it interferes with this current, resulting in the initiation of an alarm (Dungan, 2008).

Heat detectors operate by many different approaches, but the most common are fixed temperature detectors that function at a very specific, fixed activation temperature that is set by design (Dungan, 2008). They may be self-resetting or self-destructing, meaning the entire unit must be replaced once it functions.

FIRE EXTINGUISHERS

Fire extinguishers are required by almost all building and fire codes in the United States, with *NFPA 10: Standard for Portable Fire Extinguishers* (2010) providing the prevalent standards to be followed. The Occupational Safety and Health Administration regulation 1910.157 mirrors these requirements for uses in places of employment. Fire extinguishers are a tool with limited fire-suppression abilities and serve as a first line of defense for small fires. Fire extinguishers are comprised of several operating designs and are intended for use on one or more of the five classes or types of fires. Selection is based on the fuels present in a building and the most likely fire scenario, as well as the capabilities of the occupants. Fire extinguishers must be mounted to a wall on a hanger or placed inside a fire extinguisher cabinet.

Fire extinguishers are designed to suppress what is termed as "incipient" stage fires. These are fires with less than 10 inches of flame height and still confined to the original fuel items, such as a wastebasket (OSHA 1910.157). Before using the extinguisher, the occupant should first assist anyone who is close to the fire, directly being threatened by their proximity to the fire, or unable to leave the area. Second, as a part of this process, the alarm should be sounded to notify the occupants and the fire department. Only after this is done should an occupant attempt to utilize a fire extinguisher. Despite the urge to respond, it is always best to leave the building because property is not worth injury or death.

Fire extinguishers exist in three basic operating designs according to NFPA 10. They are stored pressure, cartridge operated, and self-expelling. This refers to the method utilized to expel, or spray, the agent out of the extinguisher. A stored-pressure model consists of a solid-state chemical agent. In order to expel the agent when needed, pressurized nitrogen is added into the extinguisher to a pressure meeting the requirements of the manufacturer. This information is also on the label of the extinguisher. A cartridge-operated extinguisher uses a separate sealed cartridge to expel a dry chemical agent. The agent is stored in a large cylinder, separated from the expellant gas, which mounts on the side of the extinguisher. In order to prepare this extinguisher, the gas cartridge is punctured by the user pushing down on a flat leaver, which then pressurizes the agent cylinder. A self-expelling extinguisher uses gaseous agents, stored as liquid under pressure in the extinguisher, such as carbon dioxide. When placed under pressure in the cylinder of the extinguisher, they revert to a liquid state, but once released they change to a gas state and self-expel out of the extinguisher.

NFPA 10 also classifies extinguishers by the types of fuels and fires that they are third-party-tested and approved to extinguish. It also assigns a maximum travel distance, which is the distance an occupant would have to travel to locate an extinguisher for each class. This includes the actual path traveled among desks or other obstructions. A discussion of this can be found in the NFPA's *Fire Protection Handbook* chapter on fire extinguisher use and maintenance.

A Class A extinguisher is designed for Class A fuels, such as wood, paper, and most plastics. When they are tested, they are given a Class A rating. A Class 1A rating means the extinguisher has the suppression abilities of 1¼ gallons of water. Class A extinguishers are tested on three-dimensional fires, using a wood crib, on vertically orientated fires using a wooden wall, and on horizontally orientated fires, using excelsior (similar to wooden shavings). Class A extinguishers have a maximum travel distance of 75 feet, which means an occupant must travel no more than 75 feet in any direction to reach the extinguisher. Class A extinguishing agents include water and aqueous solutions containing chemicals that enhance water's ability to suppress fires, as well as a multipurpose agent, which works on several classes of fire.

Class B extinguishers are approved for use on Class B fires, such as ignitable liquids and gases. Class B ratings come from a test utilizing a pan of heptane with a specific square footage of surface. The rating is derived by taking 40% of the square footage of the pan an expert was able to extinguish. Class B extinguishers have either a 30 or 50 feet travel distance, depending on local code and the size of the extinguisher. In areas where larger amounts of flammable liquids are stored, the extinguisher should be readily accessible but not in direct proximity of the hazards in order to allow the occupant to reach it in an emergency. Class B agents consist of water and foam mixtures, special chemical agents designed to suppress ignitable liquid fires, and multipurpose agents. Do not use water on liquid fuel fires because this will only spread the fire and make the situation much worse.

Class C is not really a class of fuel but refers to fuels that may be part of an electrically charged item or component. Class C fires present an electrocution hazard; hence water-based agents are mostly inappropriate. If at all possible, these situations should be approached by disconnecting the electricity. Transformers, capacitors, and other electronic components store electricity after the source is disconnected and may still provide a shock. Class C-rated extinguishers are tested against 100,000 volts for electric conductivity back to the users. Once de-energized, items designed to be powered by electricity may contain both Class A and Class B fuels. There are no true Class C-only fire extinguishers; instead, multipurpose extinguishers will be used to fight A, B, and C fires, because they are capable of handling all the hazards of such fires. The travel distance of Class C-rated extinguishers is therefore based on the most prevalent hazard of either Class A or B fuels in the building.

It may seem counterintuitive, but metals can produce some of the most challenging fires to fight. Metals that support combustion are called Class D fuels. Fire extinguishers that suppress Class D fuels are rare. Instead, locations that produce metal shavings, chips, filings, and ribbons from drilling are protected with agents stored in buckets. These agents are tested against magnesium only, so consult the manufacturer's recommendations on the appropriateness of an agent against fires involving specific metal fuels. Travel distance is 75 feet, but very few facilities will have large areas with these kinds of hazards. Therefore, the agent is typically kept in close proximity to the fuels and locations where the work is being performed. Only use Class D agents on metal fuel fires. Do not use water on combustible metal fires, because a violent reaction will occur, spreading molten materials and releasing flammable gases, which will only make the situation worse. In an emergency, dry sand or soil can also be used directly on the fire.

The final class of extinguisher is Class K. This refers to cooking media such as greases, fats, and other substances used or produced during cooking. Class K agents are either dry chemicals or aqueous solutions, and the initial approach to protecting these areas is through a fixed system in and around the fryers and grills of cafeterias. These agents suppress by saphonification, which is a reaction between the agent and the cooking media that results in a crusty layer on the top of the fuel, which separates the fuel from the oxygen, thereby suppressing the fire. Extinguishers will be provided to mop up any unsuppressed fire. The travel distance is 30 feet, so they will be in close proximity to the hazard (Conroy, 2008).

NFPA 10 requires three main intervals for the inspection, testing, and maintenance of fire extinguishers. First, they should be manually inspected every 30 days. This includes making sure that they are pressurized, that they have not been used and put back, and that they are not otherwise in need of service. It should be confirmed that the extinguisher is not blocked or missing and that it is located in the right location. This should be documented on a tag attached directly to the extinguisher, which should include the initials of the inspector and date of each inspection. Records should also be maintained in a separate file for documentation and can be in the form of an extinguisher inventory checklist. Also, every year, a mechanical inspection must be performed on the outside to evaluate for bent or damaged components, rust, and dents. If an extinguisher is taken out of place for servicing, one of equal rating must be put in its place.

Every extinguisher has required internal inspection and hydrostatic test intervals, depending on the type and agent inside. The internal inspection occurs between hydrostatic pressure testing times and includes a complete discharge into a collection system so that, as the description implies, the inside can be inspected for damage, rust, and solidified deposits of agent. This is verified by placing a plastic service collar or ring, just slightly larger than the neck of the cylinder, around the neck before the extinguisher is reassembled and recharged. It may also be documented, although less reliably, with a sticker placed on the side of the extinguisher opposite the instructions.

Hydrostatic testing involves pressure testing of the cylinder to ensure structural integrity has not been compromised by metal fatigue, rust, or other damage. This testing should be done by a reputable company specializing in servicing fire extinguishers. Hydrostatic testing dates vary depending on the type of container and the agents. Most multipurpose dry chemical extinguishers are hydro-tested every twelve years. The National Fire Protection Association Standard on Inspection, Testing, and Maintenance of Fire Extinguishers (NFPA 10) contains the requirements for maintaining fire extinguishers.

A few items must be addressed prior to using a fire extinguisher. First, review the instructions on use and also ensure the extinguisher is appropriate for the types of fuels that are present. Second, always check that the extinguisher's gauge shows it is charged and the tamper tag is in place holding the pin in the handle. For extinguishers with no gauge, the presence of the tamper tag is sufficient. Third, always identify an exit, and never let fire get between you and the way out. Next, employ what is called the PASS method to operate the extinguisher; the name represents the words pull, aim, squeeze, and sweep. Approach to within eight to ten feet or until you encounter any heat or smoke. Pull the pin, and then aim the extinguisher nozzle at the base of the fire or the fuel. Aiming into the flames will accomplish nothing, except to empty the extinguisher fruitlessly. Squeeze the handle and sweep the base of the fire, advancing slowly, only getting close enough to ensure the agent you are expelling is reaching the fuel. If you become exposed to heat or smoke, leave the area immediately, taking the fire extinguisher with you (Conroy, 2008).

WATER-BASED SUPPRESSION SYSTEMS

Most modern schools are required to have water-based sprinkler systems installed to provide suppression of any fire that may occur. *NFPA 13: Standard for the Installation of Sprinkler Systems* provides guidance designers follow during the installation of water-based suppression systems. This is the only standard of its kind followed in the United States, and all of the following information is taken from this document. Fire sprinklers are automatic and thermally activated by the heat energy released during a fire. This delivery of water can make a major impact on life safety of the occupancy by limiting the growth of the fire and in turn the amount of smoke produced during a fire event.

The most visible component of a sprinkler system is the sprinkler head. A sprinkler head consists of a deflector, which looks like a round disk with teeth cut out into it. The deflector is supported by two frame arms that run between the deflector and the threaded base, which is screwed into the sprinkler system piping. To keep the water from escaping, an orifice cap, which acts much like a cork in a bottle, is held in place by a thermally activated releasing mechanism. There are two systems of thermally activated releasing mechanisms used for this purpose. The first is a frangible bulb, which will appear like a small glass tube filled with a colored liquid, in the center of the sprinkler head. The color of the liquid indicates the activation temperature of the sprinkler head. When heated, the liquid begins to change from a liquid state to a gaseous one, and the expansion of the gas causes increased pressure, which will eventually overcome the structural integrity of the bulb, causing it to shatter and releasing the orifice cap. This allows water to flow out of the head and into the immediate area. The second type of releasing mechanism is called a fusible link. This consists of form-fitted metal parts held in place by solder, which acts as a glue to hold the components and the orifice cap in place. The solder has a specific melting point, and once that is reached it will melt, allowing the metal components to separate, releasing the orifice cap and allowing water to flow. There is a color code for the marking of fusible link heads, but it is not frequently followed. Instead, the activation temperature is stamped on the deflector or the body of the sprinkler itself.

The sprinkler head should never be used for any purpose other than the one it is designed to address. Never allow anyone to hang or attach anything, no matter how innocent it may seem, to the sprinkler head. This may inadvertently hold the releasing mechanism in place even after the releasing mechanism has functioned. The result is a fire that will now grow unchecked. Also, never paint or cover the sprinkler head for any reason, because this will delay or prevent activation. Finally, avoid damaging or obstructing the sprinkler head. A damaged sprinkler head requires replacement in order for the system to provide its intended coverage. A minimum of 18 inches of clearance must be maintained around the sprinkler head. Materials should never be stacked higher than the sprinkler head.

Most systems found in schools are wet pipe systems, meaning all the piping of the system contains water at all times. Starting from the outside, a connection is made to a source of water with adequate pressure and flow, rated in gallons per minute, called the water main. As part of the plumbing code, the water main is equipped with a backflow preventer, which prevents the nonsanitary water found in the sprinkler systems from returning into the potable water supply. Next, a main control valve is placed in the line to allow for the system to be turned on and off. This connection must be electronically monitored or locked in the open position.

The next component is the riser, or the vertical run between the water main and the piping that distributes the water throughout the building. This is usually located next to an outside wall in a location that is near a roadway or parking area. This is so the fire department may access it from the outside through what is called the fire department connection, which interconnects with the riser just above the riser valve. This is used by local authorities to pump additional water into an activated sprinkler system to help supplement its fire-suppression activities. Typically, in the middle of the riser is the water control valve, or the heart of the system, which acts as a hub to direct the water flow through a series of piping called system trim. This smaller series of pipes directs water through piping to allow for the activation and testing of alarms and the electronic supervision of the system and to a main drain, which allows drainage of the system.

The valve will also have two gauges located on it or piped into it that monitor pressures that are important to the reliability of system. The lower gauge is the water supply gauge, which monitors the water-supply pressure. This gauge will fluctuate as the water pressure in the system fluctuates from various levels of water usage. The upper gauge expresses pressure in the actual system and should be higher than the lower gauge.

Water-flow detection devices are also trimmed onto the valve or on the riser itself, and there are three types commonly used. The first type is a pressure sensor, which uses a diaphragm assembly attached to trim that during non-flow conditions has no water and hence no pressure bearing down on it. Once water flows, it enters this piping and places pressure on the assembly, completing a circuit that sends a signal to the fire alarm system for the building. The next type of monitor is a water-flow switch called a paddle vein detector. This is comprised of a paddle that projects through the wall of the pipe of the riser and is connected to an electronic switch. When water movement occurs, the paddle is pushed upward, which in turn completes a connection in the electronic switch assembly. The third type is a water motor gong. This consists of a water-operated mechanical bell that rings when water flows. This provides local notification only, and although the bells are quite loud, only those that are close enough to hear them will know that an emergency is occurring.

Sprinkler systems have numerous testing requirements that must be adhered to, including *NFPA 25: Standard for the Inspection, Testing, and Maintenance of Water-based Fire Protection Systems.* This testing is critical to ensure the reliability of the system. Testing should only be performed by knowledgeable individuals with proper training. None of the components should be operated on the system without the presence of a trained individual. This is to avoid false alarms or inadvertent disabling of the system. There are three main test points in each sprinkler system consisting of the main drain, the alarm test trim, and the inspector test.

The main drain is used to confirm the reliability of the water supply. Imagine the sprinkler system as a garden hose with a nozzle. Once you have finished using it, it is turned off at the valve, but there is still pressure in the hose itself until it is released. The sprinkler system is also a closed system, meaning that if someone were to turn off the valve supplying the system or a pipe were to collapse and begin leaking water into the ground, pressure would still be trapped in the system, and the supply gauge would continue to show pressure even though in reality only a minor amount of water may flow should a head activate. By operating the main drain, you can confirm that the water needed for the system to function is available.

The alarm testing trim allows for water to be introduced into the alarm trim, which in turn activates the building alarm system. This allows for testing of the water-flow monitors, which will send a signal to the alarm panel (panels will be discussed in the life safety chapter). The inspector test connection allows for simulating the activation of one sprinkler head. This is located at a remote location from the riser, and is made up of a valve leading to a short run of piping leading to a drain or to the outside. At the end of the pipe, a sprinkler head is attached, and the activation mechanism, frame arms, and deflector are removed. This simulates the same opening size as the sprinklers in the building and allows for full system testing without the need to flow water inside the building.

The fryers and cooking apparatus found in cafeterias will most likely be protected by kitchen hood systems. NFPA 17 *Standard for Dry Chemical Extinguishing Systems* and NFPA 17A *Standard for Wet Chemical Extinguishing Systems* will apply depending on the type of agent deployed by the system. Jones (2009) explains these systems and their requirements by comparing them with sprinkler systems, in that piping leading from the source is designed to cover a certain area and distribute an agent during a fire event. Kitchen systems are either a dry chemical or wet chemical agent system. They are designed specially to suppress fires involving fryers, flat top grills, and other cooking appliances. These systems may be activated by fusible links or by manual pull stations. They consist of an agent container and an expelling gas, either stored with or separately from the agent. More advanced systems may include a heat detector as well as fusible links. The discharge heads should only be covered by coverings provided by the manufacturer. Annual inspections should be performed by trained individuals and according to the manufacturer's recommendations.

CONCLUSION

The features of fire protection are essential components in the overall safety systems included in the school built environment. They must not be defeated or altered in anyway, or the risk of their failure during a fire event is high and the risk of injury or death is increased. A program of inspection, testing, maintenance, and service along the manufacturers' recommendations and local fire codes will help to ensure that these systems will operate appropriately in the event of a fire. Documentation of these activities should be maintained in order to record these activities and prove compliance with accepted standards. Visual inspections should be done periodically to check for any conditions that may adversely affect the systems described in this chapter, and trained personnel should be consulted if questions arise.

CASE STUDY

During the hottest part of the year, you find that a fire door, normally held closed with a self-closing device and no automatic closure release, is being propped open by wooden wedges, thereby defeating the function and the purpose of the door.

- Explain the process that should be used for reporting this issue in a school in terms of who would be contacted and how this issue should be approached.
- Provide a written memo explaining the violation, the possible effects during a fire, and alternate solutions to fix this issue.

EXERCISES

Perform a fire protection audit on your school building, identifying the requested features for compliance.

1. What is the fire-resistance rating of the fire doors in the hallways?
2. Are the fire doors held open by any means other than an automatic releasing mechanism?
3. Are the fire extinguishers present, and have recent visual inspections been documented on their inspection tags?
4. Are sprinkler heads clear of obstruction?
5. Are any objects hung or draped on the sprinkler heads?

REFERENCES

Conroy, T. (2008). *Fire Protection Handbook* (20th ed.). Chapter 17.5: Fire extinguisher use and maintenance. National Fire Protection Association, Quincy, MA.

Courtney. (2008).

Cowan, D. (1996). *To Sleep with the Angels: The Story of a Fire.* Elephant Paperbacks, Chicago, IL.

Davoodi, H. (2008). *Fire Protection Handbook* (20th ed.). Chapter 18.1: Confining fires in buildings. National Fire Protection Association, Quincy, MA.

Dungan, K. (2008). *Fire Protection Handbook* (20th ed.). Chapter 14.2: Automatic fire detectors. National Fire Protection Association, Quincy, MA.

Farr, R. (2008). *Fire Protection Handbook* (20th ed.). Chapter 1.5: Fire prevention and code enforcement. National Fire Protection Association, Quincy, MA.

Isman, K. (2008). *Fire Protection Handbook* (20th ed.). Chapter 16.2: Automatic sprinklers. National Fire Protection Association, Quincy, MA.

Jones, M. (2009). *Fire Protection Systems.* Delmar Cengage Learning, Clifton Park, NY.

National Fire Protection Association. (2010). *NFPA 10: Standard for Portable Fire Extinguishers* (2010 ed.). National Fire Protection Association, Quincy, MA.

National Fire Protection Association. (2010). *NFPA 13: Standard for the Installation of Sprinkler Systems* (2010 ed.). National Fire Protection Association, Quincy, MA.

National Fire Protection Association. (2010). *NFPA 72: National Fire Alarm and Signaling Code* (2010 ed.). National Fire Protection Association, Quincy, MA.

National Fire Protection Association. (2010). *NFPA 25: Standard for the Inspection, Testing, and Maintenance of Water-Based Fire Protection Systems* (2011 ed.). National Fire Protection Association, Quincy, MA.

National Fire Protection Association. (2010). *NFPA 80: Standard for Fire Doors and Other Opening Protectives* (2010 ed.). National Fire Protection Association, Quincy, MA.

National Fire Protection Association. (2012). *NFPA 1: Fire Prevention Code* (2012 ed.). National Fire Protection Association, Quincy, MA.

National Fire Protection Association. (2012). *NFPA 101: Life Safety Code* (2012 ed.). National Fire Protection Association, Quincy, MA.

14 Life Safety

William D. Hicks and Greg Gorbett

CONTENTS

On November 28, 1942, over one thousand people packed into the Coconut Grove in Boston, Massachusetts, one of the nation's premier nightclubs at the time. The club was decorated in a South Seas theme, with flammable papier-mâché and plastic wall and ceiling hangings. A fire started in a back room and quickly engulfed the entire club in smoke and superheated gases. The main entrance was served by a single revolving door, which soon was obstructed by the crush of bodies. Side doors had been bolted shut to prevent thefts and sneaking in, and inward-opening doors were held tightly shut by the crush of bodies piled high against them. Out of one thousand people present that night, 492 people died, which was 32 more than the fire marshal had listed as the maximum capacity of the building (Esposito, 2005).

Flash forward to February 20, 2003, at the Station nightclub in Warwick, Rhode Island. This time a concert by the 1980s rock group Great White is being held in an over-packed club. In attendance was a local television cameraman doing an investigation on the safety of nightclubs in response to the deaths of 21 people in a stampede (caused by pepper spray, not a fire) at the E2 Nightclub in Chicago just three days before, on February 17. There, only two doors at the bottom of a staircase served as egress, because the rest were locked. Here, pyrotechnics set acoustic foam ablaze, and in less than 5 minutes 100 people were dead, 230 were injured, and 132 made it out unscathed. The back exits were blocked by staff initially, trying to protect the backstage area. Again, the main entrance was a major problem, this time designed purposefully to create a choke point to control the flow in and out of the club, and it does its job too well. In video shot that night you can see the nearly 50 people, stacked like wood on one another, burned to death. The rest never even made it to the door, overcome by the fast moving smoke, heat, and flames (Grosshandler et al., 2005).

In terms of life safety and the importance of egress, we keep paying in lives to learn the same lesson. In order to safely evacuate a building during any emergency, you must have the proper amount of exits, and they must be free and clear, with outward-swinging doors. The term "life safety" encompasses many faculties of safety in the built environment but revolves around the notification and safe evacuation of occupants from schools during fire or other emergencies. Many of the contributing factors to our largest loss of life events are due to failure of life safety systems. Over our history, we have seen our greatest losses of life caused by an inadequate number of exits, improperly installed or maintained components of egress, failure to train those charged to direct the evacuation, and delaying the beginning of evacuation. Yet the Station nightclub fire shows we have not learned our lesson.

ALARM AND NOTIFICATION

The alarm system panel is the brain of the building's detection and notification system. The requirements for alarm system inspection, testing, and maintenance are found in NFPA 72, *The National Fire Alarm and Signaling Code*. It links the detection and suppression systems with the ever important notification devices that signal the occupant to evacuate. It also provides supervision of all devices and monitors for failure of devices on the system. When an initiating signal is received, the alarm system panel activates the strobes, horns, and speakers throughout the building. This provides for early notification of occupants and helps prevent injury or death from fire. In addition to local alarm notification, the panel will alert a monitoring point, which will in turn notify the local fire department of the situation.

The main fire alarm panel will be installed in a central location, which may be in an office or a closet where the main power panels are found. The panel will consist of several processors and should only be opened by trained and certified individuals. Panels also consist of a battery backup, which will normally be housed in the main panel itself. The battery backup systems follow a twenty-four-hour/five-minute protocol, meaning the battery must provide standby power for at least twenty-four hours and yet still be capable of operating all notification devices for five minutes (Moore, 2008).

Fire alarm panels will have remote displays called enunciator panels at the main entrances. These display the status of the alarm system and will sound a notification (but not the fire alarm) in case a supervised item has been tampered with or communication is lost with a detector. The signals will be distinct between a trouble signal and supervisory situation. The panel can also monitor for many other conditions as assigned. There are many function buttons on the panels. During an alarm or other signal, it is important that none of the function buttons are pushed, because this may result in a resetting of the memory and loss of the reason or location for the signal from the panel. No signal from the panel should be ignored, and a technician should be notified to investigate the issue.

Fire and building codes will reference NFPA 72, *The National Fire Alarm and Signaling Code*, and NFPA 70, *The National Electric Code*, for installation, inspection, testing, and maintenance requirements for fire alarm panels and their components. Alarm panels undergo specific annual testing, but every time an initiating device is tested, the alarm panel should be reviewed for normal function.

MANUAL PULL STATIONS

Manual pull stations are provided as part of the notification system, and all requirements are found in NFPA 72. These are manually activated initiating devices that will sound the audible and visible notification when pulled. The technology behind pull stations is simple: They usually incorporate either a two position switch or momentary switch wired in an open circuit. When the activation handle is pulled or slid, this in turn throws the position of the switch, which closes the circuit and signals the alarm panel to activate the systems notification devices.

There are several types currently used. The first is a single action, which simply requires a downward pull on a hinged or sliding handle that operates the switch. The next is a double action pull station, which requires two distinct actions to initiate the signal. This first action is usually in the form of a cover, either incorporated into the body of the device or placed around the outside of the station. Break-glass pull stations are mounted behind fixed glass that must be broken in order to reach the pull station. Another approach is the use of a break rod placed under the lip of the activation component that holds the pull station in place and requires extra force in order to break the rod and then allow the pull station handle to move (Jones, 2009).

NOTIFICATION

Notification of the occupants is a critical consideration for obvious reasons. If they never receive the signal to evacuate, they will never have a chance to escape. Signals come in the form of audible and visible notifications but can also include tactile, voice, or text messaging and the use of vibrating devices for the hearing impaired. There are several considerations in determining the approach to notification. First is the nature and condition of the occupants, such as if they are sleeping or awake, adults or children, or disabled in any way. There are also considerations to be made in terms of the ambient conditions (noise level) and other sounds normally present that may mask the alarm. Finally, it is necessary to determine whether the alarm will be sent out to all occupants or to a dedicated person who then is in charge of investigating alarms and initiating evacuation.

Audible notification consists of sounding devices, such as horns, sirens, buzzers, and bells. Modern devices follow a temporal three pattern, which has been adopted as a universal audible fire alarm signal. This is a series of three sound bursts evenly spaced. Audible devices are placed throughout the occupancy to insure total coverage. Audible devices are mounted on the walls or on ceilings. The applicable fire and building code will have minimum spacing and heights at which the device can be mounted above the floor and below the ceiling. This is to ensure that the device can be heard and is not obstructed by large crowds (mounted too low) nor destroyed or obstructed too soon by the smoke and heat at the ceiling (mounted too high). Audible devices should be visibly inspected to ensure nothing has been placed in a position to obstruct the device. The appropriate building or fire code, along with the manufacturer's recommendations, should be consulted for the types and frequency of inspection, testing, and maintenance (Schifiliti, 2008).

Sound pressure testing can also be performed to ensure that a minimum recommended decibel (dB) level is being met. Most manufacturers, fire and life safety codes, and research recognize 75 dB as the minimum sound level necessary for public audible notification. Care must be used if any doors create barriers between the notification device and the population. Doors will significantly lessen the dB level, lowering the effectiveness of the device. Areas with large populations or ambient noise levels above 75 dB will need to have measurements taken to ensure the device will provide correct warning (Cholin, 2008).

The use of addible alarm notifications in the form of tactile prerecorded voice announcements is becoming more prevalent. These announcements must be intelligible and must provide clear instructions for occupants. Tactile prerecorded text notification is also gaining popularity, as it can be displayed in many languages and over many devices.

Visible notification refers to the use of flashing, rotating lights or strobes to warn occupants of the need to evacuate. As with the audible devices, fire and building codes will mandate mounting location and spacing of devices to ensure visibility by occupants. The same caution must be used to avoid an obstruction of the device by decorations or any other means.

Visual inspection should occur as a regular part of building safety assessment. The appropriate building or fire code, along with the manufacturer's recommendations, should be consulted for the types and frequency of inspection, testing, and maintenance. Modern strobe lighting will flash between one and two times per second at a designed manufacturer light level measured in candela. A lumen meter can be used to measure light levels (NFPA 72, 2010). Visual devices should be visible at all points in the building to ensure effective notification.

EGRESS AND EVACUATION

Every building receives a certificate of occupancy from the authority having jurisdiction or the person in a community charged with fire and life safety inspections, such as the fire chief or fire marshal. Once in place, any changes that have the slightest potential to affect the feature of fire protection or life safety must be approved by the authority having jurisdiction immediately. Many things are assessed during this approval process. One of those is the occupant load. This is a measurement of the number of people who can safely occupy a classroom, lunchroom, or auditorium (Farr and Sawyer, 2008).

The study of human fire behavior looks at how people will react in the presence of an emergency. This, along with many incident investigations, has identified a number of interesting factors. One of which is that people will try to exit by following the path they came into the building. This is why the main entrances must be designed to handle 50% of the total occupant load. These main exits should never be altered or blocked in any way. Also, the doors of exit access pathways should always open in the path of egress to avoid pile ups should people reach the door prior to it being swung open (Bryan, 2008). Another concern is the building design through creating of choke points at exits. This was shown tragically in the video documentation of the Station nightclub fire, where an ID checkpoint had been created to control entry into the club at the main exit. Because many of those who came in this way attempted to exit here, the reverse effect occurred during the evacuations and the reduction in the width of the egress contributed to the loss of life. Another factor was the failure to train the security staff and other employees in how to properly evacuate the facility.

The first consideration in identifying the occupant load is the type of occupancy. This is a classification made based on the activities that will be occurring in a given building or room. NFPA 101, *Life Safety Code* (2012) identifies an occupant load of one person for every 20 square feet in educational occupancies. Depending on the fire and building code used in a given city, a number will be established identifying the square feet per person to identify the occupancy load of that building. The type of seating, the presence of tables, whether the seats are fixed or moveable, or lack of seats is considered to establish this number, along with the activities that will occur. This number is then divided into the total square feet of the area, resulting in the maximum number of people who can occupy the room safely.

A similar approach is used to assess the number of exits available to the occupants and can also be found in NFPA 101 (2012). An occupant load is multiplied by 0.2 inches to determine how many inches of egress are required and by 0.3 inches to assess the width of stairways. Understand that doors are currently required to be a minimum of 36 inches wide in order to accommodate those occupants utilizing wheelchairs, walkers, or crutches. For example, in rooms with occupant loads of less than fifty persons, one exit door may suffice and may swing inward or against the path of egress. Where rooms are occupied by more than 50 people, two exits are required to be provided. These exits must be remote from each other, and the doors must swing outward, or in the direction of travel.

There are three components of an egress system. These are:

- The exit access, or pathway through the building
- The exit and area around it allowing escape
- The exit discharge leading to a public area or right of way

Each component is a vital factor in the occupant's ability to escape safely. In design, we intend to keep them free from any hindrance, but as occupants, we do many things to disrupt the system, and hence place occupants in danger of not being able to escape.

The exit access is the pathway through the building that the occupants must follow in order to reach the discharge and safety. This path of travel must be clear of any obstruction that protrudes into the path. Water fountains, doors, displays, and cabinets should not restrict the width of the egress because the capacity of this pathway is essential to safe evacuation. Another important situation is the presence of dead ends that do not provide two ways of escape. These are strictly limited under NFPA 101 to 20 feet in educational occupancies without sprinklers and 50 feet in buildings with sprinklers (NFPA 101, 2012).

Strict limits should be placed on the types of decorations allowed to be placed in these hallways or in stairways. Of particular note, the placing of lightweight combustibles such as paper, plastics, and other easily ignitable materials along the walls and ceilings must be carefully controlled, because they provide an easy path of flame travel that can easily outrun occupants down a hallway, spreading fire and smoke, which would block the path of escape. Fire and building codes (NFPA 1, 2012; NFPA 101, 2012) provide flame-spread ratings on any wall coverings, permanent or temporary, of exit access and stairways to prevent such disasters. The exit access must also not lead the occupant through an area of high hazard, such as a kitchen or machine shop. Instead, it should move through the safest area or path possible.

During the design phase of the building, fire doors are sometimes placed in the middle of hallways for seemingly no reason, but they play a critical role in life safety. First and foremost, these "horizontal exits" are usually tied to the fire alarm system and will swing closed when an alarm sounds. This separates one part of the hall from another, thereby limiting or preventing the spread of fire and smoke throughout the building. These types of doors will also be found at the entries to stairwells, and they will have self-closing arms designed to keep them closed. These doors are also there to keep smoke and fire out of the stairwells, yet all too often they are propped open for ventilation purposes. This is a serious situation, because the occupants have now defeated the protective design of this door. Should a fire occur, the door will be held open, allowing smoke and heat to spread into the stairs, exposing escaping occupants to the very dangerous by-products of combustion (Lathrop, 2008).

Along the exit access, exit signs must be used to indicate the direction to the nearest exit discharge (door). Again, care should be used to avoid obscuring these signs, and they must be clearly visible from all points in the discharge. Each sign is required to be illuminated and have a backup source of power should the power fail. They require regular testing according to the manufacturer's recommendations and the applicable fire, building, and life safety code. Each exit sign is equipped with a test button to facilitate inspection of function. Additional signs may also be required and are sometimes placed along the wall at floor level in case smoke obscures the illuminated signs at ceiling level.

Stairways act as both part of the exit and as areas of refuge for those cut off from escape. The doors and wall surface materials form a one- or two-hour (depending on the code) fire-rated assembly providing protection from smoke and fire for this important escape route. Stairways consist of risers, treads, landings, and standard hand railings. Risers are the vertical part of the stair and must be closed. Treads are the part of the stair we stand upon, or tread on. Treads and risers must have little deviation in height or depth, within 3/16 of an inch. Landings are placed to break up a stair run to provide several safety factors, such as to break one's fall or to give a person a place to gather one's senses or to rest in long stairways (Lathrop, 2008). With most codes, the minimum width of stairs is 44 inches, but as the occupant load increases, so does the required width of the stairway. Depending on the applicable fire, life safety, and building codes, the top railing of a handrail must be at least 34 inches and no more than 42 inches in height, and the top rail should be between 1¼ and 2½ inches wide to allow easy grip. An intermediate rail should be provided that splits the difference between the floor and handrail and prevents persons from passing through and falling on the treads below, or worse, to the floor below. Absolutely no storage should be allowed in stairwells because this will add fuel and or obstruct the exit, hindering the egress of individuals (NFPA 101, 2012).

Once the occupant has reached the actual door that leads to safety, they have reached the exit. This exit will be separated from the building by fire-rated construction and assemblies. The exit door will be a self-closing door equipped with what is called panic hardware. Panic hardware refers to a door opening mechanism that is designed to be operated from the direction of egress. It allows for the door to be opened in case of panic, in that it consists of a bar or assembly that crosses a majority of the width of the door and should panicking occupants reach this door and pile up against it, the hardware would operate and open the door. Exit doors are self-closing and may not be locked while the building is occupied. The areas around exit doors must be kept clear of storage or debris, including ice and snow, to maintain a safe pathway.

As part of the egress system, emergency lighting will be provided. Emergency lighting consists of battery-powered lights that illuminate the egress path in case of power failure. This also includes battery backup for illumination of exit signs. Exit signs may contain low-voltage bulbs for operation during main power interruption. These will need to be tested by interrupting the 110-volt power supply and causing the battery to supply power. Emergency lighting assemblies contain a test function to ensure the system is operating. Most manufacturers and local fire and building codes require monthly testing of the functionality by operation of the test button on the assembly for thirty seconds, with a thirty-minute longevity test annually. The thirty-minute test is best achieved by operating or removing the circuit breaker or fuse (NFPA 101, 2012).

The final component, the exit discharge, refers to the portion of the egress system between the termination of an exit and a public area. This path must discharge directly to the exterior at grade or it must provide a direct path to grade and must lead occupants to the public pathway. A public pathway could be a sidewalk, street, or area open to the air and leads to an area designated for public use.

Whatever form this path takes, it must comply with all requirements for maintaining egress, such as illumination, width, and accessibility. As previously mentioned, the owner/occupant must keep this path clear of snow, ice, debris, and property of any kind. The owner should also ensure that if the path does lead parallel with or directly onto a street, the occupant has warning before reaching this point. Preplanning with the local fire department should be done to identify these areas to avoid the gathering of occupants in the same area as the fire department (Lathrop, 2008).

THE EVACUATION PROCESS

It is the responsibility of the school to have an evacuation plan in place in case of fire. The plan should be in writing, and training on the plan should be provided regularly to faculty and staff. Interaction with local fire department personnel is critical in successful operations during emergencies. Local or state laws will dictate how often the drills should occur and when in the school year they should be held. NFPA 101 (2012) states that fire drills should be held not less than once per month when the building is open for occupancy. Drills should be held on different days and at different times and use the building's fire alarm system as it is intended. Make sure to notify the fire department and your alarm monitoring company prior to holding drills.

The two most important parts of the process are maintaining an orderly evacuation and accounting for the students prior to evacuating and once outside in a safe area. These two elements will be achieved by training and practice of faculty, staff, and students. Teachers and teacher's aides should receive training prior to any actual drill, identifying the actions to be taken and the direction of egress identified by their supervisors. This should include assignments on checking of lavatories and other rooms where students may be. This should be practiced several times over the course of a year to instill good response from both the students and the staff (Szachnowicz, 2008).

First, personal items should be left behind because this creates a great amount of disorganization prior to attempting to line up and will present a trip hazard during the evacuation. The group should form at the doorway, and a staff member should first check the door for heat and then observe for smoke. A quick accountability check should then be performed, and any instructions to the students should quickly be given. Take any keys that may be necessary to open or reopen doors because you should always close doors behind yourself to hinder fire movement in the building. Then follow the predetermined paths to exit the building. More than one direction of egress should be identified for each location in case smoke or fire is encountered. Remember, most deaths in fires are caused by smoke inhalation. It is deadly to inhale the by-products of combustion. If you encounter smoke or fire, turn around. If you are surrounded, seek refuge behind a closed door and call 911 or hang a blanket out of the window and signal for assistance.

Once out of the building, move to the predetermined gathering location. This area should be away from where fire department operations will occur, as well as away from traffic or any other hazards (Szachnowicz, 2008). Once outside of the building and in a safe area, faculty and staff should immediately account for all students. Faculty and staff should maintain daily tracking of which students are in attendance and the location of any other students who are not under their direct supervision. A supervisor should check with each faculty and staff member to ensure they have accounted for everyone. It is also important to track each staff member's whereabouts to account for their well-being.

CONCLUSION

The features of life safety should never be taken for granted. Time and time again we see major loss of life caused by inappropriate egress, lack of staff training, and lack of planning for emergencies. We must be vigilant in keeping the features of life safety in place as designed and vigilant in our preparedness for fire events that will occur. All plans should be reviewed at least annually, with practice and training a continuing part of normal operations in educational occupancies. The treasures you are protecting are irreplaceable.

CASE STUDY

John has recently become the principal of South Elementary. On his first day, he is escorted throughout the school to gain a more full understanding of the location of classrooms and facilities support functions. While touring the school, he notes that two exit doors are chained with padlocks. When asking the maintenance manager about the locks, he is told that there have been concerns over theft of school property.

- What life safety issues are presented by the chained doors?
- How should John address the situation?

EXERCISES

Obtain a copy of your fire evacuation plan and perform an audit on your egress systems by following through with your plan and assessing its compliance with the components discussed in this chapter.

1. Are the components of egress free and clear of added hazards such as decorations, storage, and blocked or locked doors?
2. Does the panic hardware work properly, and do all doors swing in the direction of egress?
3. Are any exit doors locked in a manner that prevents egress?
4. Are the exit signs clearly visible from all positions in the hallways and stairways?
5. Are the audible and visible notification devices covered, obstructed, damaged, or missing?
6. Is the path from the exit discharge to a public area away from the building free and clear of any obstructions or hazards?

REFERENCES

Bryan, J. (2008). Human behavior and fire. In *Fire Protection Handbook* (20th ed.). National Fire Protection Association, Quincy, MA.

Cholin, J. (2008). Inspection, testing, and maintenance of fire alarm systems. In *Fire Protection Handbook* (20th ed.). National Fire Protection Association, Quincy, MA.

Esposito, J. (2005). *Fire in the Grove.* Decapo, Cambridge, MA.

Farr, R., and Sawyer, S.. (2008). Fire prevention and code enforcement. In *Fire Protection Handbook* (20th ed.). National Fire Protection Association, Quincy, MA.

Grosshandler, W. L., Bryner, N. P., Madrzykowski, D. N., and Kuntz, K. (2005) *Report of the Technical Investigation of the Station Nightclub Fire (NIST NCSTAR 2).* National Institute of Standards and Technology, Gaithersburg, MD.

Jones, M. (2009). *Fire Protection Systems.* Delmar Cengage Learning, Clifton Park, NY.

Lathrop, J. (2008). Concepts of egress design. In *Fire Protection Handbook* (20th ed.). National Fire Protection Association, Quincy, MA.

Moore, W. (2008). Fire alarm systems. In *Fire Protection Handbook* (20th ed.). National Fire Protection Association, Quincy, MA.

National Fire Protection Association. (2010). *NFPA 72: National Fire Alarm and Signaling Code.* National Fire Protection Association, Quincy, MA.

National Fire Protection Association. (2012). *NFPA 1: Fire Prevention Code* (2012 ed.). National Fire Protection Association, Quincy, MA.

National Fire Protection Association. (2012). *NFPA 101: Life Safety Code* (2012 ed.). National Fire Protection Association, Quincy, MA.

Schifiliti, R. (2008). Notification appliances. In *Fire Protection Handbook* (20th ed.). National Fire Protection Association, Quincy, MA.

Szachnowicz, A. (2008). Educational occupancies. In *Fire Protection Handbook* (20th ed.). National Fire Protection Association, Quincy, MA.

15 Working at Heights

Ronald Dotson

CONTENTS

School districts have employees that work from heights as often as any other industry. This includes teachers. One of the leading causes of injuries to teachers is a fall from a different level (Isaacs, 2010). Oversight of job peripherals in hazard analysis usually allows this to occur. Teachers decorate classrooms and are even required to maintain a classroom that is inviting and conducive to classroom learning. This involves reaching heights that, when proper step stools or ladders are not provided, result in the use of desks or chairs to accomplish the task. Falls are among the leading causes of death in the industry (U.S. Bureau of Labor Statistics, 2010). It is important to note that falls are not necessarily deadly because of the height. Objects that can exacerbate or increase the risk of injury when a fall occurs are often overlooked. Fall prevention plans have four core duties: identification of the hazard, proper abatement, training of personnel, and rescue planning.

The first aspect of implementing an effective program is to identify the key advocates of the program and match their roles to their capabilities. Every program has administrators, qualified persons, competent persons, authorized users, and rescuers. It is acceptable to have one person fill various roles.

Every program has administrators. Here the safety manager is likely the administrator. He or she will perform three vital tasks:

- Develop, assess, and improve the program continually
- Create and provide a vision of where the program needs to be
- Develop procedures for accomplishing the four core duties of the program (American National Standards Institute, 2007)

Continual improvement can be made by effective investigations and program evaluations for effectiveness. This is also tied to the vision. Many organizations merely talk about abating a fall hazard with personal fall arrest, and the program is as complex as using a harness whenever a worker is four feet or more off the ground. Advanced programs target hazard elimination from different techniques or procedures first. Continual improvement also means that tracking metrics for all incidents on a site must be effective. Job classification tasks that may not traditionally stand out as having a height hazard must still be examined for falls. For example, how exposed are teachers to a fall? They are certainly not the first class of worker that we would picture. We may picture roofers, framers, painters, and bricklayers as targeted groups first. Many tasks or job positions have been overlooked in relation to fall hazards. We must realize that falls are not just from different heights or from heights over four feet. Deaths and impaling occur from same-level falls frequently. Proper procedures for accomplishing the four core duties must involve all of the levels of people associated with the program communicating with each other.

An authorized person under fall protection is the user. An authorized user must be able to recognize hazards and alert others. Their duty is to recognize the hazard, report it, and then execute due diligence to utilize the prescribed fall protection system correctly. Safety managers must make an impression to achieve buy-in at this level because motivation is the key to compliance. This does not mean that authorized persons report the hazard and action is then taken. It means that they should be consulted in job hazard analysis, any planning, and any time a hazard has been encountered that has been overlooked. The degree of associate reports or attempts to report encountered hazards may be a good indication of a mature program. In other words, it says something about the culture. A fall hazard survey should be conducted before beginning a new job or project. Individual schools should perform this at least annually for teachers and staff. They will have duties that require working from height.

Qualified persons are those that have training and experience to recognize and abate the fall hazard. It should be a goal in a school setting to have authorized users become qualified persons. This will enhance the quality of any fall hazard survey. Qualified persons perform four additional roles:

- Supervise design, selection, installation, and/or inspection of fall protection systems
- Assist in investigations
- May provide training
- Serve in rescue activities and may help plan them (American National Standards Institute, 2007)

The qualified person is above an authorized user. Their training and experience allow for increased participation. Only employees with training and documented work experience can be considered "qualified."

A competent person is a person who has training and experience but also has established institutional authority to stop work until the situation is corrected. Competent persons comprise the highest level of positions. Hierarchically, the positions range from user, to qualified, to competent. A competent person for fall protection then also performs addition duties. They include:

- Conduct hazard surveys, especially early assessments when projects are being planned
- Identify hazards formally
- Stop or limit work at a site
- Supervise selection of equipment and its use
- Verify compliance and worker training levels
- Investigate
- Conduct inspections on systems and components
- Remove damaged equipment
- Supervise rescue activities (American National Standards Institute, 2007)

Rescue activities are important but are very frequently overlooked. Even fall protection equipment manufacturers sometimes fail to list personal rescue items as parts of personal fall arrest systems. The competent person must be a rescue activity supervisor. In an educational setting with many school sites, the safety professional is usually not immediately present. This means that someone near and supervising the activity must be a competent person with rescue duty as well. Rescuers also have various levels of competency. These positions may be filled by one person depending on the structure of the organization.

There are two rescuer levels as well: competent and authorized. Authorized rescuers may be part of a school first aid response team. They should be able to perform the following rescue duties:

- Perform primary rescue duties according to their training level
- Verify plans are in place
- Inspect rescue equipment

The competent rescuer:

- Develops procedures
- Verifies adequate training
- Verifies proper storage and care of equipment
- Evaluates procedures and equipment (American National Standards Institute, 2007)

The safety manager must compile the information from each site and perform the following specifically in order for the program duties to be fulfilled:

- Survey the site, preoperational and operational
- Report on site specific hazards

- Plan to involve the qualified and competent persons
- Train or verify training to match the site
- Develop rescue plans with qualified and competent persons
- Perform mock rescue exercises, even if it is a table-top drill (American National Standards Institute, 2007)

These typical duties for a safety manager for an entire site can be used as a guide for the competent supervisor to perform a fall hazard survey for specific jobs or projects. For each job with a fall hazard, the competent person can adapt his or her knowledge and experience to perform a short training session on the hazards present, the plan for abatement, and then plan for rescue activity.

FALL PROTECTION SYSTEMS

There are five fall protection system levels. They are:

- Elimination
- Passive control
- Fall restraint
- Fall arrest
- Administrative controls (American National Standards Institute, 2007)

Elimination is the best choice. It is not always possible or practical under the circumstances. It usually involves a change of process. It is difficult because many times we find that the training and experience or the financial situation of the organization does not allow for it. For example, entire sections of trusses could be hoisted by crane at one time, lowering the amount of exposure, rather than setting one truss at a time. Of course, the company must have a crane and operator or have the resources to perform this. Another example may be to assemble sections at ground level.

Passive control is the next reliable method. Here workers are separated from the hazard. An example is the use of a guardrail system.

Fall restraint systems include self-retracting lanyard systems, positioning systems, and travel restraint. With travel restraint, solid tethering may be used as long as the worker cannot access the leading edge. It must be noted that travel restraint is not allowed for workers over 310 pounds of body weight. Travel restraint may only be used on low-slope surfaces as well. Positioning systems must limit the worker to less than two feet of fall or movement and can only be used on vertical surfaces. When possible, positioning systems should be partnered with other systems and not used for workers weighing more than the 310-pound limit (American National Standards Institute, 2007).

Fall arrest systems are many times the first solution for safety professionals. However, fall arrest means that the target of elimination is not the hazard of fall but rather injury from contact with a lower surface. Fall arrest is not a first choice. Here the hierarchy is at best fourth. Workers can be injured in the fall, in the arrest, during the lapse in time from fall to rescue by "suspension trauma," and during rescue activities. Additionally, many do not understand fall distance. Too many times six-foot-long shock-absorbing lanyards are worn when the height is not great enough to justify the lanyard. In other words, the worker would still contact the lower surface.

Lastly, administrative controls such as warning line systems that rely on a monitor for verbal warning can be used. These are considered least effective because they rely on human action for correct compliance and because workers do not have the appropriate equipment erected or in place to actually prevent the fall from the edge.

RESCUE

Let us review some points of fall rescue plans that are specific to fall protection. A goal of rescuing someone from a suspended harness and lanyard should be four minutes. American National Standards Institute (ANSI) lists six minutes as the necessary response time for fall situations. However, you should note that suspension trauma can begin to set in at the one-minute-thirty-second mark, and four minutes is considered crucial because most people begin to exhibit signs of trauma at this point. The plan should address:

- Means to summon rescue personnel, company, and first responders
- Method of rescue available
- Rescue personnel availability and training
- Type of equipment available and location
- Mock exercises (American National Standards Institute, 2007)

A means to signal is probably already planned for in general safety and should not vary much, if at all. Methods and equipment available connect with personnel training level and the work environment. Mock exercises or training drills on a regular basis and table-top drills for specific sites are important.

A couple of simple points will facilitate rescue. The first is that no matter what type of fall arrest is used, a plan should be in place for retrieving the fallen worker if possible with a retrieval line. Retrieval lines are lines attached that will enable workers to quickly pull a fallen worker over to them while the connecting mechanism slides along the horizontal lifeline. A second simple tip is to issue rescue steps for all harnesses. These inexpensive steps stay attached to the side of the harness, and the fallen worker, if conscious after the fall is arrested, can deploy the step. The step hangs near the foot and allows the worker to step up and take pressure away from his legs and lower extremities. This will prevent suspension trauma.

FALL PROTECTION PLANNING

When fall arrest is the option chosen for employee protection, the competent person must consider the height hazard presented and the available options. The point is not just to have fall protection on the employee but to actually prevent the employee from falling to the level below. There are several parts to a personal fall arrest system: the harness, the lanyard, the connection mechanism, the lifeline, the anchor point, and rescue attachments such as relief steps and retrieval lines. These must combine to arrest the fall and prevent the employee from hitting the lower level.

A common mistake is to equip an employee with a six-foot shock-absorbing lanyard. Because the maximum free-fall distance is six feet, the lanyard is just short of six feet in length [29 CFR 1926.502 (d)(16)(iii)]. Any lanyard used to arrest a fall must have an energy-absorbing capability. Therefore, the lanyard uses the force of the fall to pull out or activate a deceleration device that reduces the shock of sudden stop. The shock-absorbing portion of the lanyard is three feet in length. This equates to a total length of nine feet for a deployed six-foot shock-absorbing lanyard. The height of the average employee might be six feet, and it is recommended that three feet of additional space be calculated into the total fall distance. So a six-foot lanyard is only good for a height hazard of twelve feet from the anchorage point. Lower height hazards would require that the lanyard itself be attached to a lifeline or anchor point that is above the worker's head or far enough back from the leading edge that the fall would be arrested before striking the lower level. Other lanyard options exist. One strategy is to use a self-retracting fall limiter or lanyard system that allows the worker to move but when a sudden jerk is experienced from a fall, the lanyard locks, limiting the fall to less than two feet.

When employees are working from aerial lifts, it is recommended that the preferred strategy be one of elimination. An aerial lift has a boom basket with guardrail system around it, but because of the hazard of the boom and structure catapulting a worker out of the basket when it crosses rough terrain, the guardrail system cannot be relied upon for adequate protection. The strategy is to eliminate the fall hazard by keeping the worker in the basket. This requires a self-retracting device or a solid lanyard.

Scissor lifts are also commonly used with school employees. Scissor lifts are much different than aerial lifts. Scissor lifts keep the basket with guardrail and the frame of the machine over the center of gravity. The catapult effect is not present, and the guardrail itself meets Occupational Safety and Health Administration (OSHA) requirements. Fall protection may also be used if the manufacturer allows for it. Scissor lifts may not be designed to withstand the force of arresting a fall.

LADDER SAFETY

Ladders are a vital tool for most work environments. School districts have maintenance activities, custodial activities, and even teaching duties that require their proper use. Teachers need access to a type of ladder and the training to use it properly. Classroom decor and storage necessitate the use of a step stool or stepladder. This will prevent the human behavior of taking the quickest route to accomplishing the task from facilitating the use of desks, chairs, bookshelves, or other objects to get to items just out of reach. According to the Ladder Safety Institute, approximately 160,000 people are injured each year from failing to adhere to safety precautions. Of these, three hundred are deaths. Ladder safety is a combination of five steps. Proper planning must address ladder selection, ladder inspection, ladder setup, use, and proper care (American Ladder Institute, 2011).

SELECTION

Knowing the types of ladders is the first step for proper selection. There are basically three types of ladders: stepladders, single ladders, and extension ladders.

Stepladders are ladders that have a folding section and are freestanding. They can range in various heights beginning with step stools. The stepladder has steps instead of rungs for balance. It has side rails that contain the steps; spreaders that lock down, maintaining the spread of the ladder; a top cap that is not used for standing; a top step that is not used for standing; possibly a platform protruding from the rear that serves as a rest for tools and paint; and rear rails not used for climbing. The bottoms of the side rails where the ladder meets the ground or floor are its shoes or feet. Shoes and feet are the terms for all ladders.

Extension ladders are composed of two single ladders arranged so that they extend for long reach capability. Extension ladders have a base section that stays in contact with the ground and a fly section that extends upward. The fly section will have rung locks that lock the extension upward. The fly section could also have guides that keep it in line with the base section and possibly a rope and pulley system for extending the fly section. A single ladder has shoes or feet, side rails, rungs, and is not self-supporting. It is simply a base section from an extension ladder. Extension ladders and single ladders rely on a surface to rest upon and for the user to position them properly at a 75° angle or a ratio of four feet of height to one foot of base distance on the resting surface [29 CFR 1926.1053 (b)(5)(i)].

Proper selection mainly involves understanding the hazards presented by the work environment and having the capability to overcome the hazards. Hazards presented by the base surface, the top surface, the work height, reach, and load need to be understood. For example, spur plates at the shoes of the ladder prevent softer grounds from allowing the ladder to slip. Spreaders can be attached to the rear of the ladder to provide stable top bases while working around windows or uneven surfaces.

The weather plays a part as well in the environment. Windy conditions may create unstable conditions or slippery rungs. The access area may be around doors or electrical boxes and lines. Ladders are constructed from three basic materials: wood, fiberglass, and aluminum. When work is being done in an environment that may present a hazard of becoming electrically energized, a nonconductive ladder must be used.

Another point to consider in ladder selection is the duty rating. The amount of load that a ladder can safely support is often overlooked. The weight of the load is the weight of the worker, tools, accessories, or any other item that will be stored on or located on the ladder at anytime. Ladders have ratings on legible tags on their side rails giving the load rating. The ratings are:

- Type III = 200-pound load
- Type II = 225-pound load
- Type I or Heavy Duty = 250-pound load
- Type I A or Extra Heavy Duty = 300-pound load
- Special Duty Type I AA = 375-pound load (American Ladder Institute, 2011)

INSPECTION

An inspection of the parts of the ladder must be completed prior to use, during use, and after use. Begin the inspection at the shoes and proceed upward checking the rungs, side rails, any spreading devices (stepladders), rung locking devices (extension ladders), moving parts, ropes (extension ladders), and top rails, as well as checking for missing labels. Any ladder with repairable damage can be tagged for repair and removed from the immediate work area. Those with structural damage that cannot be repaired must be discarded from the school. It would not be advisable to give the defective product to an employee for home use. This might be considered negligent and would allow grounds for a lawsuit.

Some organizations use a ladder program that rotates ladder use from month to month to relieve constant use. Ladders are inspected by safety personnel during the out-of-service time frame and repaired if required. Although this may extend the life of the ladder, the ladder must still be inspected daily prior to use, during use, and again afterward. During use, if any unserviceable condition appears, appropriate action must take place.

SETUP

Setup presents several key hazards to include ergonomic concerns. Ladders should be carried with the front end slightly elevated from the back or bottom end to allow a single carrier to maneuver the ladder around obstacles and prevent ladder damage. The bottom shoes of the ladder should be against the surface of the building, pole, or other surface that the top of the ladder will rest against (other than stepladders or articulated ladders). The user will then lift the top end of the ladder and stand it erect by walking and lifting each rung. Once the ladder is against the surface of the building or other top support, the bottom should be brought out to a 75° angle. This is explained in terms of a ratio of 4:1, or for every four feet of working height, or distance from the shoe to the top of the side rail, the bottom should be away from the vertical surface one foot. A quick guide for workers to set up the correct angle is to instruct them that when their toes are against the bottom shoes, and they are standing up straight, they should be able to reach the rung nearest shoulder height with their open palms with arms outstretched (American Ladder Institute, 2011).

The top support and bottom support are critical. Soft ground surfaces may dictate a mud plate. Sometimes homemade slip supports or 2 × 4 boards secured to a ¾-inch piece of plywood form corners for the shoes to rest against. Many times the spur can be used on softer ground to prevent slippage. Other times and especially before the top of the ladder is tied off, a second person can hold the ladder.

The top of the ladder should extend three feet above any surface that the climber will be stepping onto [29 CFR 1926.1053 (b)(5)(i)]. The top of the ladder should be secured or tied to prevent movement and slippage at the top. Someone can hold the ladder in place when the ladder is first being climbed and has not been tied or secured at the top of the ladder.

Setup should avoid doors unless they are locked, blocked, and either guarded or labeled with signage. Additionally, the area around the ladder's base should be blocked off or marked with cones or other warning devices, especially when equipment is operated near or around corners.

Clearance for workers to maneuver up and down the ladder and not have to dodge encumbrances over the rungs is important. In cases where a bump hazard does exist it may be possible to use a shield to cover corners or allow for the climber to pass by without injury.

Lastly it is important to know that the length of the ladder does not translate to working height. They are two different things. It is a general rule that workers cannot stand on the top three rungs of a straight ladder or extension ladder and cannot stand on the top rail and highest step of a stepladder. Therefore, the working height of a ladder is not the length. Extension ladders, for example, have an overlap between the base section and fly section. For ladders under forty-eight feet, the overlap is three feet, and for those from forty-eight to sixty feet, the overlap is six feet. This means that a sixty-foot extension ladder with an overlap of six feet has a working height of fifty-four feet. Because a person may not stand on the top three rungs, this results in a standing height of approximately fifty-one feet (American Ladder Institute, 2011). The worker's reach while standing plus the fifty-one feet or standing height will determine the working height possible for the ladder and worker. Ladders that are either too short or too long are unsafe. The working height of a ladder is usually listed on a label on the side rail. For example, the working height of a sixteen-foot extension ladder is generally nine feet six inches.

When ground conditions are unlevel, ladders must be utilized with what are called levelers. Levelers are mounted to the side rails near the feet and can extend each side individually so that the ladder maintains level contact with the ground surface.

PROPER USAGE

Use is where many safety professionals spend most of their time. It is important because many injuries occur from not following usage guidelines. The most common safety violation centers on reaching away from the ladder and loosing balance. As a general rule, a worker should never stretch beyond normal reach and should maintain his or her hips between the side rails. Additionally, workers should not jump or slide the ladder to the side while still on the ladder.

Another common guideline for use, and an OSHA mandated rule, is to face the ladder and maintain a three-point contact while climbing and descending the ladder [29 CFR 1926.1053 (b)(20–22)]. This refers to keeping two hands and one foot (three points) or two feet and one hand (three points) in contact with the ladder while climbing. It is a common unsafe practice for workers to climb with tools in their hands. Tools should be hoisted on lines, secured in tool belts, or handed to workers once positioned on a ladder. There is a common misconception about three-point contact. The misconception is to mandate three-point contact while working from a ladder. Workers must be able to work with two hands. One technique to help maintain balance on a ladder is to rest the shins or legs and hips against the rungs. This additional shifting of the person's weight forward and against the ladder along with another point of body contact allows the worker to maintain balance.

Working with power tools and tools with locking on devices while on a ladder presents a critical hazard. Not only can tools be dropped, striking workers below, but workers on the ladder can fall with the tools, and with the tools capable of still running, the hazard becomes much worse. Only cordless tools like saws and drills should be used on a ladder. A chain saw or drill left running may fall and strike a fallen worker. It is also important to note that areas beneath ladders are controlled access zones, and entrance is only allowed when the workers acknowledges entry and stops all work. A worker holding a ladder from beneath while work occurs from the ladder is not advised.

Some other common misuses of ladders are sources for injury as well. Leaning a stepladder against a wall or pole in the same manner as a single or extension ladder is an OSHA violation for good reason. The ladder can easily slide down the resting surface, allowing the worker to fall. Ladders should never be tied together in any manner to obtain additional reach. Another common safety violation exists when ladders are used on top of an aerial lift, scissor lift, truck bed, or other elevated surface or device. It is much more efficient to properly plan and select the correct size ladder for the project.

Experience reveals that shortcuts are taken when the best piece of equipment is not readily available. Human behavior is to select the quickest way to accomplish the task. Although ladders are expensive, having a proper inventory available at each school in the district will aid in preventing a traveling maintenance crew from taking a shortcut rather than traveling back to the shop for the proper ladder or power tool to be used from the ladder. Use cordless tools. Only buy cordless tools as a general rule and prevent the more expensive injury.

CARE

Storage of ladders is also an injury prevention issue. Ladders should be hung or chained to prevent turnover. Some organizations lock the ladders and provide a common key to personnel who have been trained to properly use the tool. This ensures that only qualified personnel have access to a ladder. Teachers should have access to ladders and step stools. They must be able to decorate classrooms, store, and retrieve items safely. Not providing them the tools and training can result in injury. They cannot and will not wait on maintenance personnel or custodians to do the job for them.

Ladders should be cared for by cleaning and lubricating them after work. All moving parts should be lubricated, and the rungs and side rails should be cleaned of any dirt, grime, or substance that can stick or create a slippery condition (American Ladder Institute, 2011).

Wood ladders cannot be painted with opaque substances that obscure the presence of cracks or factory writing and labels. However, clear preservatives may be used. Aluminum or fiberglass ladders will typically last longer with proper care.

TRAINING SOURCE

A basic and free site for introductory ladder safety training can be found through the American Ladder Institute at www.laddersafety.org. Formed in 1948, its mission is to educate the public on ladder safety, develop industry standards, and represent the interests of its membership, ladder and ladder-component manufacturers (American Ladder Institute, 2011). The American Ladder Institute currently offers online training courses on ladder safety that can be utilized by school employees.

CONCLUSION

Working from heights is one specific hazard that must be addressed with a comprehensive program. The type of program that has been suggested in this chapter comes from a combination of general safety practices, OSHA regulations, and the American National Standards Institute as utilized from experience and training. Establishing competent persons within the workforce and delegating the duties according to position and level of training and experience creates a safe culture regarding fall protection. Working from ladders is one of the most common sources of falls from a different level. Ladders are a common tool of use in a school by maintenance and grounds personnel, custodians, and teachers decorating classrooms. Providing educational service employees with the tools, equipment, training, and policy in the form of a structured system will help minimize falls from different levels.

CASE STUDY

Recently, one of the two elementary schools suffered the loss of an experienced teacher because of disability from a work-related fall. The teacher was climbing on a desk to hang decorations from the ceiling in order to have her classroom presentable for the beginning of the school year. The teacher was wearing dress shoes, and the desk was found overturned. Her back and neck struck another desk on the way to striking the floor. There was not a witness to the incident. A teaching assistant discovered the teacher minutes after the fall.

You have had some informal discussions with administrators and with the maintenance manager in which you discovered that teachers do not have access to ladders at school. The maintenance manager feels that teachers should not be using ladders. One assistant superintendent agrees with the maintenance manager. One teacher at the elementary has pointed out to you that they must decorate classrooms and reach items that are above their heads quite often.

- What is the root cause of the teacher's fall?
- How would you address the situation?

EXERCISES

1. List the varying options and responses along with the pros and cons of the options that you might enact as school principal to prevent falls from happening in the future. What recommendations would you make to the central office in regards to district-wide policy?
2. Develop a school policy as a draft to be presented to faculty and staff at your school in regards to ladder safety.

REFERENCES

American Ladder Institute. (2011). *Online Ladder Safety Training.* Accessed March 10, 2012. http://www.laddersafetytraining.org.

American National Standards Institute (ANSI). (2007). *Z 359.2-2007 Comprehensive Managed Fall Protection Programs.*

Isaacs, J. (2010). School liability issues. 2010 Kentucky School Board Association Annual Meeting, Louisville, KY.

U.S. Bureau of Labor Statistics. (2010). Fatal injuries news release. Accessed May 2, 2012. http://www.bls.gov/newsrelease/cfoi.t01.htm.

U.S. Department of Labor. (1996). Stairways and ladders. Subpart X. Title 29 Code of Federal Regulations, Part 1926.1053. Occupational Safety and Health Administration, U.S. Department of Labor, Washington, D.C.

16 Ergonomics

Paul English

CONTENTS

The term "ergonomics" is derived from the Greek words ergon, meaning "work," and nomos, which means "law." In essence, ergonomics is the study of how people interact in their work environment. When we think about how we complete routine tasks in our everyday lives, we implement ergonomic improvements on a daily basis without consciously knowing it. It could be as simple as moving a computer keyboard closer to you for increased comfort or changing the location of a job task to a table at a higher elevation. At some point, we make the conscious decision to change the environment we are in for a more favorable outcome. In more recent years, ergonomics has also become to be known as human factors (Dul and Weerdmeester, 2008).

Many experts agree that the catalyst of the modern day study of ergonomics was born in direct relation to World War II. Between 1941 and 1946, over 16 million men served and fought for the United States (Leland and Oboroceanu, 2010). The armed services would accept large volumes of individuals who volunteered for service with any of the armed forces during this era. The exceptions to these rules were those with severe health conditions such as heart ailments or a history of tuberculosis. Out of sixteen million men, the average height was 68.1 inches with a weight of 150.5 pounds (Karpinos, 1958).

All branches of the military at the time understood that to be successful, they needed to design weapons and vehicles to fit the majority of the fighting force. The government believed that designing for the masses, specifically men who were 5 foot 7 inches, would eliminate the need to provide variation in the tools and training, thus creating a lean system. Although this sounds good in theory, it did not transpose well on the battlefield. If a soldier was 3 or 4 inches above the average height and was assigned to drive a tank that was designed for someone shorter, it would make the war experience that much more unforgettable.

The good news is today, the military, along with many other organizations, recognizes the benefit and power of providing and applying good ergonomic practices in all aspects of the workplace. Not only can it give a business a competitive advantage, but it can give a business or other entity the social and moral compass to do what is right for the employees. No one ever comes to work and says, "I want to get hurt today." Unfortunately, many organizations look at the immediate hazards in the workplace and fail to identify ergonomic stressors that can lead to a cumulative trauma disorder (CTD). Ergonomic injuries will not usually show signs or symptoms immediately, but over the course of time as the accumulation of stress on the body occurs, ergonomic injuries or CTDs can be severe and disabling to employees.

SIGNS AND SYMPTOMS OF CUMULATIVE TRAUMA

Cumulative trauma can be caused by many different influences within the workplace, creating many different types of injuries. Although deeper approaches to correcting ergonomic issues are encouraged and needed, ergonomic stressors can be generalized into three categories:

- **Force:** How much force needs to be exerted to complete the job task?
- **Frequency:** How often does the job task need to be completed?
- **Posture:** What is the posture of the employee completing the task?

If one of more of these areas of force, frequency, or posture is incorrect for the employee completing the job task, it will lead to a CTD. A CTD can take on many different attributes and manifest themselves into different known categories:

- **Carpal tunnel syndrome:** Affecting the medial nerve, which runs through the body channel in the wrist called the carpal tunnel. Caused by over-exertion or twisting of the wrist, especially when excessive force is used. Symptoms can include burning, prickling, and itching sensations in the wrist, thumb, or fingers. Some patients report the feeling of the hand falling asleep. In severe cases, overall hand strength can be compromised.
- **De Quervain's disease:** Tendons to the thumb are inflamed. Scarring of this tendon can lead to restricted motion of the thumb.
- **Epicondylitis:** The inflammation of the tissues in the inner thumb side of the elbow. Sometimes referred to as "tennis elbow," because many tennis players contract the condition from excessive downward rotation of the forearm and wrist, while playing tennis.

- **Ganglion cysts:** Bumps that appear on the wrist full of synovial fluid (joint fluid). Usually occurs from excessive and repetitive use of wrist and hands. Surgical removal of cysts can be required as treatment.
- **Tendonitis:** Inflammation of a tendon that occurs when a muscle or tendon is repeatedly used and tensed. Symptoms usually appear in sore body parts such as shoulders, wrists, hands, and elbows.
- **Tenosynovitis:** Tendons in the wrist and fingers become sore and inflamed because of repetitive motion and bad body posture.
- **Thoracic outlet syndrome:** A disorder that occurs from completing overhead tasks that affect the shoulder. The loss of feeling on the pinky side of the hand and hand weakness are symptoms consistent with this condition.
- **Trigger finger syndrome:** Is a form of tendonitis caused by repetitive flexing of the fingers against vibrating resistance such as a drill or other power tools. In this condition, the tendons become inflamed, causing swelling and loss of dexterity. Possible causes can be using tools with handles that are too big or small.

Although this is by no means a definitive list of signs, symptoms, or outcomes of CTD issues, it presents a picture of how excessive force, frequency, and posture can influence workplace injuries. The key is to identify poor job design and workstations before injuries can occur and design new workstations and job tasks with ergonomics in mind.

ERGONOMIC ISSUES AND OSHA5

The Occupational Safety and Health Administration (OSHA) has published voluntary guidelines for different industries to help identify different ergonomic issues commonly found in the workplace. OSHA's mission is, "to assure safe and healthful working conditions for working men and women by setting and enforcing standards and by providing training, outreach, education and assistance" (OSHA, 1989). OSHA has the ability to levy fines against employers for violating safety and health standards. During the Clinton administration, an ergonomics regulation was published during the close of his presidency. Once Clinton left office, the first repeal that President Bush enacted was that of the ergonomics regulation. Deemed too costly for employers to comply with, it was viewed as a big-government policy that would hurt small business.

OSHA, left with no standard to enforce, turned to what is called the "general duty clause" of the Occupational Safety and Health Act. Found in section 5(a)(1) of the act, it states:

(a) Each employer:
  (1) Shall furnish to each of his employees employment and a place of employment which are free from recognized hazards that are causing or are likely to cause death or serious physical harm to his employees
  (2) Shall comply with occupational safety and health standards promulgated under this act

(b) Each employee shall comply with occupational safety and health standards and all rules, regulations, and orders issued pursuant to this act which are applicable to his own actions and conduct.

OSHA can use the general duty clause to cite workplace safety hazards in the absence of a specific regulation that addresses the safety issue at hand. OSHA used the power of the general duty clause when it investigated alleged safety and health violations at a Pepperidge Farm, Inc., facility in Downingtown, Pennsylvania. The facility was cited by OSHA for improper recordkeeping of workplace injuries. Many of the injuries that were not recorded properly were ergonomically related issues that the company identified but failed to fix. Later documents uncovered during the investigation showed that both medical staff at the facility, as well as occupational safety professionals, had identified ergonomic issues, but management failed to correct the hazards. OSHA argued that Pepperidge Farm failed to keep employees safe from a recognized hazard. As OSHA stated,

> The lifting items here involve employees lifting 100-pound bags of sugar, 68-pound blocks of butter, roll stock weighing up to 165-pounds, and cookie tins weighing up to 38-pounds. The repetitive motion items involve employees performing in quick succession assembly line tasks, such as dropping paper cups from a stack with one hand and filling them with baked cookies in the other hand (*Secretary of Labor v. Pepperidge Farm, Inc.*, 1996).

This case later became the infamous "Downingtown Case" that illustrated to industry that OSHA record keeping for workplace injuries was a very serious matter. The proposed penalty from OSHA was for $1.4 million, which was settled many years later for a fraction of the imposed penalty. What this case also did was put many companies on notice that failure to identify and recognize ergonomic issues in the workplace can lead to significant penalties and fines. Not only does it make good business sense to identify ergonomic stressors in the workplace, but it will help eliminate any preponderance of recognized hazards in the workplace under the Occupational Safety and Health (OSH) Act.

HOW TO IDENTIFY ERGONOMIC ISSUES

The easiest way to identify whether there are any ergonomic issues at a school is to look at past injury data. OSHA 300 record keeping logs, past accident investigation reports, or insurance losses can help you determine whether any issues exist. Looking at this data and asking some key questions will help you identify and locate specific job tasks or occupations that have incurred ergonomic injuries. These questions should include:

- What job tasks or occupations have the most sprain or strain injuries?
- Which job tasks or occupations have incurred lost workdays or have a high severity?

- Which job tasks or occupations have a high rate of surgical intervention? These same cases will also have the highest indemnity and medical costs associated with the workers' compensation claim.
- Which job tasks or occupations have the highest turnover rate and why? Studies have shown that turnover rates have a direct relationship to ergonomically poor job designs.

EMPLOYEE SURVEYS AND REPORTS

Employees who report signs and symptoms of CTDs should be interviewed for further information. The purpose of including ergonomics in the safety program is to identify issues before anyone becomes injured, to be proactive in safety. Employees who report signs and symptoms should be encouraged to identify specific job tasks or body parts that are creating the discomfort. More detailed information is needed to identify specific controls that can be implemented. To this end, a symptom survey can be created to identify data to be reviewed.

BASIC CTD SYMPTOM SURVEY

1. What job or occupation are you currently doing? _____
2. How many hours per week? _____ Hours per day are you completing this task? _____
3. What other job tasks have you completed in the past year for more than 2 weeks at a time? _____
4. Have you had pain or discomfort in the past year? ☐Yes ☐No If "Yes," in what body part did you experience the pain or discomfort? ☐Neck ☐Shoulder ☐Elbow/Forearm ☐Hand/Wrist ☐Fingers ☐Upper Back ☐Lower Back ☐Thigh/knee ☐Lower Leg ☐Ankle/Foot
5. What best describes the problem you are having? ☐Aching ☐Burning ☐Cramping ☐Pain ☐Loss of Color ☐Numbness ☐Swelling ☐Stiffness ☐Tingling ☐Weakness ☐Other
6. When did you first notice the problem? _____ month _____ year
7. How long does each episode last? ☐_____ hours ☐Day ☐Week ☐Month ☐>Month
8. How many episodes of this pain/discomfort have you had in the past year? _____
9. What do you think is the cause of the problem? _____
10. Have you had this problem in the last 7 days? ☐Yes ☐No
11. How would you rate this problem on a scale of 1–10, with 1 being "no issue" and 10 being "unbearable"? _____ When it was at its worst? _____
12. Have you had medical treatment for this problem? ☐Yes ☐No If no, why not? _____

13. If yes, where did you receive treatment?

_____

14. Have you lost any workdays because of the pain or discomfort? ☐Yes ☐No
15. What do you think would improve the pain/discomfort?

_____

These questions are subjective and need to be so in order to establish what the employee is experiencing as opposed to what is occurring in the workplace. Once the data are collected from the employee, it can be measured to the work environment to identify possible solutions needed to increase safety and reduce ergonomic stressors, if any actually exist. Employee input is invaluable when identifying ergonomic issues.

Safety committees or safety teams that are in place at a school can also be efficient ways to identify ergonomic issues in the workplace. If a process is already in place for a safety team or committee, take the time to train these people in how to identify ergonomic issues. Faculty and staff may be more apt to talk to a fellow coworker on a safety committee or safety improvement team than to go to the school administration with a safety issue. Talking to the employees and getting their input is invaluable when identifying ergonomic issues.

Many different tools and tips can be accessed via the Internet to help identify ergonomic issues in the workplace. The National Institute of Occupational Safety and Health (NIOSH) assists OSHA in identifying trends in workplace safety and health to create a safer workplace. NIOSH was created as a result of the Occupational Safety and Health Act of 1970 and is part of the Centers for Disease Control and Prevention, which is housed in the Department of Health and Human Services. The mission of NIOSH is to

> Generate new knowledge in the field of occupational safety and health and to transfer that knowledge into practice for the betterment of workers. To accomplish this mission, NIOSH conducts scientific research, develops guidance and authoritative recommendations, disseminates information, and responds to requests for workplace health hazard evaluations. NIOSH provides national and world leadership to prevent work-related illness, injury, disability, and death by gathering information, conducting scientific research, and translating the knowledge gained into products and services, including scientific information products, training videos, and recommendations for improving safety and health in the workplace (Centers for Disease Control and Prevention, 2011).

NIOSH has completed many studies in ergonomic hazards and how to identify possible ergonomic issues in the workplace. The General Ergonomics Risk Analysis Checklist is just one tool that can help identify ergonomic issues in the workplace (Table 16.1). Careful consideration must be given when identifying ergonomic stressors in the workplace. A "yes" answer to any of these questions would indicate further investigation. NIOSH has created additional checklists for each area identified in this general checklist. These additional checklists can be found in NIOSH Publication No. 97-117.

TABLE 16.1
General Ergonomics Risk Analysis Checklist

Question	Yes	No	Comments

Manual Material Handling

Is there lifting of loads, tools, or parts?

Is there lowering of tools, loads, or parts?

Is there overhead reaching for tools, loads, or parts?

Is there bending at the waist to handle tools, loads, or parts?

Is there twisting at the waist to handle tools, loads, or parts?

Physical Energy Demands

Do tools and parts weigh more than 10 pounds?

Is reaching greater than 20 inches?

Is bending, stooping, or squatting a primary task activity?

Is lifting or lowering loads a primary task activity?

Is walking or carrying loads a primary task activity?

Is stair or ladder climbing with loads a primary task activity?

Is pushing or pulling loads a primary task activity?

Is reaching overhead a primary task activity?

Do any of the above tasks require five or more complete work cycles to be done within a minute?

Do workers complain that rest breaks and fatigue allowances are insufficient?

Other Musculoskeletal Demands

Do manual jobs require frequent, repetitive motions?

Do work postures require frequent bending of the neck, shoulder, elbow, wrist, or finger joints?

For seated work, do reaches for tools and materials exceed 15 inches from the worker's position?

Is the worker unable to change his or her position often?

Does the work involve forceful, quick, or sudden motions?

Does the work involve shock or rapid buildup of forces?

Is finger-pinch gripping used?

Do job postures involve sustained muscle contraction of any limb?

Computer Workstations

Do operators use computer workstations for more than 4 hours a day?

Are there complaints of discomfort from those working at these stations?

Is the chair or desk nonadjustable?

Is the display monitor, keyboard, or document holder nonadjustable?

Does lighting cause glare or make the monitor screen hard to read?

Is the room temperature too hot or too cold?

Is there irritating vibration or noise?

Environment

Is the temperature too hot or too cold?

Are the worker's hands exposed to temperatures less than 70° Fahrenheit?

(continued)

TABLE 16.1   (Continued)
General Ergonomics Risk Analysis Checklist

Question	Yes	No	Comments
Is the workplace poorly lit?			
Is there glare?			
Is there excessive noise that is annoying, distracting, or producing hearing loss?			
Is there upper extremity or whole-body vibration?			
Is air circulation too high or too low?			

General Workplace

Are walkways uneven, slippery, or obstructed?
Is housekeeping poor?
Is there inadequate clearance or accessibility for performing tasks?
Are stairs cluttered or lacking railings?
Is proper footwear worn?

Tools

Is the handle too small or too large?
Does the handle shape cause the operator to bend the wrist in order to use the tool?
Is the tool hard to access?
Does the tool weigh more than 9 pounds?
Does the tool vibrate excessively?
Does the tool cause excessive kickback to the operator?
Does the tool become too hot or too cold?

Gloves

Do the gloves require the worker to use more force when performing job tasks?
Do the gloves provide inadequate protection?
Do the gloves present a hazard of catch points on the tool or in the workplace?

Administration

Is there little worker control over the work process?
Is the task highly repetitive and monotonous?
Does the job involve critical tasks with high accountability and little or no
  tolerance for error?
Are work hours and breaks poorly organized?

Source: National Institute for Occupational Safety and Health. (1997). Tool tray 5A: General ergonomic
         risk assessment checklist. Accessed December 11, 2011. http://www.ergo2.amisco.org/eptbtr5a.
         html.

QUALITATIVE ERGONOMIC ASSESSMENT TOOLS

OSHA has created the ergonomic assessment tool in Table 16.2 for video display
terminal (VDT) workstations. This checklist is from the now-rescinded OSHA
Ergonomics Safety Standard from 2000. Although the standard may not be currently in effect and enforced by OSHA, it still provides employers with a good

TABLE 16.2
Sample Ergonomic Evaluation

Working Conditions   Yes   No

The workstation is designed or arranged for doing VDT tasks so it allows the employee's

A. Head and neck to be about upright (not bent down/back)

B. Head, neck, and truck to face forward (not twisted)

C. Trunk to be about perpendicular to floor (not leaning forward/backward)

D. Shoulders and upper arms to be perpendicular to floor (not stretched forward) and relaxed.

E. Upper arms and elbows to be close to body (not extended outward)

F. Forearms, wrists, and hands to be straight and parallel to floor (not pointing up/down)

G. Wrists and hands to be straight (not bent up/down or sideways toward little finger)

H. Thighs to be about parallel to floor and lower legs to be about perpendicular to floor.

I. Feet to rest flat on floor or be supported by a stable footrest.

J. VDT to be organized in a way that allows employee to vary VDT tasks with other work activities or to take microbreaks or recovery pauses while at the VDT workstation.

Seating

1. Backrest provides support for employee's lower back (lumbar area).

2. Seat width and depth accommodate specific employees (seat pan not too big/small).

3. Seat front does not press against the back of the employee's knees or lower legs (seat pan not too long).

4. Seat has cushioning and is rounded/has "waterfall" front (no sharp edges).

5. Armrests support both forearms while employee performs VDT tasks and do not interfere with movement.

Keyboard/input device: The keyboard/input device is designed or arranged for doing VDT tasks so that:

6. Keyboard/input device platform(s) is stable and large enough to hold keyboard and input device.

7. Input device (mouse or trackball) is located right next to keyboard so it can be operated without reaching.

8. Input device is easy to activate and shape/size fits hand of specific employee (not too big/small).

9. Wrists and hands do not rest on sharp or hard edge.

Monitor: The monitor is designed or arranged for VDT tasks so that:

10. Top line of screen is at or below eye level so employee is able to read it without bending head or neck down/back. (For employees with bifocals/trifocals, see next item.)

11. Employee with bifocals/trifocals is able to read screen without bending head or neck backward.

12. Monitor distance allows employee to read screen without leaning head, neck, or trunk forward/backward.

13. Monitor position is directly in front of employee so employee does not have to twist head or neck.

14. No glare (e.g., from windows, lights) is present on the screen that might cause employee to assume an awkward posture to read screen.

(continued)

TABLE 16.2 (Continued)
Sample Ergonomic Evaluation

Working Conditions	Yes	No

Work area: The work area is designed or arranged for doing VDT tasks so that:

15. Thighs have clearance space between chair and VDT table/keyboard platform (thighs not trapped).

16. Legs and feet have clearance space under VDT table so employee is able to get close enough to keyboard/input device.

Accessories

17. Document holder, if provided, is stable and large enough to hold documents that are used.

18. Document holder, if provided, is placed at about the same height and distance as monitor screen so there is little head movement when employee looks from document to screen.

19. Wrist rest, if provided, is padded and free of sharp and square edges.

20. Wrist rest, if provided, allows employee to keep forearms, wrists, and hands straight and parallel to ground when using keyboard/input device.

21. Telephone can be used with head upright (not bent) and shoulders relaxed (not elevated) if employee does VDT tasks at the same time.

General issues

22. Workstation and equipment have sufficient adjustability so that the employee is able to be in a safe working posture and to make occasional changes in posture while performing VDT tasks.

23. VDT workstation, equipment, and accessories are maintained in serviceable condition and function properly.

Passing Score = "YES" answer on all "working postures" items (A–J) and no more than two "NO" answers on remainder of checklist (1–23).

Source: Bernard, T. E. (2009). Analysis tools for ergonomists. Accessed December 1, 2011. http://www.personal.health.usf.edu/tbernard/ergotools/index.html.

understanding of ergonomic issues that can be identified when looking at how computer workstations are designed. This assessment tool can be of particular use in schools for faculty, staff, and students. Faculty and staff can benefit from good ergonomic design of their desks and work spaces where they spend large amounts of time using computers to perform work. Students can benefit from good ergonomic design of computer work spaces in labs or in the classroom.

WASHINGTON STATE DEPARTMENT OF LABOR AND INDUSTRIES

States have the right to protect workers from workplace hazards by establishing state administered OSH programs. These programs must meet or exceed standards established by federal OSHA. Washington's is an example of a state-run OSH program. They have taken the initiative to create a number of ergonomic tools that include:

- **Lifting calculator** to help identify how much a person can lift and at what height the lift becomes a hazard.
- **Hazard zone checklist** to help observers identify bad body postures as well as excessive force and frequency of job tasks. Small pictograms are included with each step to help observers identify issues.
- **Caution zone checklist** to help identify possible issues that were not covered as a possible hazard in the Hazard Zone Checklist. Small pictograms are included with each step to help observers identify issues.

These qualitative tools for identifying ergonomic issues in the workplace are helpful to use because almost everyone can understand what to look for and how to use the tools. Training employees on how to identify ergonomic issues should always be included when performing observations. The tools provided by the Washington State OSH program serve as a good example of including pictograms to help remind observers of what to look for when identifying force, frequency, and body posture issues.

SEMIQUANTITATIVE ASSESSMENT TOOLS

Many people seem to think that ergonomics is too confusing when attempting to quantify job tasks. There are several semiquantitative tools that can be used that are very user friendly.

RULA AND REBA

The rapid upper limb assessment (RULA) and rapid entire body assessment (REBA) are methods that were developed by Dr. Lynn McAtamney and Professor E. Nigel Corlett, both ergonomists from the University of Nottingham in England. RULA is a postural targeting method for estimating the risks of work-related upper limb disorders. REBA is a postural targeting method for estimating the risks of work-related entire body disorders. Both assessments provide a quick and systematic assessment of the postural risks to a worker. The analysis can be conducted before and after an intervention to demonstrate that the intervention has worked to lower the risk of injury. Both assessments are also extremely easy to understand by observers and individuals who do not understand ergonomic stressors.

The scoring of the RULA tool involves a simple single-digit scale based on the score of the tool and the action level or severity of the job task. The higher the action level identified, the more action should be taken to eliminate the hazard. The score of the tool may identify a specific job task in a group of tasks that one employee does that is putting the employee at risk. For example, if five out of six different tasks an employee carries out scores a RULA score of 3, it would put the job at an Action Level of 2. The last job task may have a RULA score of 6, which at that point would push the Action Level up to 3 and would place the employee in a higher risk category for possible ergonomic injuries. REBA scores on a scale of 1–157 with 1 meaning

TABLE 16.3
RULA Scoring

Action Level	RULA Score	Interpretation
1	1–2	The person is working in the best posture with no risk of injury from their work posture.
2	3–4	The person is working in a posture that could present some risk of injury from their work posture, and this score most likely is the result of one part of the body being in a deviated and awkward position, so this should be investigated and corrected.
3	5–6	The person is working in a poor posture with a risk of injury from their work posture, and the reasons for this need to be investigated and changed in the near future to prevent an injury.
4	7+	The person is working in the worst posture with an immediate risk of injury from their work posture, and the reasons for this need to be investigated and changed immediately to prevent an injury

Source: Cornell University. (2011). Cornell University Ergonomics Web. Accessed December 5, 2011. http://ergo.human.cornell.edu/cutools.html.

that there is no action necessary and 15 meaning that the risk level for injury is extremely high and action must be taken immediately (Table 16.3).

LIBERTY MUTUAL MANUAL MATERIALS HANDLING TABLES

Liberty Mutual has performed several research studies to determine the best way to identify proper lifting and carrying techniques for policyholders to eliminate back claims from manual material handling. Separate tables have been created to help identify ergonomic issues for pushing/pulling tasks, carrying tasks, and lifting/lowering tasks.

RODGERS MUSCLE FATIGUE ANALYSIS

Muscle fatigue analysis was proposed by Dr. Suzanne H. Rodgers as a means to assess the amount of fatigue that accumulates in muscles during various work patterns within five minutes of work. The hypothesis was that a rapidly fatiguing muscle is more susceptible to injury and inflammation. With this in mind, if fatigue can be minimized, so should injuries and illnesses of the active muscles. This method for job analysis is most appropriate to evaluate the risk for fatigue accumulation in tasks that are performed for an hour or more and where awkward postures or frequent exertions are present. Based on the risk of fatigue, a "priority for change" can be assigned to the task (Bernard, 2010).

NIOSH LIFTING EQUATION

NIOSH created a lifting equation to help identify static lifting task issues for ergonomics. The tool considers the weight of the load, distance of the lift, distance to the floor, and frequency of the lift (Centers for Disease Prevention and Control, 1994).

CAUTION WITH ERGONOMIC ASSESSMENT TOOLS AND SOLUTIONS

This is by no means a complete list of ergonomic assessment tools. Depending on what task you are assessing, different tools will help more than others. Anyone who is performing assessments and observations to identify ergonomic issues must exercise some caution. These tools will help identify ergonomically poor job tasks; they will not give you a solution. The purpose of these tools is to give you the data and information needed to make an informed decision on how to correct an unsafe condition or unsafe act.

Be prepared to correct the issue once an assessment has been completed. It is one thing to think you might have a recognized hazard; it is completely different to verify that the hazard exists. School administrations must be prepared and provide support to correct the issues these tools will uncover. The last thing a school would want to do is to identify a hazard and do nothing to correct the issue. As stated before, OSHA has a duty to enforce regulations among employers who fail to protect employees from recognized hazards in the workplace. Once an evaluation has been completed, the hazard has been recognized.

Some problems do not need to be analyzed deeply, a practice sometimes called "analysis paralysis." Quick fixes can jumpstart an ergonomics program. It can be as simple as moving heavier items from a top shelf to a lower shelf for easier retrieval.

Some problems will be more complex than others. Do not be afraid to ask for help. Risk management or insurance carriers typically have access to safety professionals who can assist with ergonomic studies and solutions.

Ergonomic solutions, much like standard safety solutions, come in all shapes and sizes. Do not discount process changes in how a task is completed. Process changes or process sequences can be altered to produce a safer work environment.

Ergonomics should be included in the cost and planning of new facilities with employees in mind. This is your chance to eliminate the hazard and engineer it out of the process.

Although ergonomics has been around for almost seventy years, it is still an emerging field. The key to a good ergonomics program is to identify ergonomic stressors before an injury occurs. Force needed to complete a task coupled with an awkward body posture and increased frequency will lead to ergonomic injuries. Faculty and staff must be encouraged to report the signs and symptoms of possible ergonomically related injuries as soon as possible.

CASE STUDY

Stacy is currently the principal's administrative assistant and has held this position for over fifteen years. Job tasks include filing, typing, and answering phone calls. She comes to work one day with a splint on her right dominant hand because of pain in her wrist. She states that her hand has been falling asleep and that the sensation wakes her up at night. She has not received medical treatment for her condition.

Stacy purchased the splint for her wrist at the local pharmacy after doing some research on the Internet regarding her condition.

- What possible condition is Stacy suffering from given the work that she performs?
- What ergonomic assessment tools would you use to help identify ergonomic issues?
- Which ergonomic assessment tool would work best for this situation? Why?

EXERCISES

1. Does employee involvement enhance an ergonomics safety program? If so, how?
2. What are some of the cautions that should be identified and discussed when using ergonomic assessment tools?
3. How can OSHA determine whether ergonomic injuries are a "recognized hazard" in your workplace?
4. What is the difference between the ergonomic assessment tools RULA and REBA? Which is the best tool to use for sedentary (sitting) workstations?
5. Why is it important to train employees to identify ergonomic issues in the workplace?

REFERENCES

Bernard, T. E. (2009). Analysis tools for ergonomists. Accessed December 1, 2011. http://www.personal.health.usf.edu/tbernard/ergotools/index.html.

Centers for Disease Prevention and Control. (1994). *Applications Manual for the Revised NIOSH Lifting Equation.* National Institute for Occupational Safety and Health, Atlanta, GA.

Centers for Disease Control and Prevention. (2011). The National Institute for Occupational Safety and Health (NIOSH). Accessed December 4, 2011. http://www.cdc.gov/niosh/about.html.

Cornell University. (2011). Cornell University Ergonomics Web. Accessed December 5, 2011. http://ergo.human.cornell.edu/cutools.html.

Dul, J., and Weerdmeester, B. (2008). *Ergonomics for Beginners.* CRC Press, Boca Raton, FL.

Karpinos, B. D. (1958). Weight-height standards based on World War II experience. *Journal of the American Statistical Association*, 53, 415.

Leland, A., and Oboroceanu, M.-J.-J. (2010). *American War and Military Operations Casualties: Lists and Statistics.* Congressional Research Service, Washington D.C.

National Institute for Occupational Safety and Health. (1997). Tool tray 5A: General ergonomic risk assessment checklist. Accessed December 11, 2011. http://www.cdc.gov/niosh/docs/97-117/eptbtr5a.html

Occupational Safety and Health Administration. (1989). About OSHA. Accessed December 3, 2011. http://www.osha.gov/about.html.

Secretary of Labor v. Pepperidge Farm, Inc., OSHRC Docket No. 89-0265 (Occupational Safety and Health Review Commission September 20, 1996).

17 Hazard Communication

E. Scott Dunlap

CONTENTS

There are many hazards that exist in a school environment that can result in harm to faculty, staff, and students. These hazards can include falls, cumulative trauma, and natural disasters. The Occupational Safety and Health Administration (OSHA) has created a regulation that is commonly referred to as its hazard communication standard (Occupational Safety and Health Administration, 1996), but it focuses only on the use of hazardous chemicals in the workplace. OSHA addresses the general need for employers to communicate to employees the hazards of various chemicals that are used at work. These chemicals can include products used by maintenance staff, janitorial staff, contractors, visitors, and teachers. Hazardous chemicals must be identified and included within the scope of a school hazard communication program. This chapter will explore the primary components of the hazard communication standard that apply to the use of hazardous chemicals in a school. The regulation will need to be read in its full text to identify all issues, details, and exemptions that might apply.

PURPOSE

The regulation opens with a detailed statement as to the purpose for which it exists:

1910.1200(a)(1)
The purpose of this section is to ensure that the hazards of all chemicals produced or imported are evaluated, and that information concerning their hazards is transmitted to employers and employees. This transmittal of information is to be accomplished by means of comprehensive hazard

197

communication programs, which are to include container labeling and other forms of warning, material safety data sheets and employee training.

1910.1200(a)(2)

This occupational safety and health standard is intended to address comprehensively the issue of evaluating the potential hazards of chemicals, and communicating information concerning hazards and appropriate protective measures to employees, and to preempt any legal requirements of a state, or political subdivision of a state, pertaining to this subject. Evaluating the potential hazards of chemicals, and communicating information concerning hazards and appropriate protective measures to employees, may include, for example, but is not limited to, provisions for: developing and maintaining a written hazard communication program for the workplace, including lists of hazardous chemicals present; labeling of containers of chemicals in the workplace, as well as of containers of chemicals being shipped to other workplaces; preparation and distribution of material safety data sheets to employees and downstream employers; and development and implementation of employee training programs regarding hazards of chemicals and protective measures. Under Section 18 of the Act, no state or political subdivision of a state may adopt or enforce, through any court or agency, any requirement relating to the issue addressed by this Federal standard, except pursuant to a Federally-approved state plan.

This purpose statement addresses a number of key issues that are appropriate for schools. The first is the identification of hazards. Although manufacturers are responsible for identifying the hazards of specific chemicals, schools must be aware of the existence of these hazards among chemicals that are used. The second issue is communication. Once hazards are identified, a means of communication must be established through which this information is communicated to employers and employees. This is accomplished through the use of such things as training, container labels, and material safety data sheets (MSDSs). The third issue is the need for an employer to maintain a written hazard communication program.

SCOPE AND APPLICATION

The existence of an OSHA regulation does not mean that it applies to every employer. It is important to review the scope and application section of a regulation to ensure how it applies in a given environment. The hazard communication standard is unique in one respect from other regulations in that it has specific target audiences, to include manufacturers, importers, and employers (schools). The scope and application section of the hazard communication standard is contained in 29 CFR 1910.1200(b). The general information is as follows:

1910.1200(b)(1)

This section requires chemical manufacturers or importers to assess the hazards of chemicals which they produce or import, and all employers to provide information to their employees about the hazardous chemicals to which they are exposed, by means of a hazard communication program, labels and other forms of warning, material safety data sheets, and information and training. In addition, this section requires distributors to transmit the required information to employers. (Employers who do not produce or import chemicals need only focus on those parts of this rule that deal with establishing a workplace program and communicating information to their workers. Appendix E of this section is a general guide for such employers to help them determine their compliance obligations under the rule.)

1910.1200(b)(2)
This section applies to any chemical which is known to be present in the workplace in such a manner that employees may be exposed under normal conditions of use or in a foreseeable emergency.

The section goes on to address a number of specific issues in unique environments and applications. The information contained here indicates that the regulation includes schools as an "employer" and indicates the scope of activity, such as maintaining a written program and utilizing material safety data sheets.

DEFINITIONS

To more fully define the applicability of the regulation, it is important to review how OSHA defines certain words pertaining to the standard. Definitions can be found in 29 CFR 1910.1200(c). One issue is to determine what OSHA means by a "hazardous chemical." Three definitions clarify this issue:

"Hazardous chemical" means any chemical which is a physical hazard or a health hazard.

"Health hazard" means a chemical for which there is statistically significant evidence based on at least one study conducted in accordance with established scientific principles that acute or chronic health effects may occur in exposed employees. The term "health hazard" includes chemicals which are carcinogens, toxic or highly toxic agents, reproductive toxins, irritants, corrosives, sensitizers, hepatotoxins, nephrotoxins, neurotoxins, agents which act on the hematopoietic system, and agents which damage the lungs, skin, eyes, or mucous membranes. Appendix A provides further definitions and explanations of the scope of health hazards covered by this section, and Appendix B describes the criteria to be used to determine whether or not a chemical is to be considered hazardous for purposes of this standard.

"Physical hazard" means a chemical for which there is scientifically valid evidence that it is a combustible liquid, a compressed gas, explosive, flammable, an organic peroxide, an oxidizer, pyrophoric, unstable (reactive) or water-reactive.

Additional words can be explored to determine how certain things might be applied, such as what is meant by a "label": "'Label' means any written, printed, or graphic material displayed on or affixed to containers of hazardous chemicals." This definition indicates that a label can include those which are included by the manufacturer that have been professionally printed. Because of the word "any," it can also include locally printed labels that may be applied to a secondary container into which the hazardous chemical has been transferred for use, such as a spray bottle.

HAZARD DETERMINATION

An evaluation must be conducted to determine whether a chemical presents either a physical hazard or health hazard as defined by OSHA. The responsibility for doing so is as follows:

> 1910.1200(d)(1)
> Chemical manufacturers and importers shall evaluate chemicals produced in their workplaces or imported by them to determine if they are hazardous. Employers are not required to evaluate chemicals unless they choose not to rely on the evaluation performed by the chemical manufacturer or importer for the chemical to satisfy this requirement.

This information from the regulation indicates that the primary responsibility for conducting the evaluation rests with the chemical manufacturer or importer. The responsibility of the school is to review the information provided and ensure that all chemicals that are used in the workplace that are considered to be hazardous are included within the scope of the hazard communication program.

WRITTEN HAZARD COMMUNICATION PROGRAM

Employers are responsible for creating a written hazard communication program that addresses the information from the standard that applies to a given workplace. Information on how to physically create a written program is contained in a separate chapter within this text. The written program must address certain issues as delineated by the regulation:

> 1910.1200(e)(1)
> Employers shall develop, implement, and maintain at each workplace, a written hazard communication program which at least describes how the criteria specified in paragraphs (f), (g), and (h) of this section for labels and other forms of warning, material safety data sheets, and employee information and training will be met, and which also includes the following:

>> 1910.1200(e)(1)(i)
>> A list of the hazardous chemicals known to be present using an identity that is referenced on the appropriate material safety data sheet (the list may be compiled for the workplace as a whole or for individual work areas); and,

1910.1200(e)(1)(ii)
The methods the employer will use to inform employees of the hazards of non-routine tasks (for example, the cleaning of reactor vessels), and the hazards associated with chemicals contained in unlabeled pipes in their work areas.

The regulation specifies that the written program must address how the school will manage container labeling, material safety data sheets, and training. These items will be addressed in the procedures section of the written program. Procedures will specify such things as how labeling will occur, where material safety data sheets will be maintained, and what information will be reviewed in employee training sessions. Information provided in the program will help faculty and staff understand how certain tasks are to be carried out in the school. The written program should also address the assignment of responsibilities to certain individuals. This delineation of responsibility can be coupled with procedures to ensure inclusion not only of the activities but also a specification of who must be engaged in the activity. For example, a procedure might exist that delineates how a new chemical can be introduced into the school so that it can be properly included within the scope of the hazard communication program. In addition to this procedure, a delineation of responsibility can be stated as to who is responsible for actually reviewing a new chemical's material safety data sheet and either approving or denying its use in the school.

The written program will also address employee information and training. A training section in the written program can outline what information will be provided in various forms of training, the intervals at which training will be conducted, and training that will be presented to various levels of individuals within the school system.

LABELS AND OTHER FORMS OF WARNING

According to the regulation, chemical manufacturers or importers are responsible for ensuring that each hazardous chemical is properly labeled. The information that must be included on a chemical label includes the following:

1910.1200(f)(1)(i)
Identity of the hazardous chemical(s);

1910.1200(f)(1)(ii)
Appropriate hazard warnings; and

1910.1200(f)(1)(iii)
Name and address of the chemical manufacturer, importer, or other responsible party.

The identity of the hazardous chemical is often the clearest item on a manufacturer label in that it is simply the name of the material that is in the container.

"Appropriate hazard warnings" are typically listed under the primary marking on the front of the container and use words such as:

- Flammable
- Combustible
- Explosive
- Corrosive
- Carcinogen

These words indicate the primary physical or health hazard presented by the chemical. Additional information regarding the hazards of the chemical are typically listed in small print and in narrative form on the side or the back of the chemical container.

Contact information for the chemical "manufacturer, importer, or other responsible party" is included so that the user of the chemical can contact someone in the event that there are additional questions regarding the hazardous chemical. This provides a proactive and reactive avenue to gain information regarding the use of the chemical or exposure to it.

MATERIAL SAFETY DATA SHEETS

Schools must maintain copies of MSDSs for each hazardous chemical that is used by faculty or staff. An MSDS is an information summary of the hazardous chemical that provides the reader with important pieces of information regarding the chemical. An MSDS must include the following information:

1910.1200(g)(2)(i)
The identity used on the label, and, except as provided for in paragraph (i) of this section on trade secrets:

1910.1200(g)(2)(i)(A)
If the hazardous chemical is a single substance, its chemical and common name(s);

1910.1200(g)(2)(i)(B)
If the hazardous chemical is a mixture which has been tested as a whole to determine its hazards, the chemical and common name(s) of the ingredients which contribute to these known hazards, and the common name(s) of the mixture itself; or,

1910.1200(g)(2)(i)(C)
If the hazardous chemical is a mixture which has not been tested as a whole:

1910.1200(g)(2)(i)(C)(1)
The chemical and common name(s) of all ingredients which have been determined to be health hazards, and which comprise 1% or

greater of the composition, except that chemicals identified as carcinogens under paragraph (d) of this section shall be listed if the concentrations are 0.1% or greater; and,

1910.1200(g)(2)(i)(C)(2)
The chemical and common name(s) of all ingredients which have been determined to be health hazards, and which comprise less than 1% (0.1% for carcinogens) of the mixture, if there is evidence that the ingredient(s) could be released from the mixture in concentrations which would exceed an established OSHA permissible exposure limit or ACGIH Threshold Limit Value, or could present a health risk to employees; and,

1910.1200(g)(2)(i)(C)(3)
The chemical and common name(s) of all ingredients which have been determined to present a physical hazard when present in the mixture;

1910.1200(g)(2)(ii)
Physical and chemical characteristics of the hazardous chemical (such as vapor pressure, flash point);

1910.1200(g)(2)(iii)
The physical hazards of the hazardous chemical, including the potential for fire, explosion, and reactivity;

1910.1200(g)(2)(iv)
The health hazards of the hazardous chemical, including signs and symptoms of exposure, and any medical conditions which are generally recognized as being aggravated by exposure to the chemical;

1910.1200(g)(2)(v)
The primary route(s) of entry;

1910.1200(g)(2)(vi)
The OSHA permissible exposure limit, ACGIH Threshold Limit Value, and any other exposure limit used or recommended by the chemical manufacturer, importer, or employer preparing the material safety data sheet, where available;

1910.1200(g)(2)(vii)
Whether the hazardous chemical is listed in the National Toxicology Program (NTP) Annual Report on Carcinogens (latest edition) or has been found to be a potential carcinogen in the International Agency for Research on Cancer (IARC) Monographs (latest editions), or by OSHA;

1910.1200(g)(2)(viii)
Any generally applicable precautions for safe handling and use which are known to the chemical manufacturer, importer or employer preparing the material safety data sheet, including appropriate hygienic practices, protective measures during repair and maintenance of contaminated equipment, and procedures for clean-up of spills and leaks;

1910.1200(g)(2)(ix)
Any generally applicable control measures which are known to the chemical manufacturer, importer or employer preparing the material safety data sheet, such as appropriate engineering controls, work practices, or personal protective equipment;

1910.1200(g)(2)(x)
Emergency and first aid procedures;

1910.1200(g)(2)(xi)
The date of preparation of the material safety data sheet or the last change to it; and,

1910.1200(g)(2)(xii)
The name, address and telephone number of the chemical manufacturer, importer, employer or other responsible party preparing or distributing the material safety data sheet, who can provide additional information on the hazardous chemical and appropriate emergency procedures, if necessary.

A unique challenge regarding the use of material safety data sheets is that although the above information is required to be produced, there is no uniform format in which a MSDS must be generated. This causes material safety data sheets to vary in appearance. All material safety data sheets from one manufacturer will typically be uniform in appearance, but those between different manufacturers will appear different in format. Effort is being made to create a globally harmonized system that will promote more uniformity in this respect (Williams, 2009). At this time, the OSHA requirements are what direct organizations in the absence of a promulgated globally harmonized system.

Hazard communication often ranks among OSHA's most frequently cited violations. One reason for this is that it can be difficult to maintain a consistently updated list of hazardous chemicals and accompanying MSDSs where large numbers of hazardous chemicals are utilized. One strategy to address this risk is to limit the number of hazardous chemicals that are used in a school. This can be accomplished by identifying safer alternative chemicals that are not hazardous and replacing the hazardous chemicals. A second way to accomplish this is to limit the types of hazardous chemicals that are used. For example, an evaluation of hazardous chemicals might reveal that three different hazardous chemicals are being used that serve the same purpose. Two of these chemicals can be eliminated from what is permitted to be used, leaving only one to manage within the scope of the hazard communication program.

EMPLOYEE INFORMATION AND TRAINING

Employers are responsible for providing information and training to employees. Employers must "provide employees with effective information and training on hazardous chemicals in their work area at the time of their initial assignment, and whenever a new physical or health hazard the employees have not previously been trained about is introduced into their work area." In this regard, the regulation indicates the following for what must be included in information and training:

> 1910.1200(h)(2)
> "Information." Employees shall be informed of:
>
> > 1910.1200(h)(2)(i)
> > The requirements of this section;
> >
> > 1910.1200(h)(2)(ii)
> > Any operations in their work area where hazardous chemicals are present; and,
> >
> > 1910.1200(h)(2)(iii)
> > The location and availability of the written hazard communication program, including the required list(s) of hazardous chemicals, and material safety data sheets required by this section.
>
> 1910.1200(h)(3)
> "Training." Employee training shall include at least:
>
> > 1910.1200(h)(3)(i)
> > Methods and observations that may be used to detect the presence or release of a hazardous chemical in the work area (such as monitoring conducted by the employer, continuous monitoring devices, visual appearance or odor of hazardous chemicals when being released, etc.);
> >
> > 1910.1200(h)(3)(ii)
> > The physical and health hazards of the chemicals in the work area;
> >
> > 1910.1200(h)(3)(iii)
> > The measures employees can take to protect themselves from these hazards, including specific procedures the employer has implemented to protect employees from exposure to hazardous chemicals, such as appropriate work practices, emergency procedures, and personal protective equipment to be used; and,

1910.1200(h)(3)(iv)
The details of the hazard communication program developed by the employer, including an explanation of the labeling system and the material safety data sheet, and how employees can obtain and use the appropriate hazard information.

Although the regulation indicates that information and training must be provided at the time of initial assignment and when a new hazard is introduced, a best practice is to also integrate annual refresher training and nonroutine task training. Annual refresher training can be used to simply review the key aspects of the hazard communication program, risks associated with the use of hazardous chemicals, information that is available to employees (written program, material safety data sheets, and container labels), and employee protection. Nonroutine tasks can be integrated into activities where infrequent use of hazardous chemicals occurs. A review can be conducted that addresses the unique hazards of the chemical, proper procedures that must be followed, and personal protective equipment that must be worn, such as safety glasses, goggles, or gloves.

In summary, OSHA's hazard communication standard is designed to identify hazards that exist in certain chemicals. This information must then be integrated into a written hazard communication program that includes:

- A list of all hazardous chemicals that are used in the school
- Use of material safety data sheets for each of the hazardous chemicals that have been listed
- Use of container labels that communicate hazard information
- Employee training

CASE STUDY

John is the new school system maintenance manager for East County schools. The county school system is comprised of fifteen different schools. John has a staff of maintenance technicians and housekeepers that are responsible for maintaining all of the school buildings and grounds. He recently conducted an audit of all of the chemicals in each school and much to his surprise found that there are 315 hazardous chemicals that are used throughout the county. Among the 315 chemicals, he found that there are actually 125 applications, which means that multiple hazardous chemicals are being purchased and used to serve the same purpose. He also found that there is no hazard communication program present that directs activity related to the use of hazardous chemicals in the workplace.

- What should John do to address the issue of such a large volume of hazardous chemicals in the county schools?
- How should he go about implementing a hazard communication program in the school system?

EXERCISES

For the following questions, identify a single school environment in which you would like to situate your responses and answer each question accordingly.

1. What is the purpose of OSHA's hazard communication standard?
2. What is a health hazard?
3. What is a physical hazard?
4. What role does a school have in the hazard determination process?
5. What information must be included in a written hazard communication program?
6. What information must be included on a chemical label?
7. What is a material safety data sheet (MSDS)?
8. What information and training must be provided to employees?

REFERENCES

Occupational Safety and Health Administration. (1996). Hazard communication. Accessed November 17, 2011. http://www.osha.gov/pls/oshaweb/owadisp.show_document?p_table=STANDARDS&p_id=10099.

Williams, S. (2009). Web-based technology: A competitive advantage for global MSDS management. *Professional Safety*, 54(8), 20–27.

18 Environmental Hazards

Paul English and E. Scott Dunlap

CONTENTS

The environment that teachers and staff work in can have adverse effects on safety and health. Numerous issues, such as poor air quality and exposure to hazardous substances, can create an unsafe environment for all employees. In identifying and mitigating these hazards, a strong relationship must be built with buildings and grounds maintenance staff. This chapter intends to identify some environmental risk factors that are common in educational institutions.

ASBESTOS

Asbestos is the name given to a number of naturally occurring fibrous minerals with high tensile strength, the ability to be woven, and resistance to heat and most chemicals. Because of these properties, asbestos fibers have been used in a wide range of manufactured goods, including roofing shingles, ceiling and floor tiles, paper and cement products, textiles, coatings, and friction products such as the automobile clutch, brake, and transmission parts. The Toxic Substances Control Act defines asbestos as the asbestiform varieties of: chrysotile (serpentine), crocidolite (riebeckite), amosite (cummingtonite/grunerite), anthophyllite; tremolite, and actinolite (Environmental Protection Agency, 2011).

Because of the properties of asbestos, it was widely used as an insulation material for severe heat conditions in building construction. Asbestos also has fibrous qualities that allow it to be woven into other materials to increase strength of different materials. Asbestos-containing material (ACM) could be found in many different materials that were identified by the Occupational Safety and Health Administration (OSHA) and the Environmental Protection Agency (EPA). In 1989, the EPA identified these different materials as possibly being ACM:

- Building materials
  - Cement corrugated sheet

- Cement flat sheet
- Cement pipe
- Cement shingle
- Roof coatings
- Flooring felt
- Pipeline wrap
- Roofing felt
- Nonroof coatings
- Vinyl/asbestos floor tile
- Automotive parts
  - Automatic transmission components
  - Clutch facings
  - Disc brake pads
  - Drum brake linings
  - Brake blocks
- Asbestos clothing
- Commercial and industrial asbestos friction products
- Sheet and beater-add gaskets (except specialty industrial)
- Commercial, corrugated, and specialty paper
- Mill board
- Roll board (Environmental Protection Agency, 2011)

The properties of asbestos supported the development of many different materials that were incredibly strong and heat resistant. The use of asbestos in automobile brakes was the industry standard because asbestos resists heat when brakes are applied and because of its strength, which makes the brakes pads last longer. In 1972, OSHA began to regulate exposure to asbestos in general industry (U.S. Department of Labor, 1995).

As OSHA gained traction as the new enforcement agency for occupational safety, many of the different ACM were fazed out of production. Obviously, changing out ACM brake pads was a lot easier than changing out floor tiles or pipes insulated with ACM. In the mid-1970s, public backlash and outcry became the norm as many public schools and institutions were identified as "at-risk" facilities because of the amount of ACM used in the construction of the buildings. The knee-jerk reaction was to remove all ACM from all facility buildings in the name of public health and interest.

Many experts agree that not all ACM is dangerous. Many of the materials that are considered "hard ACM," such as vinyl floor tiles, generally do not pose a health hazard. ACM that is loose bound, such as spray-on insulation and soundproofing or fire-resistant material, can become "friable" and pose the greatest hazard. The term friable refers to the ACM that can unravel or crumble, allowing asbestos fibers to be released in the air (Lang, 1984). The problem with many buildings and schools that used asbestos as a building material was that the ACM was not loose, unraveling, or friable material. The national push to remove all ACM from buildings and mass panic led to many buildings being abated for ACM that posed no danger at all to health. However, once the ACM was disturbed during the abatement and removal, it was then considered friable material that did pose a health hazard.

People who are exposed to ACM are at a higher risk for lung diseases. Several specific diseases are directly related to exposure to asbestos. Three of the major health effects associated with asbestos exposure includes:

- **Asbestosis:** Asbestosis is a serious, progressive, long-term noncancer disease of the lungs. It is caused by inhaling asbestos fibers that irritate lung tissues and cause the tissues to scar. The scarring makes it hard for oxygen to get into the blood. Symptoms of asbestosis include shortness of breath and a dry, crackling sound in the lungs while inhaling. There is no effective treatment for asbestosis.
- **Lung cancer:** Lung cancer causes the largest number of deaths related to asbestos exposure. People who work in the mining, milling, and manufacturing of asbestos and those who use asbestos and its products are more likely to develop lung cancer than the general population. The most common symptoms of lung cancer are coughing and a change in breathing. Other symptoms include shortness of breath, persistent chest pains, hoarseness, and anemia.
- **Mesothelioma:** Mesothelioma is a rare form of cancer that is found in the thin lining (membrane) of the lung, chest, abdomen, and heart. Almost all cases are linked to exposure to asbestos. This disease may not show up until many years after asbestos exposure. This is why great efforts are being made to prevent school children from being exposed (Environmental Protection Agency, 2011).

What makes ACM so dangerous is the actual size and shape of the asbestos fiber. If you were to look at a fiber under a high-power microscope, you would find something similar to a fish hook with a barb on the end. The fiber is small enough to pass through all the body's natural defenses and get caught in the lung wall. Breathing in enough ACM will allow the material to accumulate and kill healthy cells, leading to lung disease.

Most of the diseases related to ACM are cumulative diseases in nature. This means that the disease will not show up in a person immediately but rather over an extended period of time. Many past cases of occupational disease forced OSHA to incorporate a medical records standard designed to address how employee exposure records are handled. It is for this reason that all employee exposure records dealing with asbestos must be kept by the employer for thirty years. Any exposure testing of employees dealing with occupational disease must be kept for the employee's duration of employment, plus thirty years. This means that if an employee starts to work at a location that monitors for ACM in 2012 and retires after twenty years of service in 2032, those exposure records need to be kept by the employer until 2062.

As stated before, not all ACM is immediately dangerous, but care should be exercised in determining whether a material contains asbestos. Once hard ACM is disturbed, the risk factors increase dramatically. If you suspect that there is ACM material in a school, there are several steps you must take to prevent exposure:

- If building construction or renovation is scheduled, be proactive and perform an asbestos survey for the area in question. Buildings constructed after 1980 will likely have very little ACM, if any at all. Any construction before 1970 should be scrutinized.
- If there is any material found that is dusty, falling apart, or otherwise questionable, you must err on the side of safety and prevent any further disturbance of the material. Keep faculty and staff away from the material and try to isolate the area as much as possible. OSHA identifies steps for building and facility owners dealing with possible ACM in the communication of hazards section of the asbestos standard, which includes the following:
  - In buildings built before 1980, treat thermal system insulation and sprayed-on and troweled-on surfacing materials as ACM, unless properly analyzed and found not to contain more than 1 percent asbestos.
  - Train employees who may be in contact with ACM to deal safely with the material.
  - Treat asphalt and vinyl flooring materials installed no later than 1980 as ACM unless properly analyzed and found to contain no more than 1 percent asbestos.
  - Inform employers of employees performing housekeeping activities of the presence and location of ACM and presumed ACM that may have contaminated the area.
- Employers shall provide training and information to all employees who may be exposed to ACM or possible ACM containing materials. Awareness training should include:
  - Where possible ACM is present or possibly present
  - Health effects of ACM
  - Signs of damage and deterioration of ACM
  - Proper response to fiber release episodes involving ACM (U.S. Department of Labor, 1995)

If a building or facility has been determined to have ACM, many different engineering and monitoring controls are needed if the decision has been made to remove the material. It is for this reason that asbestos abatement is extremely labor intensive and costly. Remove people from the hazard of friable ACM and isolate the area to prevent any exposure. Any material in question should be tested by a professional with experience in asbestos testing and abatement.

LEAD

Lead is a naturally occurring element that can be found on the periodic table of elements. Considered to be one of the first heavy metals to be discovered and widely used, it is also considered by many to be the downfall of the Roman Empire in history. It was discovered that lead was used for making plumbing pipes for the waterways in the Roman cities. Unprotected and unshielded lead pipes transporting drinking water allowed lead to poison many residents. Reviewing ancient texts also revealed that aristocrats in Rome made wine in lead vessels that were heated, creating a sweeter wine. What was actually making the wine sweeter was the lead acetate entering the wine from the lead vessels. Dr. Jerome O. Nriagu, a Canadian scientist, found that two-thirds of them, including Claudius, Caligula, and Nero, "had a predilection to lead-tainted diets and suffered from gout and other symptoms of chronic lead poisoning" (Wilford, 1983).

Fast forward to the present day and the hazards and uses of lead are heavily regulated by OSHA and the EPA to protect both employees and the environment. The Consumer Products Safety Commission has also lobbied Congress to ban certain products from being imported that do not meet the U.S. standards for lead safety to protect consumers. Many products imported from China used lead-based paint in the manufacturing process, which led to an immediate ban on certain children's toys including all-terrain vehicles and smaller motorcycles built by Honda (Motorcycle. com, 2009). The goal of the new ban was to protect children under twelve that could be exposed to lead.

Children are more susceptible to the effects of lead poisoning than adults. Lead is more dangerous to children because:

- Babies and young children often put their hands and other objects in their mouths. These objects can have lead dust on them.
- Children's growing bodies absorb more lead.
- Children's brains and nervous systems are more sensitive to the damaging effects of lead.

If not detected early, children with high levels of lead in their bodies can suffer from:

- Damage to the brain and nervous system
- Behavior and learning problems, such as hyperactivity
- Slowed growth
- Hearing problems
- Headaches (Environmental Protection Agency, 2011)

Lead use, much like asbestos use, was curtailed in the mid-1970s, because the EPA identified buildings and facilities at risk built before 1978. The majority of lead exposure can be found in the disturbance of lead-based paint. Lead-based paint was more durable than other paints offered at the time and also held color for a longer period of time. Holding the same qualities as asbestos, lead-based paint is not a danger if not disturbed, or friable. Once disturbed, however, several hazards become apparent.

In September 2011, the EPA released new precautions for dealing with lead exposure. Renovations of buildings or facilities built or remodeled before 1978 may have had lead-based paint used in the process. Again, the key to understanding unsafe exposure is to determine whether lead-containing materials (LCM) exist in the workplace. Several different tests can be completed to determine whether lead is present. If it has been determined that LCM does exist, there are several things that must be identified and completed before abatement can begin:

- Alternative facilities while abatement is occurring: If a school or classroom is being abated for lead-based paint, facilities need to be identified that can be used while the renovation is taking place. Bathrooms and dining facilities should be isolated from the area of renovation.
- Pets and animals: Biology labs, class pets, and service dogs should not be exposed to lead contaminants if possible. Much like small children, pets and animals are at a higher risk for lead exposure.
- Furniture and furnishings: These should be moved from the area if possible. Precautions should be taken during the renovation to limit the amount of dust produced. Removal of furniture is recommended instead of covering up with plastic.
- Heating, ventilation, and air conditioning: These systems should be turned off while renovations and remodeling are occurring. Leaving systems on can create a greater hazard if lead-based paint dust becomes airborne. This issue may give the administration a small window of opportunity to perform abatement work if a facility or building is located in an extreme climate or region.

There are several different renovation methods that should be incorporated into the project to limit the amount of dust created. Power tools, such as sanders, grinders, or needle guns, should be equipped with high efficiency particulate air (HEPA) attachments to capture any dust from these operations. (Environmental Protection Agency, 2011).

LCM is found not only in paint, but also in ceramic tiles and water pipes. As stated earlier regarding the historic use of lead pipes in Roman times, the entire pipe was made of lead. In many of the older buildings and institutions in the United States, pipes used in construction were made of galvanized steel and cooper. However, some of the joint materials used to join pipes together used lead-based products. This was the case that was discovered by the Seattle Public School District in 2004 (Buchanan, 2006).

Because the water supply for the school district came from the city of Seattle, the district was not required to test for lead. Upon testing the drinking water in several different schools, it was determined that the lead levels were high and that action needed to be taken. Spending a total of $3 million, the district tested all of the schools in the district, replaced old pipes, and brought in bottled water for those schools that were undergoing renovations. Further investigation from the original study determined that eleven schools had severely rusted pipes, which would have had to be replaced. This discovery and subsequent abatement of LCM in the plumbing led the school board to adopt tougher drinking water standards than the EPA setting for the permissible exposure level of lead in drinking water from 10 to 20 parts per billion (Buchanan, 2006).

Lead, much like asbestos, can be found in older buildings and facilities across the country. Undisturbed, these materials pose little to no health hazards. Once disturbed through renovation or agitation, both will create environmental stress on animals, students, and employees. The key to successful management of these issues is to know what materials to look for and where to find them. Proper planning, protection, or abatement will eliminate any possible hazards.

SICK BUILDING SYNDROME

The term "sick building syndrome" (SBS) is used to describe situations in which building occupants experience acute health and comfort effects that appear to be linked to time spent in a building, but no specific illness or cause can be identified. The complaints may be localized in a particular room or zone or may be widespread throughout the building. In contrast, the term "building-related illness" is used when symptoms of diagnosable illness are identified and can be attributed directly to airborne building contaminants. A 1984 World Health Organization Committee report suggested that up to 30% of new and remodeled buildings worldwide may be the subject of excessive complaints related to indoor air quality. Often this condition is temporary, but some buildings have long-term problems. Frequently, problems result when a building is operated or maintained in a manner that is inconsistent with its original design or prescribed operating procedures. Sometimes indoor air problems are a result of poor building design or occupant activities (Environmental Protection Agency, 2010).

SBS has become a very controversial topic in the environmental health and safety field. Because it has yet to be determined what causes SBS, only general guidelines have been established to help identify and eliminate potential causes. Because no one specific cause can be established and because different people will react to different environments, SBS can become a monumental issue in the workplace. According to the EPA, symptoms found in people claiming SBS have ranged from headache, eye, and nose irritation to muscle cramps and inability to concentrate. One theory is that as building technology increases, new buildings and large renovations make a building more energy efficient. This creates a tighter and more sealed building than ever before, referred to as "tight building syndrome" (California State University Employees Union, 2009). The EPA has cited some contributing factors that were common in SBS cases. These possible causes include:

- Inadequate ventilation (heating, ventilation, and air conditioning): Inadequate ventilation in building could be a contributing factor in SBS. In 1989, the American Society of Heating, Refrigerating and Air Conditioning Engineers (ASHRAE) increased the makeup air standard for office spaces from 5 to 15 cubic feet per minute. This was an effort to increase indoor air quality standards for all buildings.
- Chemical contamination (indoors): Chemicals that contain volatile organic compounds (VOCs) have been identified as a potential source for SBS claims. Adhesives, carpeting, and cleaning agents have the potential to contain VOC. Any type of equipment that combusts or can combust material will also give off a certain amount of VOC. This equipment can include space heaters, heat guns, gas heaters, and wood stoves.

- Chemical contamination (outdoors): Air makeup vents that pull fresh air into buildings also have the potential to pull contaminants into the building. Outside processes that were added after the building's construction may increase the air quality hazard inside the building. Hazardous waste storage locations poorly placed outside a building can pull VOCs into the building.
- Biological contamination: Mold and bacteria gained international attention with the discovery of Legionnaire's disease and Pontiac fever. Stagnant water found in humidifiers and ducts as well as water damage to building insulation, ceiling tiles, and carpet can lead to bacteria and mold growth (Environmental Protection Agency, 2010).

Because there is no specific answer for SBS, all possible risks associated with the causes of SBS must be investigated. Some possible causes will be obvious, such as the discovery of mold in an office or classroom. Other possible causes may be harder to identify, such as a new cleaning agent being used or being used improperly. As with all environmental stressors, early detection is the key to prevention. Buildings, facilities, and grounds should be inspected regularly for any possible issues affecting indoor air quality. Emergency incidents such as partial flooding from natural or man-made causes will force administrations to ensure that all moisture has been eliminated to reduce the possibility of biological growth, such as mold. Clear and constant communication with employees will help minimize SBS issues.

BLOODBORNE PATHOGENS

Although originally designed to address issues in healthcare-related occupations, OSHA's bloodborne pathogen standard (29 CFR 1910.1030) has had a far reaching effect in many workplaces. The school environment is one area that has been affected. The bloodborne pathogen standard is designed to protect employees who have "occupational exposure" to blood or body fluids in the workplace. The healthcare industry was a natural concern because of the direct contact that doctors, nurses, dentists, paramedics, and other such workers have to blood and body fluids. Exposure can have significant risks:

> Bloodborne pathogens are infectious microorganisms in human blood that can cause disease in humans. These pathogens include, but are not limited to, hepatitis B (HBV), hepatitis C (HCV) and human immunodeficiency virus (HIV). Needlesticks and other sharps-related injuries may expose workers to bloodborne pathogens. Workers in many occupations, including first aid team members, housekeeping personnel in some industries, nurses and other healthcare personnel may be at risk of exposure to bloodborne pathogens. (OSHA, n.d.)

Exposure to bloodborne pathogens can affect various employee groups within a school system. OSHA (2011) defines occupational exposure as, "reasonably anticipated skin, eye, mucous membrane, or parenteral contact with blood or other potentially infectious materials that may result from the performance of an employee's duties." The phrase "that may result from the performance of an employee's duties" refers to "occupational" exposure. An evaluation of a school system might reveal a number of groups that are within the scope of this regulation:

- Nurses
- First responders
- Custodians
- Teachers

An evaluation of all positions within the school system will need to be performed to determine who has occupational exposure as a result of job responsibilities. This evaluation might reveal individuals who might not typically be considered as indicated in the list above. Nurses and first responders might easily be identified because of their obvious contact with blood and body fluids in managing various incidents. However, other occupations might also surface. Teachers who must respond to injured children can be exposed to blood. Custodians might be exposed to the vomit of a sick child during cleaning operations.

A principle presented in the OSHA (2011) regulation is the use of "universal precautions." This refers to always making the assumption that blood or body fluid is contaminated with a bloodborne pathogen. An employee will never know who is or is not infected with AIDS or hepatitis because of the confidentiality of medical records, so a school employee who is experiencing occupational exposure to blood or body fluid should always make the assumption that the material is infected and use the appropriate precautions as outlined in the standard, such as using personal protective equipment to prevent contact with the material.

The OSHA (2011) regulation delineates training that must occur for employees who have occupational exposure to bloodborne pathogens. Training must occur:

- At the time of initial assignment
- Annually
- Whenever changes have occurred

The regulation goes on to address specific topics that must be covered in the training.

School administrators will need to deeply explore the OSHA regulation on bloodborne pathogens to gain a comprehensive understanding of what needs to be included in a bloodborne pathogen program beyond what has been presented here, to include such issues as creating exposure control plans, container labeling, and the management of waste.

CASE STUDY

Sally has been a teacher at West Elementary for thirty-five years. She has recently begun to mention that she has been having trouble breathing and is concerned that she might have been exposed to something in the school building. West Elementary was constructed in 1965 and underwent a major renovation and addition to the building in 1980, shortly after Sally was hired. The risk manager for the school system is new and cannot find records as to what safety programs were in place prior to 2000.

- What environmental issues might have impacted Sally?
- What safety precautions should have been taken during the building renovation and addition?

EXERCISES

1. In what time period was asbestos used in building materials?
2. What materials in a school could contain asbestos?
3. What physical conditions can develop as a result of exposure to asbestos?
4. What is a primary issue with the presence of lead in a school?
5. How should the potential presence of lead in a school be investigated and abated?
6. What is "sick building syndrome"?
7. How would you manage a report of a faculty or staff member who complains of exposure to an unknown substance in a school?
8. What groups within a school system might be exposed to bloodborne pathogens?

REFERENCES

Buchanan, B. (2006). The high. *American School Board Journal*, 193, 22–25.

California State University Employees Union. (2009). Health and safety sick building syndrome SBS. Accessed May 15, 2012. http://www.csun.edu/csueu/pdf/KYR/KYR-5_ Sick_Building_Syndrome.pdf?link=496&tabid=493.

Environmental Protection Agency. (2010). Indoor air facts no. 4 (revised) sick building syndrome. Accessed March 10, 2012. http://www.epa.gov/iaq/pubs/sbs.html.

Environmental Protection Agency. (2011). Asbestos. Accessed March 10, 2012. http://www. epa.gov/asbestos/pubs/help.html.

Lang, R. D. (1984). Asbestos in schools, low marks for government action. *Columbia Law School Journal of Environmnetal Law*, 26, 14–20.

Motorcycle.com. (2009). Lead toy ban could affect bikes, ATVs. Accessed December 28, 2011. http://www.motorcycle.com/news/lead-toy-ban-could-affect-bikes-atvs-87908.html.

Occupational Safety and Health Administration. (2011). Bloodborne pathogens. Accessed December 29, 2011. http://www.osha.gov/pls/oshaweb/owadisp.show_document?p_ table=STANDARDS&p_id=10051.

Occupational Safety and Health Administration. (n.d.). Bloodborne pathogens and needle stick prevention. Accessed December 29, 2011. http://www.osha.gov/SLTC/blood-bornepathogens/index.html.

U.S. Department of Labor. (1995). Occupational Safety and Health Administration. Accessed June 10, 2012. http://www.osha.gov/publications/osha3095.html.

Wilford, J. N. (1983, March 17). Roman Empire's fall is linked with gout and lead poisoning. *New York Times*. Accessed December 12, 2011. http://www.nytimes.com/1983/03/17/ us/roman-empire-s-fall-is-linked-with-gout-and-lead-poisoning.html.

19 Playground Safety

Ronald Dotson

CONTENTS

A seven-year-old second-grade student is playing on a typical playhouse cluster of attractions while on recess at school. Between the two sections of the playhouses is an elevated walkway made of chains designed for children to walk across learning balance and risk mitigation. The child falls from a foot becoming entangled in the chain. The fall results in a broken wrist.

A suit is filed against the school district and the playground equipment manufacturer. The suit alleges that the school failed to supervise the child by training the child about the hazards associated with crossing the chain walk and the equipment manufacturer was negligent with the design because the chain walk did not meet Consumer Product Safety Commission recommendations for design.

In this type of scenario, the school district may be sued for an amount well above direct injury costs. Some injuries similar to this may have a cost of approximately $6,000 or more for emergency room fees, specialist fees, and follow-up visits that may include therapy. Suits against a school district may be sought for amounts several times greater than the direct costs.

This scenario resembles a New York case from 2001. In that case, the trial court issued summary judgment for the school district and for the playground equipment manufacturer, basically absolving them from liability. However, the court of appeals reversed the decision for summary judgment on behalf of the school district. The New York Court of Appeals upheld the reversal. Therefore, the school district was held negligent for not providing education to students on the hazards encountered on the playground. The court ruled that, "A school district owes a duty to its students to exercise the same degree of care as would a parent of ordinary prudence under similar circumstances." The court referenced two cases in the decision; *Lawes versus Board of Education of the City of New York*, 16 N.Y. 2d 364 and *Merkley versus Palmyra Macedon Central School District*, 130 A.D. 2d 937. The playground manufacturer was absolved of liability because the court ruled that playground standards are voluntary, and many standards exist besides the Consumer Product Safety Commission recommendations (730 N.Y.S. 2d 132).

The mere mention of playground safety begs the notion that keeping children safe from injury while on a playground is an obvious goal. However, this may not be the case for everyone. To understand this notion, we must explore the definition of playgrounds, decide whether playground injuries are indeed a problem, and explore contemporary strategies for injury prevention.

DEFINING THE PLAYGROUND

A playground can be defined as simply an area with "specific design" for children to play. This sounds simplistic but is more complex than may first appear. The term "specific design" hints that the environment and equipment contained in the area have both psychological and physical aspects to its placement and design. Playgrounds first appeared in Germany and had more purpose than to serve as an area for the release of energy. The area served as a type of classroom meant to teach children how to play properly. Of course, a balance between play and safety must be achieved. Dr. Ellen Sandseter, a professor of psychology at Queen Maud University in Norway, views playgrounds as a learning environment for social and psychological development. She views playground injuries as necessary for a healthy adult psyche.

PLAYGROUND INJURIES

Each year approximately two hundred thousand children are seen by medical professionals at community hospitals or urgent treatment centers for injuries received on playgrounds (U.S. Consumer Product Safety Commission, 2010). School playgrounds account for most of the injuries for children of school age five to fourteen years old (Safe Kids Worldwide, 2010). Although relatively little time is spent on a school playground, playground injuries account for between 30 and 70% of all school injuries (Posner, 2000). Anywhere from 6 to 7% of elementary children will receive a playground-related injury sometime during their elementary tenure (Posner, 2000). Approximately 45% of playground injuries are severe and include amputations, internal injuries, concussions, and broken bones. Deaths do occur as well. From 1990 until 2000, 147 deaths were reported as a result of a playground injury. Approximately 70% of these deaths occurred on home playgrounds mainly involving falls from swings and strangulations from swing entanglement with clothing or strings on the child's clothing (Safe Kids Worldwide, 2010).

The Consumer Federation of America, a research group for public interest reports that female children are more likely to be injured while on the playground. Furthermore, the study of injury trends links facial injuries to younger children, usually under age five, with injuries to the hands and arms being more typical of children from ages five to fourteen (Safe Kids Worldwide, 2010).

The Arizona Department of Health Services began studying and tracking playground injuries across the State of Arizona, and it has since begun including all school injuries. The findings were consistent with other national trends identified by Safe Kids Worldwide and independent state studies such as the 1992 Pennsylvania

PTA Playground Injury Prevention Project. Several interesting trends can be pointed out:

- 2.5% of school-age children will receive medical treatment for a playground injury that is beyond first aid.
- Injuries peak during fifth, sixth, and seventh grades.
- Kindergarten through fourth grade students were four times more likely to be injured on playground equipment, whereas children in fifth, sixth, and seventh grades were injured more from rough play.
- 30% of injuries are to the head.
- 1% of playground-related injuries result in hospitalization.

The Pennsylvania PTA Project also identified some important trends, including:

- Climbing equipment is involved in 50% of cases, whereas swings are involved in 16% and slides in 11% of the cases.
- 66% of injuries occur while falling from equipment, 16% involved running into equipment, and 7% involved collisions with moving equipment.

Pennsylvania also audited thirty-five elementary playgrounds against standards. The majority of them had improper surfaces for fall mitigation, 44% of them had equipment that allowed for a child's head to become stuck, and 34% had equipment that the Consumer Product Safety Commission did not recommend for elementary playgrounds (Posner, 2000).

It is reasonable to believe that injuries among children at play will occur. However, morally speaking, one injury, especially a reasonably preventable one, is one too many. With leadership comes responsibility. Schools have a liability associated with injuries to students, visitors, and faculty or staff. In Kentucky, the principle of sovereign immunity protects school districts and individual employees from suit. In other states such as California, the legislature has allowed the waiving of sovereign immunity. In the Kentucky case, *Valesa Deck v. Tina Noble* (S.W. 3d 2011 WL 2935667), the Kentucky Court of Appeals for the eastern half of the state ruled that a teacher, Valesa Deck, could not be sued by the guardian of a minor student who was hurt while on a playground. Deck had rewarded her class with an unscheduled recess on the school playground for exceptional performance on a test. While playing on the playground at Emmalena School in Knott County, the student fell, resulting in a broken arm. The school district was relieved from the suit based upon sovereign immunity. The Nobles sued the individual teacher, Valesa Deck, on the grounds of negligent supervision. Public employees benefit from immunity only when applied to discretionary acts or functions exercising judgment in good faith and within the scope of their employment. The court in this case ruled that although it was an unscheduled recess time for a third-grade class, it was reasonable that a teacher could and would allow additional recess as a reward for performance. Therefore, the principle of sovereign immunity extended to the individual teacher.

California is one of the fifteen states that require school playgrounds meet American Society for Testing and Materials (ASTM) standards. California handles school liability as any other tort. This means that parents must prove that the school had a duty to their child's safety on the playground, that the duty was not met, that damages were incurred as a result of the school failing to meet its duty, and that the school's action or inaction was the primary cause of the damages. This also extends to the involvement of individual teachers and administrators (Tierney, 2011).

Therefore, liability for playground injuries for a school district is not as great as one might first imagine. Suits are difficult to win. However, the cost associated with the defense of a suit is a considerable liability for a school district where funds can be better utilized. In many aspects, monies spent on preventing playground injuries may be a better bargain than defending a suit. In addition to legal defense costs, student injuries are initially covered by the school's liability coverage carrier. When losses mount, the resulting premium will usually increase from that point forward. Premiums usually are not reduced. In some rare occasions, insurance companies will refund money on a year-by-year basis until the next contract period is assessed. Additionally, schools can spend monies in a reactionary mode to injuries that do occur. The strategy of spending the time and money upfront to prevent injuries is much cheaper than reacting to incidents that have already occurred, and the expense is definite.

Meeting ASTM standards is not a guarantee that playground injuries will be reduced. For Dr. David Ball, professor of risk management at Middlesex University in London, it is a matter of well-established behavioral phenomenon. He cites a study of playground injuries in England after the introduction of softened playground surfacing. The actual number of broken-arm incidents increased after softer surfacing was introduced. He offers the explanation that people will take more of a risk when they perceive the environment to be safer (Tierney, 2011).

Dr. Ellen Sandseter, a professor of psychology at Queen Maud University in Norway, views playground injuries as a necessary part of human development. Her philosophy contends that risk taking and surmounting obstacles of fear in a controlled environment allow for a more emotionally developed person later in life. The lack of risky challenge may leave adults with fears and anxieties that lead to a less productive life later on. Progressive exposure and conquering of dangers mirrors a technique used by psychologists to help adults get over phobias. The cost of limiting a child's emotional development may play out later in life. Especially, in view of limited life-threatening injuries overall. Furthermore, she cites a study where children who are exposed to a fall before the age of nine are less likely to be afraid of heights as an adult compared with those children who do not experience a traumatic fall before the age of nine (Sandseter and Kennair, 2011).

Although a straightforward examination of injury statistics and storied occurrences point toward the issue being a considerable problem that needs to be addressed from a moral stance, the liability experience in the courts and the psychological development aspect give an opposing view.

STRATEGIES FOR SAFER PLAYGROUNDS

The most popular strategy implemented is with playground equipment standards formed by the American Society of Testing and Materials. To date, fifteen states have enacted legislation requiring schools and public organizations to install playground equipment in compliance with ASTM standards. North Carolina has reported a 20% reduction in playground-related injuries since adopting the standards as law (Safe Kids Worldwide, 2010).

California has implemented a comprehensive program. The State of California has a three-tiered system for playground safety consisting of standards implementation, inspections, and educational initiatives. All school playgrounds must be inspected by a trained playground inspection official for meeting ASTM standards (Tierney, 2011). ASTM standards concentrate on equipment and surfaces, covering layout and design, types of equipment, installation and maintenance of equipment, surface materials for fall mitigation, safety zones, audit forms, and testing for entrapment hazards. The Consumer Federation of America (CFA) produces standards for playground construction in line with ASTM standards but also takes into account child development (Posner, 2000).

Using ASTM and CFA guidelines to construct new playgrounds or modify existing playgrounds is a critical first step. Building a playground correctly with the right surfaces in the beginning is the most cost-effective method of playground management. When volunteers and community supporters get involved, someone to oversee the project according to standards is necessary. Another point at this stage is to consider the maintenance cost and requirements of surface materials. For example, a community sponsor may donate the initial materials of "safety mulch," rubber chips that are used instead of wood mulch. However, continued support for future replacement materials may be in question. Another example to consider is the purchase of factory equipment made from industrial materials other than wood or the use of treated lumber to make equipment from blueprints. The exposure of children to chemicals contained in the treated lumber and the continual future treatments of the wood may justify the higher expenditure on industrial plastics.

Inspections must be completed frequently and by competent people. Competent people are those with training and experience to recognize and abate hazards but who also have the authority to correct the situation or stop usage until the correction can be made (29 CFR 1926.32f). These personnel should be very familiar with standards and stay current on playground management issues. One good suggestion is to utilize someone that is not frequently engaged in activity in or around playgrounds. Complacency sometimes allows people to overlook issues. Daily inspection prior to use is good practice. Daily inspections often help to identify newly occurring equipment problems and also identify other hazards such as a new presence of bees, snakes, or even items left from public trespass. Guns and syringes have been recovered from playgrounds.

A policy consideration is to limit the ages of children on the playground or in sections of the playground. This can assist in monitoring age-appropriate issues. Trends show that children differ in play from fourth grade to subsequent grades. Some playground equipment is designed for children of certain ages and is not appropriate for other ages. This is also a good strategy to limit the number of children that a playground monitor will have to supervise. Supervision has been shown to be an effective strategy for injury management. A supplemental strategy that facilitates supervision is to lay out playgrounds in zones. Zones for age groups and safe zones or buffers around moving equipment, especially around swings, help stop children distracted with active play and also help limit numbers and actions in areas where playground supervision is challenged from the amount of activity or the number of children.

In light of injury statistics identifying falls as the primary hazard, impact-absorbing material is paramount to injury mitigation. Safety mulch, wood chips, and even pea gravel has been used to absorb or soften the blow to a child that has fallen. Measures such as this are viewed by safety professionals as injury prevention versus accident prevention strategy.

Accident prevention as a strategy goes way back to the "investigational era" of safety management from 1915 until 1930. Education and training were considered to be the keys to preventing "accidents" from happening (Bird et al., 2003). However, with playground safety, this may be an overlooked area of meeting the duty of prudent supervision. Training is appropriate not only for supervisory personnel but for students as well. Training can center on common hazards encountered on playgrounds and on expected behaviors. Books such as *Safety at the Playground* and *Playground Safety* cover a spectrum of hazards and expected behaviors targeting the young audience as a reader. A supplemental strategy to a training session or reading initiative to young children would be the posting of educational posters that are age appropriate. Because standards are focused on engineering controls, such as design and layout, targeting behaviors with the educational initiatives can be helpful. Playground monitors must understand design criteria, injury trends, and behavior patterns.

Balancing safety and psychological development is our goal for playground safety management. Understanding the intent of playgrounds, injury trends, design standards, and behavior suggests specific management practices that fit into either injury prevention or accident prevention strategies for injury management. Peter Heseltine from the Association for Children's Play and Recreation located in Birmingham, England, cites the four main factors of playground safety to be layout, equipment design, maintenance, and behavior (Heseltine, 1993). Standards mainly address injury prevention strategies that follow Dr. William Haddon's energy exchange theory. The theory concentrates on engineering practices that prevent the exchange of energy or limit the exchange from object or condition present in the environment to the human (Byrd et al., 2003).

THE PLAYGROUND EXPERIENCE

Dr. Sandseter has identified six categories of risky play: exploring heights, experiencing high speed, handling dangerous tools, being near dangerous elements, rough and tumble play, and wandering alone (Tierney, 2011). Dr. Sandseter approached her observations from the aspect of psychological development. Using the bow-tie approach of system safety analysis for playground occurrences may yield additional categories of risky play or management practices that could help limit playground injuries to children.

The bow-tie approach is a systematic way of identifying the underlying events and conditions that align to cause a critical event. The critical event can then lead to secondary or resulting events and conditions. It utilizes both deductive and inductive reasoning to illustrate where countermeasures should be introduced to prevent future or further incidents. The model assumes that events and conditions align to cause incidents. An example using one of Dr. Sandseter's categories of risky play might reveal that the need for exploring height might align with a condition of improper surfacing to allow a fall resulting in a more serious injury than expected.

Anyone who has observed young children will notice that they desire to be close to friends and crowd each other at inopportune times. For example, a small child might be attempting to get close to a friend and the friend backs up close to a doorway. Then when the door is opened, the child's fingers are in the opening near the hinge and get pinched. Here, one might view the conditions as being the desire for close or intimate contact and the proximity of a confined area near the door. The child's hand placed in the crack of the door for support, the opening of the door itself, and the opening and subsequent closing of the door may be considered the events. All align to allow the critical incident of the fingers being trapped in the crevice. Once mapped on the model, countermeasures would be examined to be placed between the events and conditions leading to the critical event and then between any resulting conditions or resulting events.

Playground behaviors also include the desire for close contact with friends. Children like to be in a crowd with friends. Conditions such as bells or whistles indicating an end of recess can lead to children being injured. When one child trips, falls, or bumps into another child, resulting in a fall, several that are following or crowding the student may be exposed to a fall hazard as well. Another example of the desire for close contact may also result in injury around slides, swings, near climbing walls, merry-go-rounds, or other pieces of equipment.

Recommendations for the training of playground monitors can be developed from observations and investigations of playground incidents. They may include not blowing whistles to indicate an end to recess. Instead whistles might be best suited to indicate the stopping of all activity and attention for further instruction. The end of recess might be signaled from an informal gathering of children from sections of the playground. The panic effect of alarms is a documented phenomenon. Safety professionals from experience understand reactions to alarms. Additionally, educational postings might be developed that address preventing conditions or personal action that might result in an event that can lead to an injury.

School personnel may not be investigating fully or in a timely manner injuries or events on the playground. Numerous painful bumps lead to crying or fear for young children on a playground. School personnel may be numb to this and not consider these to be actual injuries needing investigative attention. They may even not refer to them as injuries unless the injury was visible and required treatment from the school nurse or use of a first aid kit. Although minor, these are injuries that require investigation. An examination of their cause, even if it is a cursory and short assessment, may identify conditions and events that can be abated in the future.

Although design criteria are addressed in the ASTM standards, the observation of playground activities can shed new light on injury management. Playground equipment pods have landings and clusters of activities that are age appropriate for the equipment. When adult access to an injured child is needed, some equipment prohibits easy rescue or response. In one instance, a child fell and got a bloody nose while on a landing approximately four feet above the ground. The landing had a tunnel at one end and a slide at the other. The guardrail sections of the landing were solid and did not allow escape or access. This meant that the playground attendant slid the child off the equipment via the slide. If the injury had been one that resulted in a concussion, this could have been a more dangerous maneuver that either created a secondary injury or exacerbated the existing injury. In other words, a normal adult would have experienced difficulty going up the slide or crawling through the tunnel to access the child. A simple addition of pins to remove the guardrails or not having limited egress from the landing would have aided in the response to the injured child.

A timely investigation may reveal information that protects the school and playground monitor. It may also point out shortcomings of preparation for such incidents. The playground monitor should be trained in first aid response. This will avoid a mere ushering of the child into the school to the school nurse with little regard for checking the status of the child for neck injury, concussion, or other condition. It may also reveal the need for phone or alarm to summon assistance, especially if the playground has a single monitor or if the recommended ratio of twenty-five children to one monitor is violated. The monitor must also be equipped and trained to handle exposure to body fluid. Any position that has a reasonable expectation of contact with body fluids is deemed to have occupational exposure according to Occupational Safety and Health Administration regulations. Additional recommendations to prevent secondary exposures to body fluid include barricades or tape to limit access to the equipment and the presence of necessary cleaning supplies and personal protective equipment to clean the blood from the equipment.

A playground management program will consist of meeting equipment and grounds standards, maintenance of the equipment and grounds to the standard, investigations of incidents, inspections, and educational initiatives. Your insurance carrier may provide an audit instrument for playground standards and an audit service from a certified playground inspector. Daily inspections prior to activities should always be practiced. School custodians or assigned maintenance workers may be best suited to perform the inspections. These individuals are probably the most familiar with physical conditions of the school and grounds. A quick cursory inspection for sharp edges, broken equipment, items lost or hidden on the playground from the previous day, and emergency response equipment can yield effective results.

Investigating all injuries and tracking or assessing minor occurrences can shed light on countermeasures that can be enacted to stop potentially injurious behaviors, correct unsafe conditions, protect employees and the school from liability, and indicate educational objectives or topics that can be covered with both the students and the staff. Observing playground behaviors is also a type of investigation. Looking for behaviors that Dr. Sandseter identified as potentially injurious and identifying trends between behaviors, incidents, and injuries, and conditions will yield many of the educational items that should be covered. Bringing this together can help make the playground a cost-effective educational environment in much the same manner as the classroom.

CASE STUDY

Two weeks ago, a three-year-old preschool child was injured on the playground. The incident occurred in the afternoon at approximately 2 p.m. The child jumped from a landing on a slide (to the side of the slide itself) and struck the ground on his side. The incident resulted in a broken collar bone and a concussion. The playground monitor was supervising only ten children at the time but had been distracted by another child and did not see the incident. One child said that the boy jumped from the landing. The boy also told the nurse that he jumped from the landing. The slide was manufactured by a reputable company and met ASTM standards for school-aged children, but not for toddlers.

- Should the school be held liable for the incident presented above? What if:
  A. The monitor was not trained in first aid and merely helped the boy up and walked him inside the building to the school nurse? Could this impact liability?
  B. A timely investigation was not completed? How would this affect the case if a suit was filed?
- Would you say that the school is liable for not training the toddler on how to use the equipment? Why or why not?
- What corrections would you implement on your elementary playground from the above situation?
- How would you manage playground safety at your elementary schools?

EXERCISES

1. What injury statistics presented in this chapter might relate to exposures that exist on your playground?
2. What organization established standards for the construction of playground equipment?
3. What surface materials might be used to protect a child against a fall?
4. How can the behavior of crowding affect injuries?

REFERENCES

Bird, F., Germain, G., and Clark, D. (2003). *Practical Loss Control Leadership* (3rd ed.). Det Norske Veritas, Duluth, GA.

Heseltine, P. (1993). Accidents on children's playgrounds. *Children's Environments Quarterly*, 2(4), 38–42. Accessed May 1, 2012. http://www.colorado.edu/journals/cye/2_4/AccidentsOnChildrensPlaygrounds_Heseltine_CEQ2_4.pdf.

Knowlton, M. (2009). *Safety at the Playground*. Crabtree Publishing Company, New York.

Pancella, P. (2005). *Playground Safety*. Heinemann Library, Chicago.

Posner, M. (2000). *Preventing School Injuries: A Comprehensive Guide for School Administrators, Teachers, and Staff*. Rutgers University Press, New Brunswick, NJ.

Safe Kids Worldwide. (2010). Playground safety. Accessed April 30, 2012. http://safekids.org/our-work/reseearch/fact-sheets/playground-safety-fact-sheet.html.

Sandseter, E. B. H., and Kennair, L. E. O. (2011). Children's risky play from an evolutionary perspective: The anti-phobic effects of thrilling experiences. *Evolutionary Psychology*, 9, 257–284.

Tierney, J. (2011). Can a playground be too safe? *New York Times.* Accessed May 1, 2012. http://www.nytimes.com/2011/07/19/science/19tierny.html.

U.S. Consumer Product Safety Commission. (2010). *Public Playground Safety Handbook* (Publication No. 325). U.S. Consumer Product Safety Commission, Bethesda, MD.

U.S. Department of Labor. (1996). Title 29 Code of Federal Regulations, Part 1926.32. Definitions. Occupational Safety and Health Administration, U.S. Department of Labor, Washington, D.C.

20 Transportation Safety

Terry Kline

CONTENTS

The U.S. Department of Transportation explains that school buses are the safest mode of transportation for getting students back and forth to schools. Students are about fifty times more likely to arrive at school alive if they take the bus than if they drive themselves or ride with friends. Most parents do not realize that a child is much safer riding the bus than being driven by a parent. Add in the environmental and financial benefits, and it is hard to find a reason to send kids to school using any other mode of transportation.

The National Association for Pupil Transportation claims the yellow school bus is the safest, most economical, most energy-responsive and most environmentally friendly means to transport our children to and from school each day. A school transportation safety program can be comprised of the following elements:

- Driver selection, training, and authorizing drivers
- Route selection
- Safe driving issues
- Preventive maintenance
- School district and personnel responsibilities

DRIVER SELECTION, TRAINING, AND AUTHORIZATION

The National Association for Pupil Transportation members are encouraged to embrace that all school bus drivers and attendants should receive relevant and appropriate training in transporting all students but particularly in the transportation of students with disabilities and special needs. Based on Pupil Transportation Safety Uniform Guidelines for State Highway Safety Programs published by the National Highway Traffic Safety Administration (NHTSA), each state entity should establish procedures to meet the personnel recommendations for operating school buses and school-chartered buses. The NHTSA Pupil Transportation Safety Guideline 17 for personnel and selection includes:

- Each state should develop a plan for selecting, training, and supervising persons whose primary duties involve transporting school children in order to ensure that such persons will attain a high degree of competence in, and knowledge of, their duties.
- Every person who drives a school bus or school-chartered bus occupied by school children should, at a minimum:
  - Have a valid state driver's license to operate such a vehicle. All drivers who operate a vehicle designed to carry sixteen or more persons (including the driver) are required by FHWA's Commercial Driver's License Standards by April 1, 1992 (49 CFR part 383) to have a valid commercial driver's license;
  - Meet all physical, mental, moral and other requirements established by the state agency having primary responsibility for pupil transportation, including requirements for drug and/or alcohol misuse or abuse; and
  - Be qualified as a driver under the Federal Motor Carrier Safety Regulations of the FHWA (49 CFR part 391) if the driver or the driver's employer is subject to those regulations (U.S. Department of Transportation, 2006)

Numerous resources are available to help the new school administer develop guidelines for school district driver selection, training, and authorization. The National Safety Council and J. J. Keller and Associates provide guidelines for developing an overall fleet management program for the local school administration. The National Association for Pupil Transportation (NAPT), the American Automobile Association, the School Bus Information Clearinghouse, and the National Association of State Directors of Pupil Transportation Services provide specific services for local and state pupil transportation agencies.

SELECTION

Fleet safety programs by Keller and Associates and the National Safety Council explain that selecting safety-conscious drivers is not as difficult, nor as time consuming, as most school administrators fear in regard to the overall program effectiveness. The question revolves around how the administrator can be sure that safe driving practices can be screened from the application and the interview process. The goal is to screen out as many potential high-risk applicants as possible because the agency does have total control over who is considered for employment and who is not considered for this position.

The administrator needs to understand what type of driver is needed to make a good candidate for consideration. The administrator needs to know the legal requirements of the position as well as the skills, abilities, personality, and behaviors that will lead to reduced-risk performance when dealing with the school bus driving environment. Any of the four basic personality temperaments may serve as an excellent school bus driver candidate, but it is crucial to work with this new driver based on the positive personality and behavior traits that candidate possesses.

The application review is the first process that contains the applicant's most important information relating to the driver's safe work and performance record. A background check on the candidate will verify the risk-taking behavior and performance of the applicant. The administrator is looking for red flags that will indicate high-risk behavior or legal records in regard to operation of a motor vehicle. The administrator must decide whether the applicant must possess a commercial driver's license (CDL) or is capable of being trained to obtain a CDL license for bus operation.

Keller and Associates recommends that an application review includes legibility, because the information contained needs to be verified. It should be understood that an illegible application is a sign that the candidate is trying to confuse or mislead the reader of the application. Driver recruiters should not take any information for granted: each item can and should be verified. Legibility of this document will also be an excellent indicator as to how other documentation such as logs, maintenance needs, and discipline reports will be completed (Driver Screening and Orientation, 2006).

Keller also points out that accuracy of the document is important, because former employers need to be contacted for verification of employment status. A credit-reporting agency used by the school district will also help to verify employment history. The driver recruiter should understand that an applicant who creates too much legwork may be trying to hide behaviors that link to poor driving records or an unsafe history of work-related activities. Applications should have a preliminary review of the application for legibility and accuracy so it can be returned to the candidate for review and correction (Driver Screening and Orientation, 2006).

National Safety Council documents point to completeness in regard to the application process. The application review process should ensure that no blank places are evident on the application. This process is especially true in the more sensitive questions in regard to criminal history. Applicants should use some determination like "NA" to indicate that they reviewed the question and have no information to include (National Safety Council, 2010).

In addition to legibility, accuracy, and completeness, the applicant review process should include obvious red flags that relate to inconsistent job performance:

- Gaps in employment
- Any frequent job shifts
- Frequent changes in residence
- Previous supervisors not with company
- Reasons for departures from previous jobs are vague

Although reviewing a position application may only require ten to fifteen minutes to identify red flags, the verification process may take some time because background and credit checks are normally completed by agencies outside of the school system. The driver interview is the next step of the process when no disqualifiers are present in the application.

The problem with candidate interviews, according to Keller Associates, is that most school agencies that recruit drivers do not have or have received little training on how to conduct an interview. The school district may have a legal incentive to have driver interviewers trained to conduct the bus driver and other fleet driver interviews within the school district. In general, interviewers must be trained to assess answers for what is said and what is not said in regard to the question. Answers that are incomplete or seem to be vague or uncertain may be indicators of the respondent trying to hide or make light of problem situations. General interviewing principles in fleet management programs include:

- Create the right environment
- Have a standard opening statement
- Keep personal inquires legal
- Keep questions performance related
- Make questions central to the hiring decision

Although general interviewing principles may be used to form the framework of the interview, there are specific inquiry methods that produce excellent results. The specific interview techniques usually relate to application statements that are relevant to the job requirements. The specific interviewing techniques include clarifying or noting:

- Gaps in employment listed
- Any job-hopping work history
- Any job change history that is repeated
- Inappropriate expressions of hostility
- Physical signs of emotional or physical abuse

Questioning tools may make the interviewers more efficient and also allow the candidate to express themselves in regard to their job needs. If the interviewer makes eye contact, listens to the candidate, takes good notes, and asks clear questions, the interview will allow for an efficient use of time and effort. The general questioning process should include:

- Asking open-ended questions
- Listening carefully to the reply without interruption
- Being objective and task-oriented
- Asking permission to take notes
- Interviewing defensively with job-related actions
- Allowing time for applicant questions

The screening process may be the most important process in developing the local district's fleet of drivers. The drivers become the face of the school district as they work with the students, parents, teachers, and the community. It is much easier to pick a driver that has reduced-risk tendencies than it is to train a driver to change the risk-taking behavior that has been part of the applicant's working experience throughout a lifetime of driving.

TRAINING

School bus driver training is one of the most important components of the school bus transportation system.

> A critical component of school bus driver training is the recognition of potential driving hazards and appropriate adjustment of driving behavior to ensure the safety of the school bus occupants. The goal of this project and report is to provide school bus drivers and substitute drivers with a list of locations/situations that should be recognized as being potentially hazardous. School bus drivers should be properly trained to deal with these potentially hazardous conditions. In addition, school bus drivers should be trained to deal with hazardous conditions that occur suddenly or are of a temporary nature. Constant dialogue between school bus drivers and route planners is critical to ensure the continued safe transportation of students in school buses (U.S. Department of Transportation, 1998).

Training and assessment are tied together because the novice school bus driver will need to develop as he or she now changes from novice to experienced to model school bus driver throughout years of service to the local school district and the community. The training aspects of drivers move from an orientation process, driver skill development, driver perceptual development, and driver audits to driver assessment and retraining efforts. All training efforts can be accomplished through local orientation, local skill training, and perceptual training developed by professional agencies. Professional pupil training agencies provide curriculum structure that can be used for state-wide and local training efforts.

Orientation of drivers should be a local process that introduces the novice school bus driver to the specific vehicles, policies, emergency procedures, and route or collision reporting procedures of the local school district or provider agency. A school district still has responsibility for driver actions and errors when contracting with a private agency to perform the school transportation services. It is incumbent on the school district administer to review policies and procedures for this orientation process for the local school entity.

Basic driver skill training, for local district bus units, is best performed by local trainers and the model drivers of the local district or contracted agency. Curriculum and instructor training is often available and controlled by state agencies and professional pupil transportation associations. Recommended skills training should at least include mirror usage, backing skills, determining vehicle space needs, parking at curbs, use of lights and devices, as well as specific braking and steering skills.

Because driving a school bus has a great deal of built-in distractions, drivers must be trained to use divided attention techniques to demonstrate reduced-risk choices.

The perceptual training techniques are often developed with simulation and perceptual training techniques and then practiced in various environments. The final task in training is distraction accommodation with simulated in-bus activities on off-street practice facilities. Video technology in school bus operation may be used for assessment reviews and further remedial training as required.

Driver audits and assessment may be performed using trainer observation ride-along procedures when students are not on the school bus unit. Standard routes for assessment will provide for most accurate driver comparisons and behavior responses. Video technology may be used for assessment after the initial trainer observation/assessment to assess any needed behavior adjustments. Remedial training can be assigned or developed as a result of yearly training needs for the experienced and model drivers.

AUTHORIZATION

Each school district should have a policy for authorizing drivers to perform the task of transporting students within the pupil transportation system. CDL licensing processes have licensing authorization rules that need to be enforced within the local school district or the contracted provider. States or local school entities may have rules based on medical authorization on yearly basis or drug-use testing polices or protocol. School administrators will need to maintain records of authorization on a yearly basis for pupil transportation drivers. School transportation administrators should engage in continuous professional education and should be certified in industry practices and knowledge.

ROUTE SELECTION

School transportation providers should utilize computer-based routing and scheduling systems to attain maximum efficiency in their operations and safety for all the children. It is preferable that all children ride on yellow school buses, but in those areas where this is not possible, children should be provided with the safest possible transit systems and equipment or safely-constructed and well-equipped infrastructure to get them to school. The NHTSA Pupil Transportation Safety Guideline 17 for route planning indicates that:

- Each State should enact legislation that provides for uniform procedures regarding school buses stopping on public highways for loading and discharge of children. Public information campaigns should be conducted on a regular basis to ensure that the driving public fully understands the implications of school bus warning signals and requirements to stop for school buses that are loading or discharging school children.
- Each State should develop plans for minimizing highway use hazards to school bus and school-chartered bus occupants, other highway users, pedestrians, bicycle riders, and property. They should include, but not be limited to:
  - Careful planning and annual review of routes for safety hazards.

- Planning routes to ensure maximum use of school buses and school chartered buses and to ensure that passengers are not standing while these vehicles are in operation.
- Providing loading and unloading zones off the main traveled part of highways, whenever it is practical to do so.
- Establishing restricted loading and unloading areas for school buses and school-chartered buses at or near schools.
- Ensuring that school bus operators, when stopping on a highway to take on or discharge children, adhere to State regulations for loading and discharging including the use of signal lamps as specified in section B.1.f. of this guideline.
- Prohibiting, by legislation or regulation, operation of any school bus unless it meets the equipment and identification recommendations of this guideline. (U.S. Department of Transportation, 2006).

Developing safe routes is a tedious and flexible process that often is adjusted yearly based on student population adjustments. Transportation staffs must constantly balance safety and efficiency, and nowhere is this more critical than on the school bus route in identifying safe bus loading zones. Bus stop placement, route hazards, and bus stop safety must be addressed by the school district while they use GPS and routing software to plan school bus routes and schedules.

As the number of school buses operating on the roadways increases, inevitable problems increased with the number of units on the roadway. Several serious tragedies occurred involving school buses causing school officials to think seriously about developing safety guidelines for school buses (U.S. Department of Transportation, 2006).

DISTRACTIONS

Distracted driving is a dangerous epidemic on America's roadways. In 2009 alone, nearly 5,500 people were killed, and 450,000 more were injured in distracted driving crashes. The U.S. Department of Transportation is leading the effort to stop texting and cell phone use behind the wheel. Since 2009, the department has held two national distracted driving summits, banned texting and cell phone use for commercial drivers, encouraged states to adopt tough laws, and launched several campaigns to raise public awareness about the issue. Distracted driving is any activity that could divert a person's attention away from the primary task of driving. All distractions endanger driver, passenger, and bystander safety. The types of distractions involved with vehicle collisions in no particular order include:

- Texting
- Using a cell phone or smartphone
- Eating and drinking
- Talking to passengers
- Grooming
- Reading (including looking at maps)

- Using a navigation system
- Watching a video
- Adjusting a radio, CD player, or MP3 player

Because text messaging requires visual, manual, and cognitive attention from the driver, it is by far the most alarming distraction.

The best way to end distracted driving is to educate school bus drivers about the danger it poses. The facts and statistics are powerfully persuasive. If school administrators and drivers do not already think distracted driving is a safety problem, please encourage staff to take a moment to learn more. The Distraction.gov website shares facts, media support, and curriculum support for agencies interested in reducing distracted driving in the pupil transportation system.

WEATHER

School administrators must deal with weather-related school delays and dismissals on a yearly basis. Weather conditions are a primary issue for pupil transportation safety because of conditions possibly worsening while the routes are in service. Policies and procedures need to be directed around safety of the students in regard to transporting students into dangerous road conditions. School districts must have plans in place to deal with emergency weather and roadway situations. Recommendations from the National School Safety and Security Services related to emergency situations include:

- Establish guidelines related to safety and emergency planning, including emergency communications procedures, for all field trips.
- Establish emergency preparedness guidelines from an "all hazards" approach, covering both natural disasters (weather related, for example) and manmade acts of crime and violence.
- Develop emergency plans with your school district, neighboring districts, and the broader community in mind. How would you mobilize buses in a major community emergency? What role do buses have in emergency management for cities and counties? What happens if public safety and emergency management officials commandeer your buses? How would an emergency impact gas supplies? Who could and would be able to drive school buses if the regular drivers were not available?
- Train school bus drivers and transportation supervisors on terrorism-related issues, bomb threats and suspicious devices, inspecting buses, heightened awareness at bus stops and while driving, increased observations skills while coming and going at schools, sharpening skills in reporting incidents, etc.
- Include school transportation supervisors and school bus drivers in district and building emergency planning processes and meetings.
- Establish mechanisms for mobilizing transportation services during irregular transportation department operations times, such as mid-day when drivers are not normally scheduled to work. Consider establishing mutual aid

agreements with neighboring school districts for rapid mass mobilization in an emergency.

- Train school bus drivers on interacting with public safety officials aboard buses, at accident scenes, in on-road emergencies, and when emergency situations exist at schools. Include protocols for dealing with school evacuations, student release procedures, family reunification issues, and associated matters.
- Have student rosters, emergency contact numbers, first aid kits, and other necessary emergency information and equipment aboard all buses.
- Put identifiers (numbers, district initials, etc.) on top of all school buses that could be used to identify specific buses from police helicopters overhead in an emergency.
- Hold periodic meetings during the school year between bus drivers and school administrators to discuss discipline procedures, safety practices, and associated issues (National School Safety and Security Services, 2007).

Although weather-related emergencies are common events in a local school district, a general emergency plan should be put into place. This school-wide emergency plan should have pupil transportation input to gain access to drivers and their responsibilities in emergency and disaster training efforts.

DRIVER CONDITION

The nature of part-time employment may lead to an older than normal driver population that may possess health issues that are yet to be determined. Part-time employment may lead to younger drivers looking for part-time employment and various methods of gaining money. Various part-time jobs may lead to fatigue events that become an issue with driver condition. It is not easy to decide to report a trusted driver, relative, or friend, but concern for the driver's safety and the safety of those being transported is usually the deciding factor for pupil transportation professionals.

The school district is responsible for a driver's functional ability to safely operate a motor vehicle. Decisions about impaired drivers are based on individual signs, symptoms, behaviors, and the observations of others, rather than the type of condition or a diagnosis. The issue is whether or not a medical condition affects a driver's ability to drive safely.

The basic signs of impairment that are available to the adult monitor riding on a school bus are listed below:

- Confusion
- Disorientation
- Memory loss
- Impaired judgment
- Extreme exhaustion
- Difficulty making simple decisions
- Chronic drowsiness
- Impaired response/reaction time

- Inability to concentrate
- Impulsive behaviors
- Severe shortness of breath
- Episodes of impaired or altered consciousness

The conditions that may cause the impairments listed above are usually medical disorders that should be the first area of concern:

- Alzheimer's and other types of dementia
- Diabetes, if frequent episodes of low blood sugar (hypoglycemia) occur
- Neurological conditions, such as seizure disorders
- Sleep disorders
- Behavioral or mental disorders
- Respiratory (lung) diseases
- Cardiovascular (heart) disease
- Visual impairments

The impairments may be due to physical impairments and, although secondary to the medical conditions above for emergency status, may need to have medical intervention in regard to fatigue, consuming intoxicating drugs and alcohol, and emotional distractions.

A driver condition-monitoring apparatus for vehicles is being developed and is available for commercial vehicles. The device and program software determine the condition of the driver while driving the automotive vehicle with voice information from the driver. The driver's voice is recognized, and the voice information for the driver is generated depending on the result of determination of the condition of the driver. The oral response of the driver is judged based on a result of recognition of the voice. Based on how the driver judges and responds to the voice information, the device can give a warning to the driver, advise the driver to take a rest, or control operation of the automotive vehicle.

Driver condition is the driving force for regular physical and drug testing evaluations. Many states and local agencies have rules and regulations in place for regular monitoring of these conditions to avoid potential collisions that may involve injuries.

OVERSIGHT

Oversight of any school district program is subject to administrative planning and supervision. Communication becomes the most critical issue in operating a fleet and planning for pupil transportation needs. The administrator is responsible for authorizing drivers to perform the task of moving school district staff and children from location to location with the utmost respect for the safety of the occupants. Competent drivers that are considerate to the public need for school safety is critical for any program effectiveness. Leaving children unattended on school buses must not be condoned in any way because it exposes our children to risk, is a frightening experience for the children, and erodes public confidence in the safety and security of the yellow school bus. Drivers should be responsible models of driver behavior because students are very watchful of drivers as they grow and expand their interests. School bus drivers should not engage in text messaging or similar activities while operating a school bus in order to show safe driver actions to those in his or her charge. School bus drivers should always utilize their lap-shoulder seat restraints on the school bus. The driver then is the front line representative of the efforts of the school district administration to provide a safe pupil transportation experience.

PREVENTIVE MAINTENANCE

The National Highway Traffic Safety Administration is in the process of completing research concerning the use of passenger crash protection systems built into the yellow school bus prior to issuing any additional requirements related to such systems. The NHTSA Pupil Transportation Safety Guideline 17 for vehicle maintenance indicates that each state should establish procedures to meet the following recommendations for maintaining buses used to carry school children:

- School buses should be maintained in safe operating condition through a systematic preventive maintenance program.
- All school buses should be inspected at least semiannually. In addition, school buses and school-chartered buses subject to the Federal Motor Carrier Safety Regulations of FHWA should be inspected and maintained in accordance with those regulations (49 CFR Parts 393 and 396).
- School bus drivers should be required to perform daily pretrip inspections of their vehicles and the safety equipment thereon (especially fire extinguishers), and to report promptly and in writing any problems discovered that may affect the safety of the vehicle's operation or result in its mechanical breakdown. Pretrip inspection and condition reports for school buses and school-chartered buses subject to the Federal Motor Carrier Safety Regulations of FHWA should be performed in accordance with those regulations (49 CFR 392.7, 392.8, and 396; U.S. Department of Transportation, 2006).

Perhaps the most important safety aspect that school districts and school bus companies can provide students is ensuring that the nation's 480,000 school buses that are in regular service nationwide remain in top operating condition. This requires a tried and true preventative maintenance schedule.

We all have preventative maintenance performed on our personal automobiles. Many of us take our car into a quick oil change shop somewhere between 3,000 and 7,000 miles and wait while they change the oil and filter, lube the steering and suspension, place air in the tires, vacuum the inside, and check the transmission, power steering, anti-freeze, and washer fluid levels. They will check the belt, and pull the air filter for inspection. This is the normal preventive maintenance (PM) schedule for most cars on the road today.

A breakdown in another part of the county can cost time and energy for a school district, as well as initiate many parent complaints. Pupil transportation staff should study the life of various components and be in the practice for changing parts before they fail, preventing a breakdown. The school bus industry operates in an area between a standard automobile and the over-the-road trucking companies. School bus technicians try to find potential problems before they turn into costly repairs that remove the vehicle from service for extended periods of time. There are a number of ways we do this. The daily pretrip inspection performed by the driver is the most common. The scheduled services or inspection performed by mechanics is another. Most states have an annual inspection requirement, which may be performed by a state official.

Preventative maintenance is the key to a safe fleet and an economically operated fleet. School buses must be inspected on a scheduled basis by a qualified mechanic. Lubrication is necessary to extend wear points. The amount of wear in many areas can be measured. Replacement of component parts can be done before failure occurs.

There is no universal PM schedule. Manufacturers recommend a schedule for their vehicles. States often mandate schedules, and each fleet develops a schedule that they believe is best for them. Some are based on days, miles, or hours of operation. Regardless, these schedules should be continuously reviewed. To determine ways in which some schedules may not be ideal, one can start by looking at the number and type of vehicle breakdowns. For example, if an operation is making service calls for items that could be found during normal PM inspections, this may indicate that the period between inspections should be shortened.

CASE STUDY

You are an administration member at a medium-sized school district responsible for the district pupil transportation program. The bus transportation coordinator calls to report a collision involving a district school bus and a passenger vehicle. Three students are treated for injuries and released, whereas the driver of the passenger vehicle sustains serious injuries including fractures. The onboard camera was in operation because the bus was being used to transport students home from after-school activities without a bus monitor. The bus was hit at 5:45 p.m. on a rural road with a 6-foot shoulder while stopped.

The sequence of events was sent to your office computer for evaluation and internal actions. The video finds two students in the rear of the vehicle having a heated argument that moments later had the attention of other passengers in the rear. The driver asks the students to calm down and not distract the driver. Punches are thrown by the occupant in the left rear seat with retaliation from the student in the right rear seat. Several other passengers stand up and turn to watch the activity, blocking the driver's view of the ensuing problem. The noise from the commotion alerts the driver of the problem in the rear of the bus.

The driver pulls the vehicle off the road with about 12 inches of the school bus in the travel lane, but well onto the limits of the roadway shoulder. The driver unbuckles as students fill the walkway of the school bus. The driver is approaching the students in the walkway when the bus is struck by a passenger vehicle from the rear. The students in the walkway fall into seated areas and into other students. The fighting students in the rear have injuries as a result of being thrown into the rear emergency door and window bars. Three of the standing students, including two that were fighting, have bleeding injuries that need treatment. Four of the other students complained of bumps and bruises but are not transported to the hospital.

The driver of the other vehicle is transported to the hospital with bleeding injuries and supporting casts on one arm and one leg. The hospital report listed a leg and ankle fracture. The school bus was able to complete the route with damage to the spare tire and the rear bumper. The passenger vehicle was towed from the scene.

- What are the school's responsibilities in this collision?
- What administrative actions should be used to correct the issues presented in this school bus collision with a privately owned vehicle?

EXERCISES

1. Enter the http://www.nhtsa.gov/School-Buses website and review comments about the items below.
   - Take a position on seat-belt usage in school buses and then defend your position.
   - Take a position on requiring the use of pupil transportation instead of allowing teens to drive to school and defend your position.
   - Take a position on school bus drivers using cell phones or district communication devices and then defend your position.
2. Enter the http://naptonline.org website and review comments about the items below.
   - Review a NAPT news brief and then provide a brief topical review.
   - Review the online PDS file under "Education." Explain the education offerings that are available.
   - Review one item located in the "Resources" file. Explain how the resource can be used in your pupil transportation program.

REFERENCES

Driver Screening and Orientation. (2006). The Transport Safety Pro Advisor. J. J. Keller & Associates. DSO 1–22.

National Safety Council. (2010). *Fleet Safety Program Guide*. National Safety Council, Chicago, IL.

National School Safety and Security Services. (2007). School bus transportation security. Accessed December 28, 2011. http://www.schoolsecurity.org/resources/school_bus_ security.html.

U.S. Department of Transportation. (1998). *Identification and Evaluation of School Bus Route and Hazard Marking Systems*. National Association of State Directors of Pupil Transportation Services. National Highway Traffic Safety Administration, Dover, DE.

U.S. Department of Transportation. (2006). *Highway Safety Program Guideline Numbers and Titles*. Guideline 17. Pupil Transportation Safety Uniform Guidelines for State Highway Safety Programs. National Highway Traffic Safety Administration, Dover, DE.

21 School Lab Safety Management

Ronald Dotson

CONTENTS

Lab safety is a function of managing hazardous materials and behavior-based safety and the control of hazardous energy. Like any safety initiative, it requires comprehensive program management. A comprehensive program has seven key elements: it has measurable goals, it utilizes a hierarchical level of positions with assigned responsibilities, it tracks organizational experience, it investigates incidents, it utilizes metrics for benchmarking, it involves all levels of the organization, and it has a commitment to continual improvement. Comprehensive management is important to lab safety because of the experimental nature of the work involved.

On January 7, 2010, a graduate student at Texas Tech University was severely injured. His injuries included the loss of three fingers, burns to his hands and face, and an eye injury. The researcher was working with a nickel hydrazine perchlorate derivative. The project was investigating the nature of new potentially energetic materials as research to be used by the Department of Homeland Security. Tests on the substances required an amount of approximately 10 grams of material. A procedure of using only 100 milligrams was established, and the immediate researchers did not receive this communication. Rather than produce several batches of the substance, they decided to make a large batch in order to preserve the consistency and reproducibility of the experiments. The injured researcher had begun to mix a large batch of the substance. In small quantities, hexane or water prevented the substance from igniting. The researcher assumed this would hold true for larger amounts as well. Once the batch had been mixed, clumps were noticed, so the researcher transferred about half of the mixture into a mortar, added hexane, and began using a pestle to break up the clumps. He initially wore his goggles properly, but when he began mixing the substance in the mortar, he had his goggles on his forehead. The substance ignited, resulting in his injuries (Chemical Safety Board, 2011).

The Chemical Safety Board (CSB) investigated the incident and produced recommendations. Texas Tech is a premier research institution and had many controls and policies in place for lab safety, but the CSB found the following:

1. Physical hazards were not properly assessed and abated.
2. There was a lack of safety management accountability and oversight.
3. Previous incidents were not documented, tracked, and shared.
4. Communication was not thorough to all levels (Chemical Safety Board, 2011).

These findings indicate that the program was not comprehensively managed.

Committee oversight and the establishment of stages of experimental implementation must be carried out. In industry, a philosophy of system safety analysis is adhered to when any new machine or process is introduced or when a machine or process is being modified. A cross-functional team meets at all stages of the process and uncovers hazards and abates the hazards before the process continues to the next stage. Stages might be organized as follows: concept stage, initial design, prototype, experimental, initial production, and production. Lab experiments and projects may also be organized into stages. Committee oversight should approve the continuation into a progressive stage. In the Texas Tech incident, the conceptual stage should have involved a certain amount of research into similar substances, previous organizational experience, and the experiences of other labs. Furthermore, an assumption was made regarding small amounts versus large amounts that could have been tested in a more controlled environment. Committee oversight uses the assumption that more review of different perspectives may prevent oversight of a hazard. The involvement of experienced researchers is a must.

Safety is one business management or leadership principle that requires bottom-up involvement. The decision that the immediate researchers made to produce one large batch instead of several smaller ones is completely reasonable within their area of work practice, but because they were not represented in the decision-making process, communication failed to reach the bottom level. Failure to reach the action or user level is a common failure of any system. Because safety directly relies on the correct implementation of practices at the user level, it must involve them in a collaborative leadership style.

MANAGING HAZARDOUS MATERIALS

Managing hazardous materials involves the communication of hazards to the user, establishing correct handling and storage procedures, inventory control, and correct response planning. The material safety data sheet (MSDS) that must be provided by the manufacturer is the starting point for identifying the hazards and correct storage and handling procedures. MSDSs for all hazardous materials in the organization must be kept and made available to employees and student users. These sheets should be available in open-access format to the employees in the area where the chemicals are utilized. There should also a system where the hazards are communicated along with the required personal protective equipment. The typical system is called a hazardous material identification system. Every substance must be labeled, and the system utilizes a numerical code for communication of health hazard of the substance, flammability of the substance, reactivity, any special hazards, and then a code or symbols showing proper personal protective equipment to be worn during handling. This system or a system similar to it should be used in a lab environment as well.

It is required that employees be trained on the hazards of the substances to be used, how to handle the substance, and what to do in case of an emergency. This should also be true for student users. Perhaps it would be best to incorporate this information into pre-experiment assignments. Students could utilize MSDS and other sources, such as chemical desk references, the computer-aided management of emergency operations (CAMEO) download from the web site of the Environmental Protection Agency (EPA), and other data bases to identify the hazards, proper personal protective equipment to be worn, and emergency response procedures. It might also be a good group practice to allow groups to review safety precautions for this information prior to beginning lab experiments.

Storing the materials should also be done properly and in conjunction with good inventory practices. The safest storage technique is to not store it at all, meaning timely ordering and obtaining only the needed amounts will avoid long-term storage of substances that can break down over time or break down the container, allowing seepage and spills to occur. Flammable and combustible substances should also be stored in metal cabinets designed for their storage, and amounts must be limited to Occupational Safety and Health Administration (OSHA) requirements.

There are requirements for entire rooms as well. Inside storage rooms have design criteria that must be met. Inside storage room design and amounts may be found in 29 Code of Federal Regulations (CFR) 1910.106. For example, storage rooms must have a ventilation system that accomplishes six air exchanges per hour (29 CFR 1910.106(d)(4)(iv)). This may also go beyond OSHA requirements and be based upon the fire suppression capability of the protective system. Your liability insurance company will inspect and provide requirements usually based upon National Fire Protection Association or factory mutual standards.

Another strategy is to find alternative chemicals or substances that can yield similar results but present fewer hazards to the user and for storage or substances that are less reactive to other needed substances. This requires constant research and effort on the chemical manager's part. Storage of chemicals requires planning for the chemicals that have a possibility of mixing upon spill or leakage. Separation is again the best method. Spill-containing methods like catch bins, spill pallets, and separate lockers will prevent any leakage from causing a chemical reaction. The CAMEO program can be downloaded free from the EPA and allows for the user to "mix" chemicals and obtain a report about the expected reaction. This further allows for response planning and knowing what chemicals in your inventory must be kept separate.

Hazardous energy present in the lab environment might include electricity, compressed gases, flammable gas, flame, and even water. Energy should be controlled by establishing two levels of user: authorized and permissive. Authorized users would be faculty and staff researchers that have a level of training and experience that allows for their usage of the energy. Permissive users would be those that must use the energy only after access has been granted by an authorized user and certain precautions have taken place or been planned for in advance. It might be reasonable for permissive users to fill out a permit that requires them to identify the hazards and plan for their abatement. It may also require an authorized user be present and supervising the activities. The access to energy must be controlled with only certain authorized users allowed to manage the access.

BEHAVIOR-BASED SAFETY MANAGEMENT

Behavior-based safety management is basically the management of safety taking into account human behaviors. A good example is the supply of protective equipment at or near the point of exposure in order to facilitate the compliance of obtaining and wearing the equipment. Behavior-based safety requires doing behavior observations and performance observations. In lab safety, this means the behavior and performance of the various levels of researcher and teacher. Observation allows you to:

- Identify practices that could cause incidents like injury or damage
- Determine the training needs
- Learn more about the habits of personnel
- Assess procedures and instructions
- Correct deficiencies immediately
- Reward desired behavior (Bird et al., 2003)

Observation requires reporting, documenting, and assessment or investigation in order to produce a response to the undesired behavior. The results must also then be shared and communicated to all levels. In other words, the organization must learn from its mistakes and undesired behaviors prior to the occurrence of an incident.

There are critical behaviors that can be identified for general safety practices in the lab, including:

- Proper selection and use of personal protective equipment
- Proper physical positioning
- Limiting of physical exposure with controls
- Proper warning or securing of materials and energy sources
- Use of safety devices
- Proper work pace
- Following proper work instructions
- Proper use of equipment
- Attention to aisle ways and work areas
- Following systematic procedures

Authorized users supervising lab activities should grade or observe students for these behaviors. Incorporating safety behaviors into lab course grades is an important strategy for user safety. Observation of organizational practice can also take place in the same manner. However, giving feedback is important. Debriefing of lab exercises should occur with not only the results but with safety behaviors as well. Group work in a lab might require one person in the group be assigned to the "safety" of the group. In a school setting, this might be rotated to provide all students with the experience.

The documentation and reporting of observed behaviors over time points to important trends that might be addressed from a systematic viewpoint. This allows for committees to produce safety countermeasures or improve system details in order to increase safe behaviors and decrease at-risk behaviors. Safety incidents usually have job factors and management system factors that can be traced as root causes of incidents (Bird et al., 2003). From the early days of scientific management, safety managers began tracking the number of near-miss incidents, minor injuries, and major injuries (Heinrich et al., 1980). In relation to behaviors, these categories translate hierarchically to near misses, damage incidents, minor injuries, and major injuries (Bird et al., 2003). The tracking of behaviors then allows for an early identification of trends before incidents are allowed to blossom into more costly ones.

EXPOSURE AND HYGIENE PLANS

OSHA has covered laboratory safety for employee exposures to hazards. The requirements can be used in a school environment to provide safety for students as well. The OSHA standards are covered under 29 CFR 1910.1450. Two strategies other than what has been previously covered are mandated: exposure determinations and hygiene plans.

Exposure determinations according to the standard concern exposure limits that are published as permissible exposure limits. These limits are the amounts that a person can be exposed to in an eight-hour time-weighted average and are listed in 1910.1450. A determination of the exposure amount for the researcher should be ascertained. For example, air sampling can be used as a method for determining exposure to air contaminants. Initial monitoring when a new substance or process is introduced is required when the exposure can be reasonably expected to reach the permissible exposure limit or exceed it. Periodic monitoring of the process should then continue to occur at intervals of no longer than six months if the exposure limit is approached or exceeded. In a school environment, we might wish to continue periodic monitoring of standard experiments to document exposure levels. At a minimum, initial monitoring should occur to document even low levels of any exposure of a listed substance in 29 CFR 1910.1000 on toxic and hazardous substances.

Other sampling may occur as well. Swipes of metal shavings, for example, might document low levels of metal exposure. However, all monitoring results should be maintained and shared with students, parents, and other faculty upon request to facilitate an atmosphere of transparency and legitimate concern for the student's welfare.

CHEMICAL HYGIENE PLANS

Schools can also use the chemical hygiene plan as guidance for a lab safety plan of their own. The chemical hygiene plan has two goals: to protect employees from health hazards associated with work procedures and hazardous chemicals in the lab and to keep personal exposure to below the established limits. The plan should be the core of educational and training efforts for the lab and be readily available to employees, students, parents, and anyone upon request (29 CFR 1910.1450).

The hygiene plan established out by OSHA for employers to follow has the following elements:

- Standard operating procedures relevant to safety
- Controls and practices aimed at reducing exposure to chemicals
- A schedule of checks and maintenance of equipment that controls exposure
- A list of circumstances where prior approval from a higher authority is required
- Provisions for training and information
- Provisions for medical consultation and examinations
- Assignment of responsibility for plan implementation
- Provisions for additional protection for work with highly hazardous substances like select carcinogens, reproductive toxins, or acute toxins to include:
  - Work in designated areas
  - Use of containment devices and protective equipment
  - Procedures for handling and removal of contaminated wastes
  - Decontamination procedures
  - Annual review of the plan (29 CFR 1910.1450).

Appendix A of 29 CFR 1910.1450 provides an outline and general guidance for the hygiene plan. The first section attempts to establish general rules and principles that guide all laboratory actions. Rather than adopt handling procedures for different substances or classes of substance, general protective procedures should be relied upon. Examples might include the establishment of minimum personal protective equipment required upon lab entry rather than only if certain hazards are identified. This also prevents some chemicals from being viewed as "safe." All substances should be treated as though they are toxic to skin contact (29 CFR 1910.1450 Appendix A). Other standard operating procedures might include the establishment of group responsibilities, safety talks before daily activities begin, and safety debriefings after the project. Once safety is implemented as a standard practice, it begins to become cultural.

The "responsibilities" section addresses the need to delineate responsibilities for individuals associated with the program. OSHA recommends a structure with titles that may be changed or adapted to a school:

- The chief executive officer (principal) has the ultimate responsibility for chemical hygiene. He or she must work with other administrators to provide various types of support and ensure levels of oversight by cross-functional teams.
- The supervisor of the unit, as described by OSHA, could be translated as the person who is in charge of plan implementation at the school level or at the single lab (science program coordinator).
- The chemical hygiene officer (CHO) is an essential role. This should be a person who is involved with direct oversight and perhaps participation in

the project itself (teacher). They will have the frontline leadership for safety. Their duties are as follows:

- Monitor procurement, use, and disposal of chemicals used in the lab
- See that appropriate audits are maintained
- Lead development of precautions, procedures, and facilities
- Know and educate all in the project on regulated substances and exposure limits
- Oversee the annual review process for continual improvement
- The laboratory supervisor, according to OSHA, has the responsibility of placing the plan in practice at the lab level. Their duties include:
  - Ensuring lab participants know and follow hygiene rules
  - Ensuring that protective equipment is available and operational
  - Providing regular housekeeping and other routine inspections
  - Knowing the current requirements of regulated substances
  - Determining the required levels of personal protective equipment
  - Ensuring that training takes place
- The laboratory worker has the responsibility of following all rules and procedures established in the chemical hygiene plan and reporting all deficiencies (29 CFR 1910.1450 Appendix A).

In a school setting, these roles might be combined. For simple laboratory projects, the CHO and the supervisor positions might be combined and assigned to the teacher conducting the lab. Project directors might actually be student participants. It is important that education begin to incorporate the practices and procedures that are foundational to industry. Simply stated, it prepares students to perform more efficiently in the world after graduation. Giving students progressive responsibilities where the benefits are of the utmost in intrinsic reward motivates many to take personal responsibility and accountability seriously.

The facility section covers the overall design and features for conducting safe projects, storage of chemicals, and the emergency response features. Generally, four sections can cover the needed material:

- The design section covers the general description of ventilation information, storage room design and features, the location of hoods, sinks, energy type locations and controls, any other physical equipment, eyewash and emergency equipment locations, suppression systems, and waste disposal equipment and arrangements.
- Equipment requires maintenance and maintenance inspections. The CHO and supervisor are responsible for ensuring that daily inspections and scheduled inspections are completed.
- Usage guidelines should also be determined. The number of participants and the types of projects should match the facility. The scale of the project is also a determining factor.
- The ventilation section itself covers more detailed information;
  - Turnover rates and capacities
  - Input areas

- Stacks for release
- Local hoods and ventilation
- Performance of hoods, snorkels, and other devices
- Isolation rooms
- Procedures for modification approvals
- Quality monitoring programs
- Periodic inspections (29 CFR 1910.1450 Appendix A)

The facility section of the plan is an important and detailed component. When drafting a plan, one should also consider listing contractors for repair and replacement issues, third-party inspectors, and facility experience. The documentation of failures, incidents, damage, modifications, and performance contribute to continual improvement.

The chemical procurement, distribution, and storage section of the plan should establish the process and procedures for procuring chemicals, storing them, distributing them, and using them (29 CFR 1910.1450 Appendix A). Simple establishment of procedures such as requesting an MSDS when ordering any hazardous material and the automatic update of the MSDS list is vital. MSDS management is often overlooked and a common source of violation. These sheets provide the source for hazard information, hazard mitigation, exposure levels, and emergency procedure and requirements.

The environmental monitoring section is where the schedule and types of sampling will be identified and documented. It might also be tracked by individual project because each project might have different needs. This section should also include the tracking and documentation of any environmental management project, such as amounts and types of wastes or regulatory agency reporting.

The housekeeping and inspection section of the chemical hygiene plan should establish cleaning procedures and acceptable methods and cleaning chemicals. It should identify basic inspection criteria for all lab participants prior to and during lab use. It should also include a training portion on these matters for school custodians.

The medical program for a school lab should account for first-aid training, the identification of school personnel trained to perform this training, and how to summon or enact an emergency protocol. This will include much of the emergency response plan for the entire school. Most of the time, students will not be required to handle substances requiring or be involved in long-term projects that mandate medical monitoring. However, teachers and staff might be another matter. Simple medical monitoring for exposure to substances can be incorporated into annual physicals.

In Appendix A of 29 CFR 1910.1450, OSHA recommends inclusion of a section on protective apparel and equipment in any chemical safety plan. This might be handled in other portions of the plan, but this section could be a place to incorporate a formal analysis for the required personal protective equipment (PPE). The formal analysis could be handled in the same format as the job hazard analysis. Instead of critical steps for job duties, chemical and energy sources could be identified, along with the protection needs and requirements, and then the list of specific PPE to mitigate the hazard.

A records section is also suggested. Record keeping might be handled in all of the other portions of the plan as well. However, it may be useful to document and gather original papers such as work orders, receipts, or medical records and experience. It is important to note that the confidential nature of some of this information might require that this portion of the plan is kept as private information, separate from the published plan.

Signage and labels are important components to educational and safety motivation. Signage instructs as to proper usage of materials and reminds users of proper procedures. Most equipment will contain warning labels and instructional labels. It may be an issue of liability to keep these labels in place and legible. Photos and documentation of all signage and label placement is important. Labels and signage should be incorporated into scheduled audits.

The spills and accidents section of the plan addresses emergency response. Proper spill response planning is critical. There is a distinction between a spill, which is at the threshold level for reporting or above, and incidental spills that do not meet reporting requirements. Spill kits should be located in the lab. Procedures should be established for who can respond to spills and how material is to be collected and disposed of.

Waste disposal is a required section as well. It may even be necessary to have a separate vendor for such disposal, depending on lab projects. Some materials may be capable of being recycled. This would be an important portion of environmental management. Batteries might be one example of lab materials to be recycled. Perhaps the school and district have an environmental management program that can be integrated by the lab. The plan must identify containment and waste locations and cover the frequency of disposal, method of disposal, and general rules.

The training section ideally should be a formal training needs assessment and storage of records. It could also be used as a central location for the listing of training topics and specific objectives for each section of the chemical hygiene plan.

Labs present a special hazard for schools. They require a comprehensive management approach and a cultural commitment from all levels to achieve a safe environment. Labs are an important educational setting. They are one of the best environments for students to become familiar with personal safety and how personal safety equates to the safety of the team.

CASE STUDY

Your high school has added a new science lab. All of the equipment is state-of-the-art and is modeled from the best technology that is available in educational settings. It is two weeks before the beginning of the school year. All of the equipment is installed, and signs are posted. In a meeting, the principal indicates that a safety plan for the lab is not necessary because everything is new and must meet governmental requirements.

- How would you respond to the principal?
- What additional safety issues need to be considered?

EXERCISES

1. Design a lesson plan for high school students on basic lab safety. You might cover basic personal protective equipment, identifying hazards and exposure levels, lab equipment usage, emergency action, and disposal.
2. What management systems might you put in place to oversee and ensure multiple levels of oversight of lab safety?
3. As a science teacher, what behaviors might you observe during lab projects? How would you use the information on these behaviors to ensure that safety protocols are effective?
4. What elements should be included in a chemical hygiene plan?

REFERENCES

Bird, F., Germain, G., and Clark, D. (2003). *Practical Loss Control Leadership* (3rd ed.). Det Norske Veritas, Duluth, GA.

Heinrich, H. W., Peterson, D., and Roos, N. (1980). *Industrial Accident Prevention: A Safety Management Approach* (5th ed.). McGraw-Hill, New York.

U.S. Chemical Safety Board. (2011). Case study: Texas Tech incident, January 7, 2010. U.S. Chemical Safety and Hazard Investigation Board, Washington, DC.

U.S. Department of Labor. (1996). Appendix A: National Research Council recommendations concerning chemical hygiene in laboratories. Title 29 Code of Federal Regulations, Part 1910.1450. Occupational Safety and Health Administration, U.S. Department of Labor, Washington, D.C.

U.S. Department of Labor. (1996). Title 29 Code of Federal Regulations, Part 1910.106. Flammable and combustible liquids. Occupational Safety and Health Administration, U.S. Department of Labor, Washington, D.C.

U.S. Department of Labor. (1996). Occupational exposure to hazardous chemicals in laboratory environments. Title 29 Code of Federal Regulations, Part 1910.1450. Occupational Safety and Health Administration, U.S. Department of Labor, Washington, D.C.

U.S. Department of Labor. (1996).Toxic and hazardous substances. Title 29 Code of Federal Regulations, Part 1910.1000. Occupational Safety and Health Administration, U.S. Department of Labor, Washington, D.C.

22 Food Safety

Sheila Pressley

CONTENTS

Promoting food safety is an important part of the fundamental mission of schools that includes giving young people the skills to become healthy young adults. Food safety at school matters because on a typical school day, millions of children eat school lunches prepared by school personnel. Children may also consume foods purchased to support school or team fundraisers or purchased at concessions during sporting events. Food safety in schools should always be a concern for a number of reasons including lost learning and teaching time, potential liability to the school or to the school board, decreased family and community confidence in schools, and, most importantly, the possibility of serious or even fatal illnesses. The Centers for Disease Control and Prevention (CDC) estimates that approximately 48 million cases of food-borne illness occur in the United States each year, resulting in 128,000 hospitalizations and 3,000 deaths (CDC, 2011). In a study conducted in July 2007, an analysis of 816 worker-associated outbreaks from 1927 to 2006 found that although 61% of the outbreaks started at food service facilities and catered events, another 11% were traced to schools, day-care centers, and health care institutions (Greig et al., 2007).

Food-borne illnesses can be caused by viruses and parasites, but the greatest number of food-borne illness outbreaks are caused by bacteria. An estimated 65% of all food-borne illnesses in the United States are caused by the ingestion of foods containing large doses of bacteria or bacterial toxins (Beck et al., 2010). Some of the most common bacterial food-borne illnesses include salmonellosis, campylobacteriosis, and sickness caused by *Escherichia coli* 0157:H7.

SALMONELLOSIS

The number of reported cases for salmonellosis is about forty thousand annually in the United States (CDC, 2011). Because many cases are milder and are not diagnosed or reported, the number is probably thirty or more times greater. The CDC estimates that approximately four hundred persons die each year with acute salmonellosis, and children are the most likely victims. There are hundreds of serotypes that can cause food infections of varying symptoms and severity, but the two most common serotypes are *Salmonella enteritidis* and *Salmonella typhimurium*. These bacteria can be destroyed by temperatures of 145°F for whole meats, 160°F for ground meats, and 165°F for all poultry. The foods implicated with salmonella are typically raw meats, poultry, eggs, milk and dairy, fish, pork, salad dressing, chocolate, peanut butter, and the list continues to grow. Symptoms include diarrhea, abdominal cramps, fever, and dehydration. Some serotypes may cause nausea and vomiting. The illness usually subsides in two to four days. However, some individuals continue to carry the bacteria in their bodies. Salmonellosis can be prevented with good personal hygiene, proper cooking temperatures, proper cleaning and sanitizing of food contact surfaces, and the use of only pasteurized eggs. The use of any recipe using raw egg, such as homemade ice cream or hollandaise sauce, should be prohibited. It is also recommended that handwashing with soap takes place after handling reptiles, birds, baby chicks, and pet feces.

If a salmonellosis outbreak occurs, a local or state health official will most likely conduct an investigation. School personnel should be prepared to address the following questions:

- Did the persons preparing the foods wash their hands properly before and after preparing foods? Do they appear healthy?
- How and where were chicken and meat menu items prepared? Were there possible cross-contamination risks?
- Were the implicated foods properly thawed? Are there occasions when frozen chicken and meats are thawed at room temperature?
- Were shell eggs broken and "pooled" as opposed to being cracked singly?
- Were there any occasions when raw milk was used?

CAMPYLOBACTERIOSIS

Campylobacteriosis is one of the most common of all food-borne illnesses, causing over 2.4 million cases per year in the United States (CDC, 2011). Although campylobacteriosis does not commonly cause death, it has been estimated by CDC that 124 persons with Campylobacter infections die each year. *Campylobacter jejuni* is a bacterial microbe that is sensitive to drying, heating, disinfectants, and acidic conditions. The foods implicated in campylobacteriosis are raw or uncooked meat or chicken, raw milk, and nonchlorinated water supplies. The symptoms include diarrhea, fever, abdominal pain, headache, and muscle pain occurring two to five days after ingestion of contaminated food or water. Illness usually lasts for seven to ten days, and relapses are possible. To prevent campylobacteriosis, cook all poultry products to a minimum internal temperature of 165°F, wash hands with soap after handling raw foods of animal origin before touching anything else, avoid consuming unpasteurized milk and untreated surface water, and wash hands with soap after contact with pet feces.

If a campylobacteriosis outbreak occurs, a local or state health official will most likely conduct an investigation. School personnel should be prepared to address the following:

- The extended recovery period of seven to ten days may be an indicator.
- Be prepared to show the chicken and meat menu items prepared.
- Explain whether the practice of prompt refrigeration and cooling of hot foods took place.
- Are refrigeration and freezer units maintaining proper temperatures?

ESCHERICHIA COLI

E. coli belong to a large group of bacteria in which most strains are harmless, but some strains can make you very sick. Some kinds of *E. coli* make a toxin called shiga toxin. The most common shiga toxin-producing *E. coli* (STEC) is *E. coli* 0157. When you read or hear about an *E. coli* outbreak, it is usually referring to *E. coli* 0157 or what is also called STEC 0157. The other kinds of *E. coli* are the non-STEC 0157. About 265,000 STEC infections occur each year in the United States, and about 36% of these are STEC (CDC, 2011). The remaining infections are non-STEC 0157.

Foods involved with *E. coli* outbreaks may include undercooked or raw hamburger, cheeses, lettuce, and unpasteurized milk or apple cider. Symptoms of *E. coli* infections include severe cramping and diarrhea that is initially watery and may become bloody. About 5–10% of those infected develop a potentially life-threatening complication known as hemolytic uremic syndrome (HUS). Children and the elderly are especially vulnerable to HUS, which can cause kidney failure, permanent damage, or death. The recovery period is an average of eight days. To prevent exposure to *E. coli*, cook all ground meat hamburgers to an internal temperature of 160°F and emphasize good hygiene and proper handwashing.

If an *E. coli* outbreak occurs within a school or district, a local or state health official will most likely conduct an investigation. School personnel should be prepared to address the following:

- How are raw meats stored in refrigeration units? Raw meats should be stored on the bottom shelves to prevent drippings from the meat from touching ready-to-eat products.
- Explain how hamburgers are prepared. The cook and the manager should have similar answers.
- Determine the source and/or vendor of the meat products and vegetables.
- Handwashing practices and other personal hygiene habits.
- Determine if there were any occasions when raw milk was used.

In addition to the food-borne illnesses mentioned in this chapter, Figure 22.1 shows the trends of other pathogens using preliminary data from the CDC FoodNet surveillance system. The figure shows the changes in incidence of laboratory-confirmed bacterial infections in the United States in 2010 compared with those from 1996 to 1998.

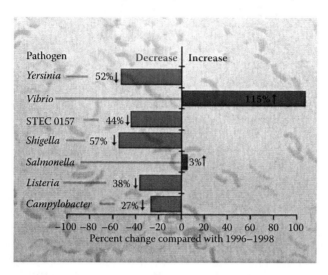

FIGURE 22.1   Changes in bacterial infections.

FOOD SAFETY REGULATIONS

Although there are many laws related to food safety at the local and state levels, the key federal regulations and guidelines that determine food safety in the
United States and dictate local laws include the Food and Drug Administration
(FDA) Food Code, the Hazard Analysis and Critical Control Points (HACCP)
System, and the Food Safety and Modernization Act. The FDA Food Code is a
model code published by the U.S. Food and Drug Administration to assist food
control jurisdictions at local, state, and tribal levels of government with technical
and legal guidance. This guidance is used to regulate the food service industries
and establishments such as schools and nursing homes. The FDA Food Code is
also a model for local, state, federal, and tribal regulators to use to develop and
update their own food safety rules and be consistent with national food regulatory policy. Beginning with the 1993 edition, there are five versions of the FDA
Food Code. The most recent version is the 2009 FDA Food Code, and it is the
first full edition to be published since the 2005 FDA Food Code was released.
The majority of U.S. states and territories have adopted one of the five versions
of the food code.

HACCP is a management system that addresses food safety through the analysis
and control of chemical, physical, and biological hazards from raw material production, procurement, and handling to manufacturing, distribution, and consumption of
the finished product. In the 1960s, the Pillsbury corporation developed the Hazard
Analysis and Critical Control Points (HACCP) system with NASA to ensure food
safety for the first manned space missions. The FDA went on to use the HACCP
principles in developing regulations for low-acid food in canning in the 1970s. Since
that time, the FDA has implemented the mandatory use of HACCP with fish and
seafood products in 1995 and with juice processing and packaging plants in 2001.
A voluntary HACCP program was introduced for Grade "A" fluid milk and milk

products in 2001. HACCP is a systematic approach to the identification, evaluation, and control of food safety hazards based on the following seven principles (U.S. FDA, 2011):

- Principle 1: Conduct a hazard analysis
- Principle 2: Determine the critical control points
- Principle 3: Establish critical limits
- Principle 4: Establish monitoring procedures
- Principle 5: Establish corrective actions
- Principle 6: Establish verification procedures
- Principle 7: Establish record-keeping and documentation procedures

The Child Nutrition and Women Infants and Children Reauthorization Act of 2004 addresses food safety in school nutrition programs and requires some schools to implement food safety based on the HACCP principles listed in this chapter. If your school is using HACCP or planning to implement a HACCP program, the sample HACCP-based standard operating procedures (SOPs) recommended by the U.S. Department of Agriculture are located at the end of this chapter (U.S. Department of Agriculture, 2005).

The Food Safety and Modernization Act was signed into law by President Obama on January 4, 2011. The main goal of this law is to ensure the safety of the U.S. food supply by shifting the focus from reacting to contamination to preventing contamination. This is the most significant change in the FDA's food safety laws in several decades, and a number of rules will be issued. With such a significant number of changes, the FDA will issue regulations over a number of months and years. Some of the enforcement measures of the new law include mandatory recalls by the FDA, as opposed to waiting for firms to voluntarily recall products, and suspension of registration. Suspension of registration gives the FDA the regulatory power to suspend a facility if their food products present a serious health hazard. Preventive measures include records inspection, expanded administrative detention, and the authority to deny entry of foods from facilities in other countries that do not allow the FDA to perform inspections.

ESTABLISHING A SCHOOL FOOD SAFETY PROGRAM

Food safety programs in school settings should include a review of current and pertinent regulations such as those mentioned in this chapter, and they must also encompass a series of training for school cafeteria employees, as well as for teachers, staff, and students. In addition to written policies and presentations, there should also be time allocated to hands-on training and the distribution of policies and procedures to employees and students. To begin a food safety program at a school, there must first be a team or a group of individuals representing the best interests of the school. The team should be diverse and should have representatives from the school and the community. Members of the team could include the school food service director or supervisor, the principal or another administrator, a classroom teacher, a parent, the school nurse or another health professional, representatives from the local heath department and the cooperative extension, and representatives from the school food vendors or suppliers.

Once the team is assembled and a leader has been elected, the first task of the team is to conduct a needs assessment. The needs assessment will include a review of current food policies and procedures so that gaps and areas requiring improvement may be identified. Once the strengths and weaknesses are identified, the team will need to lay out a plan for action. Implementing a plan will depend on several factors, such as the budget and other resources available to the school; however, many of the items may be implemented over time and may be of little or no cost. While improving school food safety reduces liability, it can also reduce absenteeism and lower other costs such as hiring substitute teachers.

Although the plan for action may be very different for each school, the protocols and policies developed as a result of the plan must be written and shared with teachers, staff, students, and parents. The way in which the plan is shared is a critical part of the success of the program. To introduce the plan and any supporting materials, consider using school events and other venues to share the information and recruit supporters for the plan. Events such as parent-teacher organization meetings, school assemblies, sporting events, festivals, open houses, and family literacy nights are conducive to sharing the food safety plan and the vision and mission of the food safety team.

Maintaining a plan and keeping staff trained and up to date will be the next phase of the plan. Be sure to call on local and state health departments, other schools or districts, local universities, and professionals with expertise in food- and water-borne disease control and other environmental health issues. Invite guest speakers to address the food service workers and employ the use of posters, films, and videos to train and educate employees. Whenever possible, recognize and reward food service employees for completing or attending training programs or exceeding expectations.

For students, the food safety plan must be explained with role-playing and plenty of time for questions and practice. Telling students to wash their hands is the first step, but the way to get them involved is to let them practice. Have a contest to see which students wash their hands more thoroughly. Using a product like Glo Germ, you can determine from the use of an ultraviolet flashlight whether students have washed their hands thoroughly. To implement the school-wide culture of clean hands and to motivate employees and students, adequate supplies for handwashing such as towels, soap, and warm, running water must be readily available. To imbed the connection and understanding of good personal hygiene and food-borne illness, the school could present a play with everyone having a role and showcasing their knowledge about food-borne diseases. Students should also be encouraged to do science projects or other required activities on the value of handwashing and sanitation as they relate to food-borne illness.

As more recalls occur in the United States and in other countries, and as school food service programs face more biosecurity concerns, food safety programs for K–12 schools will be necessary and perhaps mandatory. With the rising costs of health care services and programs, school districts and officials will have to devote more resources to preventing food-borne illness and to educating and training employees, students, and families on the value of food safety. Many school systems across the nation have voluntarily established food safety programs using local and state resources. The future health and well-being of their students depends on the strength and their ability to establish and maintain programs such as these.

CASE STUDY

As the school cafeteria prepares to open, a report of about forty thousand pounds of frozen beef patties being recalled catches your attention online. As the school principal, you are aware that the lunch menu has cheeseburgers as the main entrée for the day. Although no one has complained about the patties before and you have not heard anything from the school district, you want to be sure that your school food is safe and that your staff understands the importance of verifying this information.

- What should be done first to address your concerns about serving tainted beef?
- How will you share this news with students and parents?

EXERCISES

For the following questions, explain how these situations could be addressed in the most professional and informative manner.

1. An employee reports to work on time for his job at the school cafeteria, but he is clearly ill and has already confided in you that he has diarrhea. As the food service manager, what should you do?
2. A parent has volunteered each year in your classroom, and in the past, she has provided homemade snacks and treats for your students. The school now has a policy that prohibits homemade items, but there have been no problems in the past with her snacks and treats. What will you do when she brings them with her on the day she volunteers?
3. Each time you visit the elementary school that your child attends, you notice that soap and paper towels are missing from the restroom. You have complained to your child's teacher and to the school guidance counselor. However, you continue to see the same problem. Who will you speak to next, and how will you approach them?
4. When it comes to food safety, there are multiple laws and agencies that are involved. State and federal regulations may change frequently depending on the scientific discovery and advancement of food-borne illnesses. As a principal, how will you ensure that you and your staff have the latest information and updates related to food safety?

APPENDIX

HACCP-Based SOPs

(Taken from the U.S. Department of Agriculture at http://sop.nfsmi.org/
HACCPBasedSOPs.php)
Cleaning and Sanitizing Food Contact Surfaces (Sample SOP)

PURPOSE: To prevent food-borne illness by ensuring that all food contact surfaces are properly cleaned and sanitized

SCOPE: This procedure applies to food service employees involved in cleaning and sanitizing food contact surfaces

KEY WORDS: Food contact surface, cleaning, sanitizing

INSTRUCTIONS:
1. Train food service employees on using the procedures in this SOP
2. Follow state or local health department requirements
3. Follow manufacturer's instructions regarding the use and maintenance of equipment and use of chemicals for cleaning and sanitizing food contact surfaces; refer to Storing and Using Poisonous or Toxic Chemicals SOP
4. If state or local requirements are based on the *2001 FDA Food Code*, wash, rinse, and sanitize food contact surfaces of sinks, tables, equipment, utensils, thermometers, carts, and equipment:
   • Before each use
   • Between uses when preparing different types of raw animal foods, such as eggs, fish, meat, and poultry
   • Between uses when preparing ready-to-eat foods and raw animal foods, such as eggs, fish, meat, and poultry
   • Any time contamination occurs or is suspected
5. Wash, rinse, and sanitize food contact surfaces of sinks, tables, equipment, utensils, thermometers, carts, and equipment using the following procedure:
   • Wash surface with detergent solution
   • Rinse surface with clean water
   • Sanitize surface using a sanitizing solution mixed at a concentration specified on the manufacturer's label
   • Place wet items in a manner to allow air drying
6. If a three-compartment sink is used, set up and use the sink in the following manner:
   • In the first compartment, wash with a clean detergent solution at or above 110°F or at the temperature specified by the detergent manufacturer
   • In the second compartment, rinse with clean water
   • In the third compartment, sanitize with a sanitizing solution mixed at a concentration specified on the manufacturer's label or by immersing in hot water at or above 171°F for thirty seconds; test the chemical sanitizer concentration by using an appropriate test kit
7. If a dish machine is used:
   • Check with the dish machine manufacturer to verify that the information on the data plate is correct
   • Refer to the information on the data plate for determining wash, rinse, and sanitization (final) rinse temperatures; sanitizing solution concentrations; and water pressures, if applicable

- Follow manufacturer's instructions for use
- Ensure that food contact surfaces reach a surface temperature of 160°F or above if using hot water to sanitize

MONITORING:

Food service employees will:

1. During all hours of operation, visually and physically inspect food contact surfaces of equipment and utensils to ensure that the surfaces are clean
2. In a three-compartment sink, on a daily basis:
   - Visually monitor that the water in each compartment is clean
   - Take the water temperature in the first compartment of the sink by using a calibrated thermometer
   - If using chemicals to sanitize, test the sanitizer concentration by using the appropriate test kit for the chemical
   - If using hot water to sanitize, use a calibrated thermometer to measure the water temperature; refer to Using and Calibrating Thermometers SOPs
3. In a dish machine, on a daily basis:
   - Visually monitor that the water and the interior parts of the machine are clean and free of debris
   - Continually monitor the temperature and pressure gauges, if applicable, to ensure that the machine is operating according to the data plate
   - For hot water sanitizing dish machine, ensure that food contact surfaces are reaching the appropriate temperature by placing a piece of heat-sensitive tape on a smallware item or a maximum registering thermometer on a rack and running the item or rack through the dish machine
   - For chemical sanitizing dish machine, check the sanitizer concentration on a recently washed food-contact surface using an appropriate test kit

CORRECTIVE ACTION:

1. Retrain any food service employee found not following the procedures in this SOP
2. Wash, rinse, and sanitize dirty food contact surfaces; sanitize food contact surfaces if it is discovered that the surfaces were not properly sanitized; discard food that comes in contact with food contact surfaces that have not been sanitized properly
3. In a three-compartment sink:
   - Drain and refill compartments periodically and as needed to keep the water clean
   - Adjust the water temperature by adding hot water until the desired temperature is reached
   - Add more sanitizer or water, as appropriate, until the proper concentration is achieved
4. In a dish machine:
   - Drain and refill the machine periodically and as needed to keep the water clean

- Contact the appropriate individual(s) to have the machine repaired if the machine is not reaching the proper wash temperature indicated on the data plate
- For a hot water sanitizing dish machine, retest by running the machine again; if the appropriate surface temperature is still not achieved on the second run, contact the appropriate individual(s) to have the machine repaired; wash, rinse, and sanitize in the three-compartment sink until the machine is repaired or use disposable single service/single-use items if a three-compartment sink is not available
- For a chemical sanitizing dish machine, check the level of sanitizer remaining in bulk container; fill, if needed; "prime" the machine according to the manufacturer's instructions to ensure that the sanitizer is being pumped through the machine; retest; if the proper sanitizer concentration level is not achieved, stop using the machine and contact the appropriate individual(s) to have it repaired; use a three-compartment sink to wash, rinse, and sanitize until the machine is repaired

VERIFICATION AND RECORD KEEPING:

Food service employees will record monitoring activities and any corrective action taken on the Food Contact Surfaces Cleaning and Sanitizing Log. The food service manager will verify that food service employees have taken the required temperatures and tested the sanitizer concentration by visually monitoring food service employees during the shift and reviewing, initialing, and dating the Food Contact Surfaces Cleaning and Sanitizing Log. The log will be kept on file for at least one year. The food service manager will complete the Food Safety Checklist daily. The Food Safety Checklist is to be kept on file for a minimum of one year.

DATE IMPLEMENTED: _____ **BY:** _____

DATE REVIEWED: _____ **BY:** _____

DATE REVISED: _____ **BY:** _____

REFERENCES

Beck, J. B., Barnett, D. B, Johnson, W. J., and Pressley, S. D. (2010). *Fundamentals of Environmental Health Field Practice*. Kendall Hunt, Dubuque, IA.

Centers for Disease Control and Prevention. (2011). CDC estimates: Findings. Accessed March 13, 2012. http://www.cdc.gov/foodborneburden/2011-foodborne-estimates.html.

Greig, J. D., Todd, E. D., Bartleson, C. A., and Michaels, B. S. (2007). Outbreaks where food workers have been implicated in the spread of foodborne disease: Part 1. Description of the problem, methods, and agents involved. *Journal of Food Protection*, 70(7), 1752–1761.

U.S. Department of Agriculture. (2005). National Food Service Management Institute United States Department of Agriculture HACCP-based SOPs. Accessed March 13, 2012. http://sop.nfsmi.org/HACCPBasedSOPs.php.

U.S. Food and Drug Administration. (2011). Hazard Analysis and Critical Control Point Principles and Application Guidelines. Accessed March 13, 2012. http://www.fda.gov/Food/FoodSafety/HazardAnalysisCriticalControlPointsHACCP/HACCPPrinciplesApplicationGuidelines/default.htm#princ.

Section III

Emergency Management

23 Emergency Response Procedures

Ronald Dotson

CONTENTS

Emergency response is an example of a management program that must be living. It must be continually reviewed and adapted to current trends, best practices, legal concerns, and a plethora of issues. It probably is a dominant safety planning issue for an educator. Generally, emergency response centers on the protection of the students, but it must also include all personnel and visitors on the school grounds. It must facilitate the first responders and other agencies that may respond to an incident, and it must be viewed as reasonable and effective by the school's community.

FOUNDATIONAL PLANNING

The most basic tool for response planning is the labeling or mapping of the main building or buildings. It is important to be able to communicate quickly and effectively between both school associates and first responders. For example, when an ambulance is responding to the facility, how would you communicate to them the best entry door or location to come to on school property? Obviously, one strategy is to have an attendant stationed outside awaiting their arrival direct them to the nearest location to the emergency. Another example to consider is the response of a police agency and the establishment of a command and control center that must communicate and direct activity. How do responders communicate the location of entry or sightings? Labeling of the facility requires knowledge of tactical operations.

This process begins with the labeling of the sides of the structure. The main entrance area or front is labeled side 1. The labeling progresses either to the left or right and uses succeeding numbers. A basic four-sided structure would label its front as side 1, the side to the right as viewed from a facing person as side 2, the rear as side 3, and the remaining side as side 4. The labeling continues with doors and windows. The first door going from left to right is labeled as door 1, side 1. If a second door is visible, then it would be referred to as door 2, side 1. Windows are labeled in the same manner: for example, window 1, side 1. The labeling would also continue from side to side. When windows and doors are present on multiple floors or levels, the labeling must include the level. The first window on side 1 on the second floor would be labeled as window 1, level 2, side 1. Tactical teams that work and practice together may communicate quickly by referring to the window as window 1,2,1. Some doors are normal style entrance doors and some are roll-up or bay doors. This can be handled simply as referring to the roll-up door as a bay door.

This labeling of the outside features of the building plays a vital role in communications during an emergency. Schools, like any facility, should have an emergency response team. Each member should have a clear role to play. One such member is the driveway tender. This position awaits responding agencies and helps direct them to the nearest response point. The most important duty during any emergency is personnel accountability. The first thing first responders will need to know is whether or not all people are out or accounted for by school administrators. Roll call is a complex duty to perform correctly. It begins with evacuation route planning, shelter location planning, training, and drills. Each section or pod of a school should have an emergency coordinator that assists the others and is the last one out. Once the sheltering event or evacuation is accomplished, knowing that all of the employees and students are accounted for is critical. The emergency coordinators for each section can then report to a central person with the overall responsibility for roll call.

Another basic position is the person that can have the blueprints, alarm panel labeling information, and grounds and facility mapping. In a school setting, this person may also be the one to shut off critical operations such as gas flow for certain emergencies. Facility mapping and other information may also be prestaged in evacuation kits. The facility may need one person to contact the media and possibly parents or other key people from the community. The principal may elect to fill one or more roles mentioned and is the person of central decision making. Establishing the response team and its roles allows for more efficient response. It is also a good consideration to make the roles redundant. More than one person should be trained for each position in the event that one is absent.

Educators are adept at announcing public information to relevant media sources, such as communications regarding snow days. Occasionally, an incident will occur that cancels school after it is too late to be effective in disseminating public information. A digital message board may save a school employee or principal from standing outside and informing arriving employees and students of the situation. Employees are much easier to keep informed because call chains of communications and texting can be utilized. Planning the dissemination of current information will also facilitate response.

Determining the hazards that require planning is a matter of historical research for weather experience, natural disaster experience, school experience, state law, and experiences from around the country. The test is whether or not a "reasonable person" would have fear or concern for the possible occurrence and whether or not other schools are planning or making preparations for response or prevention. It does not stop there. One important and overlooked source of information is the Tier II and emergency action planning that local businesses must perform or update annually and submit to the local emergency planning committee (LEPC). The LEPC is a coordinated partnership between local first responders, municipalities, key organizations, and local industry. Schools should be represented on this committee and form partnerships with local agencies that could or will respond to an emergency. Seeking active participation from local agencies will allow for more effective planning for the school and will also allow for a team approach to any incident between the responding agency and the school.

The school should form a relationship and request information on the Tier II report and the community emergency action plan. This will allow for the school to understand the possible hazards that may occur and impact the school if a nearby industrial accident occurs. A school may possibly be asked to shelter in place during a chemical leak or other catastrophe. The Tier II report lists the hazardous chemicals or highly hazardous chemicals on site at the factory if the amount is ten thousand pounds or more. The emergency action plan or "Tab Q 7," as it is titled, will contain a mapped radius for worse-case evacuation radius based on a given chemical inventory. Active participation with an LEPC also allows for the school to be involved in table-top exercises and community drills. This is not just for physical use of school grounds but also for school personnel to become more experienced for handling an actual event.

The school may also have an inventory of chemicals or other hazardous materials that require management. Chemicals must also be examined for spillage or mixture possibilities and the hazards associated with a mix occurring. The Environmental Protection Agency (EPA) has chemical management software available free that can be downloaded. Use of a software program called computer-assisted management of emergency operations (CAMEO) allows for a profile of the facility to be developed. This software is also relied upon by many safety managers in industry and with first responders. The necessity for the management of hazardous materials also includes the need for a fire prevention plan for the school.

FIRE PREVENTION PLANNING

The Occupational Safety and Health Administration (OSHA) requires a fire prevention plan for the protection of employees. The Code of Federal Regulations under 1910.39 lists the minimum requirements for such a prevention plan. It is required for all general industry workplaces except for mobile stations such as vessels and vehicles. All OSHA state-plan states must cover government and municipal employees. There may also be state law in a federal-plan state that adopts these regulations for public workers or develops its own standards. Regardless, it is a best management practice for any facility to incorporate these minimum requirements.

OSHA does recognize orally communicated plans for employers with ten or fewer employees, but as a general rule, written plans will typically be required of schools because of the population of employees. Written plans are documented, are more formal, and are taken more seriously by the employees and in turn are more effective.

The first requirement is to create a list of all major fire hazards (29 CFR 1910.39). Major fire hazards would consist of the storage of large quantities of combustibles such as paper, heaters, boilers, electrical panels, flammable liquids, ovens, stoves, natural gas lines, utility entry points to the building, and storage of chemicals.

The second requirement is to create a list of all hazardous materials [29 CFR 1910.39(a–d)], which should also include the material safety data sheets (MSDSs) as an appendix to this fire prevention plan. MSDSs can be requested from the manufacturer and must be provided by law. The sheets provide the make-up of the substance, the hazards involved with it, the first aid and response to an incident involving the substance. A hazardous material is any substance, or combination of substances that has the potential to harm a person, animal, or the environment. Because it is also a combination of substances, proper handling and storage for each is required to be written. The MSDS will contain information meeting these requirements. You can also download CAMEO from the EPA website. This software will allow you to compare mixtures. When a mixture is volatile, you can plan for separate storage and must consider spill control.

A list of potential ignition sources and controls for them must also be listed (29 CFR 1910.39). Ignition sources are open flames such as a pilot light, source of heat such as an oven or dryer, and even standard light switches when in a storage area for chemicals. In this list, you must also list the sensors, alarms, or controls that alert or automatically shut down the source when the heat or a fault occurs.

For each of these special hazards, the required fire suppression system must be identified (29 CFR 1910.39). Common examples are the suppression systems installed above commercial cooking grills and stoves and the systems protecting computer rooms and assets. This fire suppression system would also include basic fire extinguishers.

There must also be procedures for controlling the accumulation of combustible materials, combustible wastes, and flammable materials (29 CFR 1910.39). This is more than timely emptying of trash receptacles. It will also include good inventory practices such as timely ordering to avoid over-storage of paper products or chemicals, or storage requirements such as keeping any item at least 18 inches below sprinkler heads, limiting the amount of student work exhibited on walls, and purchasing fire-resistant paints and waxes for the walls and floors.

There must also be a list and schedule of the regular maintenance for the safeguards, sensors, and controls for heat-producing equipment (29 CFR 1910.39). This list must also identify the person responsible for the regular maintenance (29 CFR 1910.39). This is a key component of controlling ignition sources.

There must also be a person designated to control fuel-source hazards, which include combustible and flammable substances (29 CFR 1910.39). This person must audit the grounds and schedule waste removal and storage activities.

HAZARD ASSESSMENT

The key to any emergency response is the knowledge of what to cover. This assessment is a matter of weather history, natural disaster history, educational practices, cultural trends, illness trends, and school experience. Examining each category will reveal a comprehensive set of scenarios that may be preplanned. State law may list required emergency events that need to be planned for. However, this is a minimum starting point.

The National Weather Service has much information for your area about the pattern or history of weather events such as tornadoes, floods, annual snowfall, and rainfall. There must also be a certain amount of planning for any possible event. You can access many statistics of weather concern through the National Weather Service at www.nws.noaa.gov/om/hazstats.shtml. Generally, most districts will cover plans for tornadoes, floods, and snow. Certain areas will have hazards common to them as well, such as hurricanes for coastal areas. Planning for these weather events for a school means that transportation routes, sheltering, scheduling, and announcements must be addressed. Schools may also be an important part of community response efforts for large weather events. Sheltering, for example, may be planned in case students could not be returned to home for flooding and require shelter at the school temporarily. The school may also become a community evacuation shelter for such cases as flood, tornado, or massive power outages associated with snow and ice storms.

Natural disasters such as earthquakes and wildfires must also be considered as to their possibility and history. The U.S. Geological Survey can provide real time assessments of earthquakes and provide historical data as to likelihood and severity. In reality, we make a general plan based upon the possibility that it could occur. Earthquakes are a possibility in many regions and should be standard components in response planning.

The emergency response plans from other districts in the region are also a helpful place to initially help in identifying topics to be covered. If another district addresses the issue, then it is an educational practice and should be considered regardless of actual history. Today many schools plan for lockdowns in case of escaped convicts, violent crimes, or violent actions on any part of the school grounds. This was not a standard practice years ago, but it shows a trend that has become an actual concern across the country.

Many schools deal with pandemic prevention. The quick spread of illnesses occurs because of numerous cross-exposures in a school environment. Schools are increasing efforts to prevent widespread outbreaks in order to keep enough students and employees at the school in order to make academic continuity feasible. A number of strategies can be utilized to counter an outbreak. Medical intervention is one strategy, such as providing vaccinations to employees and students or at least advertising them in partnership with the local health department. At certain times sick leave policies or attendance policies should be adapted or suspended in order to prevent a sick employee or child from attending. Educational efforts must accompany any such endeavor. Issues such as teaching children and employees to stay at home if they have a fever are important.

Other educational efforts must be commonplace as well. These efforts might include covering the mouth and nose when coughing or sneezing, regular washing of hands, regular usage of hand sanitizer, and other common methods to limit exposures. Housekeeping can include the spray of products that will kill viruses and bacteria. Regular washing and disinfecting of tables, desks, door knobs, and phones can lower exposure. Key personnel can also work from home with computers and technologies available through the Internet. Alternate schedules could also be utilized. When outbreaks of an illness are occurring and a closing is considered, education can still continue with homework packets that can be completed away from school. Whereas this situation would certainly not count as attendance days for public schools, the educational impact might be minimized.

School experience is also a consideration. It would certainly be negligent to ignore reports or stories of past events or conditions. Criminal acts such as bomb threats are certainly commonplace. The pulling of fire alarms under false pretenses, criminal mischief, assaults, or any other type incident that is a crime, security breach, or liability exposure must be tracked. This allows for the emergency response and security procedures and plans to be living documents.

PLAN OUTLINE

School districts are capable of producing simple, easily understood plans that can be adapted for dissemination to parents and students. The remainder of this chapter is dedicated to giving some additional thought and practices to certain topics. It can be used to help develop or improve an existing plan template.

RESPONSIBILITIES

The responsibilities section is the first section of the plan. The responsibilities section specifies each position by job title and lists in a quick reference format the specific duties of each position. The reasoning behind such a section is to summarize the duties for a person who wants to know what they have to do in case of an emergency. It can also serve as a training tool and as an objective gauge for grading performance in exercises or drills.

COMMUNICATION

Establishing an emergency action board or establishing a rank structure for such duties and authority is a good practice. Many duties must be performed simultaneously, and having an established structure helps organize the incident. It requires that a policy be established that determines when the board will be activated. The superintendent or other administrators should also be included in the plan. The emergency action board becomes the central reporting and control center. This concept somewhat mimics the structure of the response effort to a natural disaster established by the Federal Emergency Management Agency in the National Incident Management System.

Schools might have a site-based council. This council should be included as part of the emergency action board. This would not preclude the principal from being in charge or making decisions until the entire board is assembled. The board is established in order to help organize and assemble information and to disseminate information to the community.

The plan will also have to establish communication procedures. Communications can be person to person, with radios, or possibly with text messages or cell phone communications. This may be needed as the building evacuates, critical machinery or processes are shut down, or students and personnel are being treated or accounted for by school administrators.

There may also need to be special communication procedures for certain situations. A special code in front of a communication can signal a particular situation or needed response without warning in plain language. For example, police officers will have a special code at the beginning of a radio communication to signal that they are being held hostage by an aggressor. Special codes may also exist in department stores where security needs to lock down exit doors, and additional persons should respond to the exits. This is a standard practice with a missing child or possible abduction.

The national threat advisory system or the color-coded signal that indicates threat levels is another example of special communications. Many see this kind of system as not effective. However, this allows an organization to tighten security. When credible threats occur, key personnel can be assigned additional security checks and increased frequency of checks, and protocol can be adjusted. Credible threats might include terminations, threats from an aggressor in the community, bomb threats, or criminal acts nearby relayed by law enforcement.

EVACUATIONS

Schools are usually very efficient at fire drills or other evacuations. The routes for these drills must be mapped and posted, and the drills should be orderly and have adults assigned to assist and lead students and other adults in their groups or pods to safety. A primary route as well as a secondary route should be established and practiced. The design of the school will also play a large part in evacuation.

Exit signage should be visible from any position in the hallways or courtyards [29 CFR 1910.39(a–e)]. It is important to have emergency lighting and signage or reflective signage at a low level to aid in an evacuation during fire. Fires have an additional hazard that some overlook, and that is the toxic nature of the smoke. Although it is not usually thought that evacuations of a school may involve a thick buildup of smoke, we must prepare for worst-case scenarios. Furthermore, the reasonable expectation of an attack on a school is increasing.

The evacuation point or rally point is the location that all evacuees will move toward and assemble in. The wind direction, security, and access must also be considered when choosing a rally point. Consider the predominant direction of the wind. If a fire or spill necessitates evacuation, the wind could carry the toxic smoke or dust toward the people. It is also a concern to keep the evacuees secure and safe from traffic, leaving without notice, or being attacked. Parking lots are not a good first choice. The area should also be accessible and avoid being in the way of responding agencies. If the event is ongoing, and dismissal is necessary, parent traffic should be controllable and not impede responding agencies. Consider having a large enough and safe enough area to land a helicopter for the seriously injured. It may be difficult or impossible to have the perfect rally point, so it must be possible to move or migrate it as the situation demands.

Additional hazards may be present and should be avoided as possible rally points. Any location near where natural gas lines cross should be avoided. Fires or explosions can cause rupture or explosions down the line. Furthermore, if the building is evacuated for an earthquake, the vibration may cause a line rupture, and the gas will travel along the utility line or path of least resistance until the gas can escape or becomes ignited. Electrical lines are another utility to avoid. Falling lines or poles are a hazard to avoid.

BOMB THREATS/SUSPICIOUS PACKAGING

Bomb threats do occur at schools. They are usually false alarms. However, they must be taken seriously. The strategy here should be one of receiving the threat correctly, assembling a team for suspicious package search, increasing security, and providing the safest option for the students depending on the situation (Department of Homeland Security, 2003). Bomb threats may become law enforcement scenes because of the crime involved with the threat. Responding agencies will rely heavily on the school's team.

Those seated at phones that have access to outside lines should have a checklist available at the phone for the person who receives the threat to reference for questions to ask the caller and notes for what to pay attention to when the threat is communicated. Additionally, each person who is the primary user of the phone should have training on receiving such a threat and how to initiate an organizational response (Department of Homeland Security, 2003).

The threat level communication can be vital in such circumstances. For example, if school officials had credible intelligence that threats could occur ahead of time, then security could be tightened, helping assure that the threat is minimal. Student walk-outs or teacher strikes might be times when this could be expected. If security was tight and comprehensive enough, then evacuation and major disruption may not be needed. Otherwise, evacuation would have to be considered. However, it is not an automatic decision.

A bomb threat can be a diversion for the real intent, even if the real intent is business interruption. It could be a diversion for an elaborate plan to abduct someone or cause confusion for an unauthorized pickup. This does need to be considered and is a factor in determining the rally point.

As a general rule, evacuation must occur until the threat is not considered credible. If security could realistically rule out a planted device, then evacuation would not be required. This is not usually the case in a school environment. A team or teams that are very familiar with the school facilities must be assigned to search or aid in the search for anything suspicious. The details of the threat may determine that a search is not safe if a timed deadline is too close. It may have to wait until the time passes. Teams that search their assigned areas of the facility should also be aware of and insist on routine enforcement of policies that deal with proper placement and storage of personal items, lunch items, and equipment. Having an orderly building with employee lunches kept in designated areas only, personal items for teachers and staff kept in designated places, student items in their normal places, proper storage of equipment and chemicals, proper labeling of stored items and containers, and maintaining good housekeeping practices all play a specific role in security and emergency response.

In case of an evacuation for a bomb threat, it advisable to turn off any two-way radios and prohibit cell phone usage. These devices may inadvertently cause a detonation for some devices. This is a standard practice when explosives are in use on construction sites or during military training.

A suspicious package is any item that is out of place. There are some signs to look for, such as an odor from the package, any leakage, or noise, unusual handwriting, or cut-out-type letters for words. Packages left in restrooms or discovered in trash cans and unlabeled boxes or substances are good clues. Package delivery or mail delivery are two vulnerable entry points for delivery of an item of ill intent. Vendors must be checked in and compared with a schedule of their arrival. If a vendor arrives unscheduled, it requires a call to the company for verification.

Mail requires a delivery point that is not an exposure point to numerous people. One person should have the duty of retrieving mail and performing an initial inspection prior to exposing others to it. There are several signs that mail packages may contain harmful substances. Mail that has not been stamped or processed via the mail service is one such sign. It should be immediately isolated and reported to authorities. It could have been placed in the box before official mail arrival. Mail that has excessive postage, an abundance of misspelled wording, typing that is of several fonts, cut-out letters glued to the package to form address information, the absence of a return address, unrecognized names, bulky items, packages with excessive tape, or packages with leakage, noise, or odors should be treated as suspicious. Although these signs are typical for suspicious packages, any one alone may not mean the package is harmful (DOCJT, 2003).

The single mail contact for the organization would likely be required to undergo quarantine after exposure to a suspicious package until any substances in the package could be identified. However, this is a much better option than having to treat or assume exposure to several or all building occupants from haphazardly bringing in such a package.

OTHER RESPONSE TOPICS

There are other situations for which schools must plan. They include dismissal and parent/guardian pick-up procedures. Usually only custodial parents and guardians are allowed to pick up a child from school. However, most schools allow the custodial parent/guardian to list other authorized adults for pickup. Some schools even issue vehicle placards to help determine whether the vehicle for parent pickups at the end of the day is driven by an authorized person. Another good practice is to ask students and train them to indicate whether the adult they are being released to is the right person. School buses also now may have radios connected to a central communication point. This helps when parents are not visible at a drop off point. Today many districts require elementary children to have a parent meet the child at the bus stop. School bus radios also aid in times of mechanical failure, vehicle accidents, and unexpected delays.

Tornado reaction is also a topic that must be covered. The emergency response plan must list the evacuation points, document drills, and determine when shelter will be sought and how the school will be alerted to an advisory. Usually, a weather alert radio will be in the front office and in another redundant location to ensure that someone will be within sensory presence of the radio. School computers can also have homepages connected to a local airport or weather site for up-to-date information.

A list of media contacts and media stations for mass school announcements must also be disclosed in the plan and to parents. For example, deciding to close because of weather requires a county-wide team approach. Who will be utilized to decide on travel safety? How will this be decided? Will the bus drivers be called on for this information? Can the county and state road crews be contacted to help in this decision? Will those individuals asked their opinion on the roads be trained to some extent? These are all questions that must be answered in order to properly plan. Alternate bus routes or schedules might also be required.

CASE STUDY

You are the new principal for an elementary school in a small community. Last year, a nearby bank had been robbed in the early morning. A suspect entered the school building posing as a parent in attempt to avoid police. He was not armed at the time he came into the front office. He sat in a chair waiting for someone to assist him. An officer who came to the school to inform the principal saw the suspect and apprehended him without incident. However, the incident painted the school as ill prepared for handling emergency issues. Part of the reason for the dismissal of the previous principal was this incident.

- Could such an incident be prevented? If so, how?
- What issues would you address in designing an emergency response plan associated with such an incident?

EXERCISES

Consider the following additional circumstances:

1. The superintendent has asked that you assess the school's emergency preparedness status. Present an audit or set of assessment criteria that you would consider in order to assess the school's preparedness.
2. The superintendent has also asked you to address the concerns that parents and the community at large have at the next board meeting. Draft a letter outlining the actions you will be undertaking in order to address emergency planning and response. Be specific in terms of access control and situational awareness.
3. In Kentucky, state laws exist covering minimal procedures for tornadoes, earthquakes, and lockdowns. Research your state's laws or access Kentucky Revised Statutes for lockdown procedures. Draft a school policy for general lockdown procedures.

4. What items would you list to be included in emergency management kits, where would you locate them, and which personnel would be assigned to maintain and have the kits present?

REFERENCES

U.S. Department of Criminal Justice Training. (2003). *Non-Explosive Threats*. U.S. Department of Criminal Justice Training, Richmond, KY.

U.S. Department of Homeland Security. (2003). *Bomb Threat Response Study Guide*. Federal Law Enforcement Training Center, Glynco, GA.

U.S. Department of Labor. (1996). Fire prevention plans. Title 29 Code of Federal Regulations, Part 1910.37. Occupational Safety and Health Administration, U.S. Department of Labor, Washington, D.C.

U.S. Department of Labor. (1996). Maintenance, safeguards, and operational features for exit routes. Title 29 Code of Federal Regulations, Part 1910.37. Occupational Safety and Health Administration, U.S. Department of Labor, Washington, D.C.

24 Emergency Response and Situational Awareness

James P. Stephens

CONTENTS

He is most free from danger, who, even when safe, is on his guard.

Publius Syrus
First century BC

Each day, schools across this nation must be prepared for potential hazards and situations that may arise. Preparing an emergency response requires many actions in various areas. These actions of planning and preparation are vital to the success of any emergency response in which you will engage. Each school district must develop an emergency response plan that is individualized to their specific location and identified hazards, as well as those hazards that exist for all entities.

For example, a school in Kentucky does not have a great need to prepare for a response to a volcanic eruption. However, a school in Mt. Rainier, Washington, has a great need for such a safety plan. With the adoption of an all-hazards response plan, the National Incident Management System (NIMS), and a school (or student) emergency response team, the pathway to an emergency response is paved. The success of your emergency response is greatly increased by implementing these preparedness plans.

279

Within the U.S. Department of Education's guidelines for emergency management, there are four phases that must be addressed: mitigation/prevention, preparedness, response, and recovery (U.S. Department of Education, 2007). Our focus for this chapter is the third phase: the emergency response to an incident at or around your facility. We will not focus on individual responses to specific emergencies but will center on various aspects of emergency response preparation and steps taken in an emergency response. One specific area for emergency response preparation that we will discuss in length is situational awareness (SA) training for all employees. Improving the awareness level of all stakeholders will increase the likelihood of a positive emergency response.

TIP-OFF

Imagine going into a championship basketball game without prior scouting and team preparation. In order to appropriately prepare a winning defensive and offensive strategy, a good coach will watch video or personally watch the future opponent. We can relate this scouting effort to the risk identification actions taken in preparing an emergency response plan.

Following the receipt of an opponent's scouting report (hazard identification), the coach must either create a new "game plan" or polish one that already exists. Furthermore, he may have to blend an old plan with a new plan based upon the strengths and weaknesses of the opponent. In preparing emergency responses to your identified hazards and risks, you may find yourself creating a new plan or improving a current plan for your establishment. The path you choose is dependent on the identified hazard or risk and the individual needs of your facility.

When the coaching staff has reviewed the opponent (hazard or risk) and created a game plan (emergency action plan), they begin the process of implementing the plan through various practice techniques and methods (drills). They understand that in order to properly carry out their plan, they must mentally and physically rehearse the actions they will take. In preparing an emergency response in your setting, you must also engage those within your facility in mental and physical rehearsal and practice. This type of preparation can be achieved through mental rehearsal through situational awareness training for your employees. SA training is comprised of the following steps:

- Perception: accurately seeing things around you, such as people, objects, and vehicles
- Comprehension: understanding the elements of the situation as you observe them and as they relate to the overall situation
- Projection: thinking ahead and formulating a plan in response to the things you see and comprehend in the situation; this also relates to the ability to foresee the outcome of the situation

One's ability to avoid dangerous situations or respond effectively is often inhibited or enhanced by one's ability to maintain vigilance and perceive a risk. Failure to do either effectively adds to the chances of the interjection of human error into an already dangerous situation (Pantic, 2009). We will discuss this type of training in depth later in this chapter.

Another proven method of preparing a successful emergency response is through table-top exercises, which is a cost-effective method in emergency response preparedness. This process typically includes: orientation, table-top exercise, critique, and conclusion (Holloway, 2007). This is similar to an intersquad scrimmage game that the basketball coach implements to observe how well the players understand and execute their actions. The goal of any preparation exercise or training is preparing those who may be involved in understanding their roles and actions necessary to properly respond to the hazard or risk, before those actions are required.

RESPONSE

"Taking action to effectively contain and resolve an emergency through the implementation of the emergency management plan" is a general definition of emergency response (U.S. Department of Education, 2007). A key to this definition is that preparation must be provided for actualizing an appropriate and effective response. Looking at and implementing this definition of response requires a comprehensive emergency management plan and practicing and coordinating that plan. Emergency response must be undertaken in individual and team approaches as the situation dictates. It is imperative to recognize that every emergency situation is different, and your response must remain flexible to allow for those surprises that may occur. Therefore, using the words "never" and "always" within the emergency management and response context is unwarranted, and both words must be absent from your preparation and response mindset (Basic Tenets of Emergency Response, 2011).

EMERGENCY ASSESSMENT: UNDERSTANDING THE HAZARD

Before emergency response is initiated, the situation must be recognized as a hazard and comprehended correctly in order to choose the correct response procedure. If someone incorrectly perceives a risk to warrant an evacuation of the facility, when it is actually a hazardous material situation outside the facility, it would create a more dangerous situation instead of reducing the impact of the emergency. Therefore, all stakeholders in your school must be properly trained to perceive and comprehend the hazards and risks to which they are vulnerable.

In an effort to provide more educational training on an employee level for mitigating emergency situations in their facility, an improvement in the overall thought process with the employee is necessary. One must increase preparedness and be cautious not to generate paranoia. The biggest distinction between the two is that preparedness asserts a heightened state of readiness, and paranoia is unreasonable suspicion with an element of fear. We do not want to create fear; we should seek to create a state of full readiness and vigilance. If we can reduce the likelihood of the employee being surprised by a violent event or other emergency situation, we increase the chances for that individual to react in a more controlled manner. Situational awareness reduces the reaction time of the individual, thus allowing them a greater chance of mitigating the threat or hazard they face (Gonzales, 2004).

One such method of empowering your employees to mitigate any emergency situation is through situational awareness. SA in general terms is the training of the mental capacities and thought processes to become in tune with the individual's environment and surroundings. Although this concept is certainly not new, it has failed to reach the masses within most places of employment. Situational awareness is an important skill in various professions such as military, fire, police, medical, and aeronautics and is required for the survival of those who work in these fields. Not only is this mentality and skill set utilized while at their workstation or area of responsibility, it becomes a way of thinking and behavior each and every day. The benefits of situational awareness are observed in emergency situations and will play a large role in reducing overall exposure to injuries during the response to an emergency. In creating a more alert and aware employee, you grant them the tools for recognizing various workplace hazards and implement a successful emergency response.

It would be naïve to believe that any training or safety program will guard against all emergency situations all of the time. However, improving one's awareness of the environment will increase the probability of mitigating and surviving an emergency condition through quick perception and comprehension.

HISTORICAL BACKGROUND OF SITUATIONAL AWARENESS

Nineteenth-century philosopher-psychologist William James provided early insight into the processes of the human mind. His early work into the ability of the human thought process to engage in "selective attention" is significant within the situational awareness process. This is a key to understanding situational awareness and the ability to train the mind to draw attention to various items within the environment. With many other philosophers and psychologists studying the thought process within the human brain, understanding of situational awareness continued to grow (Smith, 2003).

During the 1930s and 1940s, the need to improve the cognitive abilities in military aircraft personnel surfaced. However, the concept of situational awareness as we recognize it today was developed in the 1970s with primary usage within the military community. Although there are many definitions of what situational awareness is, there are two definitions that should be used in an attempt to maintain a simple working understanding:

- SA is the ability to "maintain the 'big picture' and think ahead" (Dennehy and Deighton, 1997).
- SA is defined as "the 'operational space' within which personal and environmental factors affect performance" (Dennehy and Deighton, 1997).

These definitions provide a basic understanding in addition to numerous philosophical and psychological definitions and studies in the world of situational awareness that are very complex. The crux of situational awareness training with military pilots is the enhancement of the pilot's ability to cognitively process large amounts of information from their surrounding environment, hence reducing the time needed for them to formulate and actualize a successful response to any situation.

Although the majority of sociological and psychological research on situational awareness relates to aeronautics, the ideology of situational awareness is integrated in several professions. Having a heightened sense of awareness to their environment is also a major asset in the daily life of fire, police, and other emergency service providers. In the life of a law enforcement officer, situational awareness is so embedded within their psyche that it is often practiced without conscious thought. Failure to keep a heightened state of awareness for the officer can easily lead to a situation with life and death consequences.

IMPLEMENTING SITUATIONAL AWARENESS

Employees can be assisted in advancing their abilities to become more aware of their surroundings. Techniques can be taught that will change the way they go about their daily activities, and what they observe in their workplace can improve their response in an emergency. A mindset to maintain a "relaxed" situational awareness state can be developed that is in equilibrium between paranoia and preparedness. The following information presented to faculty and staff regarding situational awareness can answer many questions about SA.

As mentioned in the beginning of this chapter, it is formally recognized that the mind has the ability to draw attention to specific elements or information during any situation. We can train our mind to look for specific warning signs of danger and have a formulated plan of action for various scenarios or situations we encounter. Failure to perceive danger is one of the most common problems in a hazardous situation and often leads to injury or death. The situation-aware employee will have a better ability to perceive an emergency and have a formulated plan of action already in place. This will lead that employee to transition their immediate response to the implemented emergency management plan for their location.

One of the first steps in achieving a situational awareness mentality is to realize and accept that hazards and risks surround us. Again, a state of constant fear and paranoia is not the goal and is counterproductive to personal safety and emergency response. However, it is often difficult to get individuals to realize that they must maintain "relaxed awareness" that allows them to enjoy their daily life yet remain prepared for those unexpected events that may come our way. Complacency, apathy, and denial will prohibit an individual from becoming attuned to their surrounding and being able to recognize an emergency situation in a quick, responsive manner (Burton and Stewart, 2007).

At the other end of the spectrum from complacency, denial, and apathy is a state of heightened awareness. The body is not designed to remain in a constant state of high vigilance and preparedness, and this condition is not sustainable. The ideal condition for a "relaxed awareness" level is referred to as condition yellow (Grossman, 2004). In the book *On Combat*, Lieutenant Colonel David Grossman goes into great detail regarding the various physiological states that the body experiences. He contends that the optimal state of preparedness and readiness is condition yellow. At this level, the heart rate is just above a normal resting heart rate of 60 to 80 beats per minute, yet below 115 per minute, which is the level at which a deterioration of fine motor skills begins to occur. When an individual is faced with a highly stressful situation, certain physiological changes take place. A person who is not in a state of readiness will often freeze in an emergency. When the body goes from a completely relaxed state to a highly vigilant state in an instant, it causes the unprepared to lose their ability to respond.

PERCEPTION TRAINING

One basic principle to implement into the mindset of faculty and staff is: "If it doesn't look or feel right, then it probably isn't." It is imperative to encourage and convince employees, students, visitors, and all individuals to eliminate their reluctance to report vital information to the authorities.

The practice of accurately perceiving the elements and actions in your environment is largely dependent on an individual's knowledge base of what is abnormal or destructive behavior or circumstances. It is important for us to realize that one cannot recognize something that is out of the ordinary if one does not know what ordinary is. Knowing what is normal or ordinary in your school is the basis for situational awareness. Recognizing out-of-the-ordinary behavior in a coworker, student, or other associate is vitally essential in preventing personal violence. What we want to avoid in the safety management profession is for employees to say, "I didn't see it coming" or "I have no idea how that happened." We want employees to work in a heightened state of readiness for added safety in the workplace.

You should provide each employee the opportunity to do a thorough walkthrough of the school with the safety manager or other designee to get a good understanding of what the environment should look like in a normal state. As we just discussed, one cannot recognize or perceive abnormality without first knowing normality. During this walkthrough, the employee should identify and point out any specific areas where a greater potential for danger may exist and what the normal state looks, sounds, or feels like. During the inspection, make sure to reinforce the fact that anything that does not fit the norm should raise their awareness level and should be reported. This educational and training time should be conducted during an initiation setting for any new employees and immediately for existing employees, with an annual update for all employees. Updated informational sessions are also necessary if any changes occur in the workplace.

Building a baseline of normalcy is also necessary when interacting with coworkers, students, or other individuals. Supervisors and coworkers must pay close attention to a person's behavior at the onset of interaction with them. Although identifying risk factors in personal behavior is imperative, it is helpful to know their normal behavior in order to perceive a change in that behavior in a timely manner. "If you do not recognize the hazard, you cannot control the hazard. If you cannot control the hazard, you cannot prevent the injury" (Logsdon, 2008).

COMPREHENSION TRAINING

Strengthening an individual's ability to comprehend what they have perceived is the next step in acquiring a stronger situational awareness mentality. Some of the information that we just discussed regarding recognition blends in with the comprehension aspect of situational awareness. Keep in mind that when talking of these different levels, there are no clear end points and starting points when moving between levels. The transition from one to another resembles a smooth blending of behavior and activity.

Children are normally born with little understanding or concept of danger. Just as a child will have to learn through experience or education what danger is and what produces a negative result, faculty and staff must learn this as well in the context of the school setting. However, learning from experience is not what we want or encourage in the workplace. Learning to comprehend perceived danger, leading to the formulation of a viable and effective response, is our ambition.

In the workplace, we will build upon any experiences employees may have combined with additional educational efforts. Through these efforts and the enhancement of comprehension, the employee is preparing an effective projection of the outcome of the situation and formulation of a plan for response.

PROJECTION: PLAN FORMULATION

As an individual perceives or recognizes data or stimuli and then conceives the input as a danger, they must move to the projection of the outcome and formulation of an action plan. However, one should not wait until this time to begin mental and/or physical preparation for a response. Although you cannot train for every situation that you may face, you can formulate and prepare a plan for the most common dangers and hazards you may face. This is also true in the school environment, where faculty and staff should know the highest risk locations, activities, and situations they may face each day.

Mental preparation for a quick and effective response in an emergency is often the formative aspect of the situational result. Developing a "when-then" attitude will enable you to mentally develop hazardous scenarios, with a correlating response. The popular ideology of "if-then" can no longer suffice in the day in which we live. It is critical in situational awareness for one not only to recognize and comprehend a danger; they must also have a mental plan formulated. We all know that action is always swifter than reaction. One method of reducing your reactionary gap is through mental and physical rehearsal of your actions in preparation for certain high stress situations.

We briefly spoke of implementing table-top exercise training into your overall safety and emergency response training. These training opportunities implement the type of situational awareness enhancement we discussed. This training introduces the stakeholders to the chosen emergency and allows them to formulate a plan of action. These low-cost training opportunities allow you to develop a scenario of your choosing that is relevant to the location of your school, conduct the exercise, and then conduct a critique on each participant's proposed emergency response (Holloway,

2007). Now that an emergency situation is perceived and comprehended, the emergency response must move to the next phase, which is emergency notification.

EMERGENCY NOTIFICATION

In order to begin the proper emergency response to any given situation, quick and accurate notification to all stakeholders is imperative. An example of how quick and accurate notification to an ongoing emergency can save lives comes from the response on September 28, 2010, on the University of Texas campus in Austin. Police cite a quick response and notification process from the faculty of the university in saving lives of those in the area (Mulvaney and Garrett, 2010). Within fifteen minutes of first notifying the police of the active shooting situation, multiple avenues of notification such as sirens, e-mails, the University of Texas website, and other outlets warned the student body of the impending danger. This incident highlights the need for multiple methods for emergency notification.

With a growing dependence upon electronic communications, one may be tempted to rely solely upon this method of emergency notification. However, it is important to develop and utilize other means of communication and notification in the event of an emergency. It is very hard to imagine someone not owning a cell phone in today's society, but not everyone owns or carries a cell phone to receive instant messages or emergency notifications. In developing your notification system, you must not forget individuals such as low-income students who may not have this access (Galuszka, 2008). In a campus setting, there are also those who will not take the time to sign up for the instant emergency messaging service offered by the institution. Therefore, even if instant messaging is used in an emergency situation, many would still not receive the alert. This would likely include visitors to your campus who would not necessarily have a need to register their cell phone for your emergency alerts.

It is important to note that whatever method you choose for emergency notification, all stakeholders must be aware of the means of notification and what they must do when notified. This is a primary reason that the Federal Emergency Management Agency (FEMA) recommends plain language in lieu of codes when making your emergency notifications and communications. This allows a clear understanding of the emergency to all within your facility who may not know the codes you are using (FEMA, 2006).

Furthermore, you must have a distinct notification for various emergency situations and not a general all-hazard warning system. You may have an initial siren or tone that you utilize for various situations; however, after the warning signal is given it must have follow-up with further instructions and information regarding the emergency situation.

According to 29 CFR 1910.38(d) within the Occupational Safety and Health Administration's regulations, "[A]n employer must have and maintain an employee alarm system." The employee alarm system must use a distinctive signal for each purpose and comply with the requirements in 1910.165 (U.S. Department of Labor, 2011b); 29 CFR 1910.165 specifically deals with employee alarm system requirements for fire emergency response (U.S. Department of Labor, 2011a). It is imperative that all facilities and districts follow these requirements for emergency notification.

Having a reliable and viable method of notification and communication is only one part of a successful notification process. You must also have a clear, streamlined policy in place that begins the notification process. When an emergency is observed, time is of the essence to warn others of impending danger and begin their emergency actions.

In the full report of the Virginia Tech shooting on April 16, 2007, an area of critique was the delay in emergency notification on the part of the university. On the day of the shooting, it was the policy of Virginia Tech to assemble a policy group to determine whether an emergency message was necessary and then determine what to say. The Virginia Tech Review Panel (2007) also states that their emergency action planning did not account for an active shooter situation.

Regardless of the emergency you are faced with, the timely and accurate notification of those within your school and other stakeholders is crucial to a successful emergency response. Take the time to evaluate your emergency notification and communication plans and process. Ensure that everyone knows what your alarms identify as an emergency condition, and then move into the necessary measures to eliminate or mitigate the hazard.

TAKE ACTION

Once an emergency situation is identified and notification is made, it is time for individual and collective action. All of the efforts put forth during the days, weeks, months, and years of emergency planning come down to this moment. Your actions will be based upon the information that you have at that specific time; therefore, your response may change or evolve as more information is received.

Upon observation or notification of an emergency, it is essential to make a report to the appropriate emergency responders to reduce any delay in their response. Once that notification is made, the appropriate actions outlined within your emergency action plan must ensue. Generally, the following procedures will be implemented:

- Evacuation: the movement of people from a place of danger to a safe location. This may be used in incidents such as a fire, structural damage, or a hazardous materials situation.
- Reverse evacuation: moving people back into the building for shelter.
- Lockdown: a process of moving from a location of possible danger to a known safe location or remaining in a safe location. This may be used for dangerous situations taking place outside of the facility, as well as a dangerous situation within your school such as an active shooter event.
- Shelter-in-place: used when moving outside the building may be harmful or if there is no time for evacuation. During this response, it is common to seal the doors and windows and shut off the air handling devices. This response is normal during a hazardous material emergency outside of your school.
- Drop, cover, and hold: dropping to the floor, covering your head, and holding your position until the danger passes. A time for this emergency action is during a tornado or severe weather emergency.

Whatever emergency response is undertaken at your school, it must be carried out in a calm and effective manner. During this time, leadership is a key element to success. Visitors, students, and others within your school will be looking for guidance in the actions they need to take, and that leadership must come from the authority figures within your school system. This fact must be underscored to all employees throughout the year and reinforced in the multiple training exercises in which they participate. The actions taken in the first few moments of an emergency situation will greatly impact the end result of the emergency response.

ACCOUNTABILITY

Regardless of the emergency situation and the chosen response, the safety and accountability of all within your school is the primary focus. Knowing who is present on your premises is vital to ensuring the safety of everyone. You must know who is in your area of responsibility in order to properly account for everyone in an emergency.

Although schools built after January 26, 1992, are required to comply with the accessibility requirements within the Americans with Disabilities Act, the designers may not have considered the challenges of evacuating individuals with disabilities. Proper planning for protecting students and faculty who have special needs is essential. When planning for emergency situations, you must consider the various types of challenges these individuals may face such as mental, physical, motor, and other developmental limitations (National Clearinghouse for Educational Facilities, 2008).

Various issues must be addressed and taken into consideration as you plan. Keep in mind that you will likely have individuals with temporary mobility issues caused by things such as sports injuries that you must prepare for as well as those with permanent conditions (National Clearinghouse for Educational Facilities, 2008).

All of these issues must be considered and thoroughly addressed prior to an emergency response. While planning your emergency management plan, ensure that this area of concern is discussed and implemented into the final plan. Conduct a careful assessment of your school, and conduct drills to ensure you are capable of assisting these individuals.

CONCLUSION

Throughout this chapter, we discussed the importance of planning for emergency response. In an emergency situation, people will react based upon their knowledge, abilities, and experiences. This is the reason that training such as situational awareness, table-top exercises, and emergency drills are vital to your response. The degree of success in response is generally indicative of the preparation involved, just as the final score in a basketball game is indicative of the many facets of preparation prior to tip-off.

As you move forward in your emergency response preparation, take time to evaluate your current action plan to ensure you have identified the hazards and risks that are specific to your locations. Ensure that you are preparing yourself and all stakeholders to enact the response plans developed for your school. Conduct a thorough review of your premises, policy, and practice for assisting those with special needs. Make certain these individuals are given the necessary means to remain safe in any emergency response.

CASE STUDY

Central High School is located in Minneapolis, Minnesota. The school is situated on a campus that occupies six city blocks and has a population of two thousand students. It is among the newer schools in the Minneapolis school system in that it was constructed in 2009. A flour mill is operated two blocks away from Central High School.

- What primary risks would Central High School administration need to consider when developing an emergency response plan?
- What are some of the unique issues that Central High School will face compared to other high schools throughout the country?

EXERCISES

1. What opportunity should you provide stakeholders in your facility in order to learn the "normal" environment?
2. What is the first step to achieving situational awareness?
3. What are some benefits to conducting table-top exercises?
4. What system of accountability would you establish in your school to ensure that everyone is accounted for during an emergency?

REFERENCES

Basic Tenets of Emergency Response. (2011). Professional Safety, 56(11), 32–33.

Burton, F., and Stewart, S. (2007). Threats, situational awareness and perspective. Accessed July 19, 2009. http://www.stratfor.com/threats_situational_awareness_and_perspective.

Dennehy, K., and Deighton, C. (1997). Development of an interactionist framework for operationalising situation awareness. In *Engineering Psychology and Cognitive Ergonomics: Volume 1. Transportation Systems*, edited by D. Harris. Ashgate, Aldershot, UK.

Federal Emergency Management Agency. (2006). NIMS and the Use of Plain Language. Accessed August 14, 2011. http://www.fema.gov/pdf/emergency/nims/plain_lang.pdf.

Galuszka, P. (2008). Emergency notification in an instant. *Diverse: Issues in Higher Education*, 25(2), 14–17.

Gonzales, J. (2004). Up close and personal: Situational awareness. Accessed July 25, 2009. http://www.scribd.com/doc/4737625/200401Situational-Awareness.

Grossman, D. (2004). *On Combat: The Psychology and Physiology of Deadly Conflict in War and in Peace*. PPCT Research Publications, Belleville, IL.

Holloway, L. G. (2007). Emergency preparedness: Tabletop exercise improves readiness. *Professional Safety*, 52(8), 48–51.

Logsdon, R. (2008). Take off the blindfold: Be aware of your surroundings. Accessed April 26, 2012. http://www.rockproducts.com/index.php/features/51-archives/7145.html.

Mulvaney, E. and Garrett, R. T. (2010). Experts credit UT, police with sparing lives. *Dallas Morning News*. Accessed March 14, 2012. http://www.dallasnews.com/news/education/headlines/20100929-Experts-credit-UT-police-with-6008.ece.

National Clearinghouse for Educational Facilities. (2008). An investigation of best practices for evacuating and sheltering individuals with special needs and disabilities. Accessed July 19, 2011. http://www.ncef.org/pubs/evacuating_special_needs.pdf.

Pantic, D. (2009). Situational safety awareness. Accessed April 26, 2012. http://esvc000491.wic041u.server-web.com/docs/Situational_Safety_Awareness_Feb_09.pdf.

Smith, D. (2003). Situational Awareness in effective command and control. Accessed April 26, 2012. http://www.smithsrisca.co.uk/situational-awareness.html.

U.S. Department of Education. (2007). *Practical Information on Crisis Planning: A Guide for Schools and Communities*. Office of Safe and Drug-Free Schools, Washington, D.C.

U.S. Department of Labor. (2011a). Fire protection. Title 29 Code of Federal Regulations, Part 1910.165. Occupational Safety and Health Administration, U.S. Department of Labor, Washington, D.C.

U.S. Department of Labor. (2011b). Means of egress. Title 29 Code of Federal Regulations, Part 1910.38. Occupational Safety and Health Administration, U.S. Department of Labor, Washington, D.C.

Virginia Tech Review Panel. (2007). Report of the Virginia Tech Review Panel. Accessed November June 16, 2011. http://www.governor.virginia.gov/TempContent/techPanel-Report.cfm.

25 Readiness and Emergency Management for Schools

Amy C. Hughes

CONTENTS

There are many challenges for school systems in providing an environment where students can learn and grow. With the implementation of smart classrooms and new technology, the nearly 50 million students enrolled in schools across the United States (National Center for Education Statistics, 2011) have tremendous learning opportunities; but students face many negative social, cultural, and criminal influences, coupled with the threat of natural and man-made disasters, that compromise what has traditionally been viewed as a safe and healthy environment. The new reality in the school environment is one that requires a coordinated emergency planning and response effort: "Schools can no longer assume safety. They must plan for safety" (Time Select/Families, 1999).

School-related policy and issues primarily fall in the domain of the U.S. Department of Education (DOE). Through the Elementary and Secondary Education Act of 1965 as amended by the landmark and controversial No Child Left Behind Act of 2001 (NCLB), the DOE is authorized to manage many national level programs to address crime, school violence, and student safety. Included among them is Readiness and Emergency Management for Schools (REMS), a program established in 2003 designed to strengthen and improve school emergency response and crisis plans and training for school personnel (Skinner and McCallion, 2008). If developed appropriately, these plans can provide critical guidance and procedures for handling a multitude of situations.

The EF-5 tornado that devastated Joplin, Missouri, on May 22, 2011, killed 161 citizens (Cune, 2011) and completely destroyed much of the community, including Joplin High School, a public school with a student population of approximately 2,200 (Missouri Department of Elementary and Secondary Education, 2011). Among those killed were 7 students and 1 staff member. The school facility was a total loss (Joplin Schools, 2011). Although most school systems will not experience an incident of this size and magnitude, the widespread impact to the local school system highlights the need for well-designed, -exercised, and -executed emergency management plans developed on a district and individual school level—plans that can address incidents of varying sources, scales, and severity.

Although no federal laws exist requiring school districts to have emergency management plans, most states (thirty-two of them) and school districts surveyed in a 2008 government study reported having requirements for school emergency management planning (U.S. Government Accountability Office, 2007). However, funding was limited to initiate planning efforts, continually update these plans, maintain staff readiness through training, and purchase equipment to support the response effort.

The U.S. Department of Homeland Security (DHS), DOE, state governments, and individual school districts provide various sources of funding for emergency management planning in schools. Some of these grant programs are based on a competitive process, whereby not all applicants will receive funds; others tie funding to the inclusion of emergency plan development as a part of a broader effort to strengthen school preparedness. Some federal funding that is administered exclusively by the state or local jurisdiction (such as a large metropolitan area designated as an urban area security initiative grant recipient) can be provided to school districts for emergency planning, but only at its discretion and where it is in alignment with the jurisdiction's strategic plan and priorities.

The REMS program, also known as the *Emergency Response and Crisis Management Plan* grant in earlier years, is one of the few grant programs intended to specifically address school emergency planning and preparedness. Since its inception, the program has helped hundreds of school districts, individual schools, and regional educational partnerships in developing or improving plans that address all four phases of emergency management—prevention/mitigation, preparedness, response, and recovery—support homeland security response doctrines, and address emerging issues for the school environment. However, the future of this program is uncertain as the policy outlook shifts to focus more on the school "climate" and as government discretionary grants overall are scrutinized in tough financial times.

BACKGROUND

The REMS program was first established by the Office of Safe and Drug-Free Schools (OSDFS), a division within DOE formed in 2002 pursuant to the Safe and Drug-Free Schools and Communities Act component of the NCLB. The mission of the OSDFS is to administer, coordinate, and recommend policy for improving the quality and excellence of programs and activities that are designed to:

- Provide financial assistance for drug and violence prevention activities and activities that promote the health and well-being of students in elementary and secondary schools and institutions of higher education.
- Participate in the formulation and development of DOE program policy and legislative proposals and in overall administration policies related to violence and drug prevention.
- Participate in interagency committees, groups, and partnerships related to drug and violence prevention.
- Participate with other federal agencies in the development of a national research agenda for drug and violence prevention.
- Administer the department's programs relating to citizenship and civics education (U.S. Department of Education, 2011b).

The OSDFS housed many programs focusing on issues of importance for the school environment, including environmental health, mental health and physical education programs, drug-violence prevention, character and civic education, and policy programs. The mechanism for distribution of these programs is often through discretionary or formula grants to state and local education agencies or directly to school districts.

As of September 2011, the OSDFS was subsumed into a new Office of Safe and Healthy Students (OSHS), which integrates programs for:

- Safe and supportive schools
- Health, mental health, environmental health, and physical education
- Drug and violence prevention
- Character and civic education
- Homeland security, emergency management, and school preparedness

In addition to the REMS program, OSHS administers eighteen discretionary grants and two formula grant programs, as well as other programs. Within OSHS, the REMS program is now organized under the Center for School Preparedness, which "administers programs that promote the ability of schools to prepare for and respond to crisis and disasters (natural and man-made)," as well as homeland security (U.S. Department of Education, 2011a). The center groups several similar programs focused on emergency management and responses to violence in the K–12 and higher education environments.

REMS PROGRAM

The REMS grant program supports efforts by local educational agencies (LEAs) to create, strengthen, and improve emergency management plans at the district and school-building levels, including training school personnel on emergency management procedures; communicating with parents about emergency plans and procedures; and coordinating with local law enforcement, public safety or emergency management, public health, and mental health agencies and local government.

Local educational agencies are defined by the No Child Left Behind Act of 2001 as follows (Elementary and Secondary Education Act, 2011):

> IN GENERAL—The term local educational agency means a public board of education or other public authority legally constituted within a State for either administrative control or direction of, or to perform a service function for, public elementary schools or secondary schools in a city, county, township, school district, or other political subdivision of a State, or of or for a combination of school districts or counties that is recognized in a State as an administrative agency for its public elementary schools or secondary schools.

School systems in tribal areas governed by the U.S. Bureau of Indian Affairs are eligible for funding with certain size restrictions. Private schools are not eligible for REMS funding; however, LEAs are required to include private schools in their efforts if such schools exist in their district or jurisdiction.

Grant funds may be used for the following activities: reviewing and revising emergency management plans; training school staff, conducting building and facilities audits; communicating emergency response policies to parents and guardians; implementing the National Incident Management System (NIMS); developing an infectious disease plan; developing or revising food defense plans; purchasing school safety equipment (to a limited extent); conducting drills and table-top simulation exercises; and preparing and distributing copies of emergency management plans.

In reviewing and improving their plans, districts are required to work with community partners including local law enforcement, public safety or emergency management, public health, and mental health agencies and local government. Plans must include training for school staff, a plan to sustain local partnerships after the period of federal assistance, a plan for communicating emergency management policies and reunification procedures to parents, and a written plan for improving LEA capacity to sustain the emergency management process through ongoing training and continual review of policies and procedures. In addition, LEAs must agree to support the implementation of the NIMS and commit to developing plans that take into consideration special needs populations within the LEA. Lastly, LEAs must agree to develop a written food defense plan and an infectious disease plan designed to prepare the LEA for possible infectious disease outbreak.

Typical activities included in grantee programs include reviewing and revising existing emergency management plans, conducting vulnerability assessments of schools and other district facilities, providing training, organizing table-top exercises, procuring emergency supplies, and engaging in crisis simulation drills.

Technical assistance is provided to recipients of the REMS grants through the Readiness and Emergency Management for Schools Technical Assistance Center (TAC), established in 2004 by OSDFS. The division provides assistance for schools, school districts, and institutions of higher education on emergency management issues and questions. The TAC is also moved to the OSHS under the Center for School Preparedness.

AWARDS

In fiscal year 2010, the REMS program awarded ninety-six local educational agencies of various sizes approximately $29 million in DOE funds for activities related to reviewing and improving their emergency management plans. The awards ranged from $103,976 for the Vidor Independent School District in Vidor, Texas, a small community outside of Beaumont with a student population of approximately 4,900, to $710,053 for the Los Angeles Unified School District, which serves approximately 678,500 students in grades K–12. For the first time, awardees were given two years for full completion of the proposed projects, versus one-and-a-half years for previous awards.

In comparison, during the first year of the program in fiscal year 2003, the OSDFS provided grants to 132 local educational agencies, with awards ranging from $68,875 for the Amber Charter School (a K–5 school in New York City with a student population of 425) to $1,000,000 for the Hillsborough County Public School District in Tampa, Florida (with a student population of 194,737).

POLICY SHIFTS

A shift in priority under the Obama administration has resulted in changes throughout federal school safety initiatives and, thus, the funding programs that support them. In fiscal year 2012, greater priority is placed on the $365 million "Successful, Safe, and Healthy Students" program, which is intended to:

- Increase the capacity of state educational agencies (SEAs), high-need LEAs, and their partners to develop and implement programs and activities that improve conditions for learning so that students are safe, healthy, and successful. Programs and activities supported by this program would include those that reduce or prevent drug use, alcohol use, bullying, harassment, or violence and promote and support the physical and mental well-being of students.
- Improve conditions for learning and student outcomes, including activities aimed at preventing and reducing substance use, violence, harassment or bullying; promoting student mental, behavioral, and emotional health;

strengthening family and student engagement in school; reducing out-of-school suspensions; implementing positive behavioral interventions and supports; and implementing programs designed to improve students' physical health and well-being, including their physical activity, nutrition, and fitness (U.S. Department of Education, 2010a).

The program would consolidate or eliminate several existing programs that the administration defined as "narrowly focused" and "too fragmented." According to the budget request, the program would build upon the competitive grants awarded through the Safe and Supportive Schools program created in 2010 within the OSDFS. Effective September 2011, what programs remained were consolidated under the new Office of Safe and Healthy Students (McCallion, 2008).

Funding would be awarded to state agencies instead of directly to school districts—an about-face from fiscal year 2007, where no funding was provided for state grant programs. In 2006, the administration determined that the state grant component of the OSDFS was found to be "ineffective" because the grants were too thinly distributed to truly support the desired outcomes (McCallion, 2008). Yet state educational agencies will now be responsible for distribution of all Readiness and Emergency Management in Schools funds under the Safe and Supportive Schools program.

These moves represent a shift in attitude away from the building and sustainment of the organizational infrastructure for emergency preparedness and response and places a greater emphasis on prevention measures through direct engagement with students and families on social, cultural, and personal health issues that affect student achievement and well-being in the school environment. Up to and including the fiscal year 2010 grant cycle, prevention programs were not allowable under the REMS, because they are not directly relatable to the development of an emergency management plan.

Additionally, by specifically addressing "high-need local educational agencies," grant applicants must undertake an effort to identify those districts with the greatest need for funding. That effort in itself forces state educational agencies to define metrics, conduct surveys, and develop application processes to administer the funds available. As with many federal grant programs where states serve as the administrative and financial agent, these types of activities have a cost and thus reduce the overall amount of funding from the outset. Under previous programs within the OSDFS office, states could retain up to 5% of their award to offset administrative, reporting, training, and technical assistance costs (McCallion, 2008).

Beginning with fiscal year 2011, discretionary funding for the Readiness and Emergency Management in Schools was transferred away from grants directly to LEAs in favor of awards to SEAs. In fiscal year 2012, funding streams to both are nonexistent.

GRANT APPLICATION PROCESS AND PROJECT REQUIREMENTS

REMS funding is awarded through a competitive process. Applications are developed and submitted by the LEA or an agent working on their behalf. LEAs who received funding in previous years and that are still actively fulfilling the obligations of that award are not eligible to apply for new funding.

The guidance indicates that projects should include training for school personnel in emergency management procedures; coordination with local community partners; and plans to improve local capacity to sustain emergency management efforts after funding ends (U.S. Department of Education, 2010b). Projects considered for funding are also required to include (U.S. Department of Education, 2010b):

- Partner agreements: With their applications, LEAs must provide written and signed partner agreements with five community-based partners including local government (e.g., mayor, city manager, county executive), law enforcement, public safety or emergency management, public health, and mental health agencies. "The agreements must describe the roles and responsibilities of each entity in improving and strengthening emergency management plans and a description of each partner's commitment to continuous improvement of the plans." This requires coordination with these partners in advance of the application submission. Likely, these schools will already have a relationship with the partners through other initiatives, but this may serve to strengthen their partnership.

- Coordination with state or local homeland security plans: In their applications, LEAs must commit to ensuring a level of integration with their local jurisdiction or state homeland security plans. This is particularly important for determining external hazards, linking procedures for both shelter-in-place and evacuation actions, and requesting assistance from outside entities.

- Infectious disease plan: LEAs must include a plan for possible infectious disease outbreak, including components for disease surveillance, school-closure decision making, business continuity, and continuation of educational services.

- Food defense plan: Applicants are required to develop a food defense plan; that is, a plan that "protects against intentional contamination [by] the introduction of chemical or biological hazards into food, water, or facilities by individuals seeking to harm students or staff" (U.S. Department of Education, 2006).

- Individuals with disabilities: Plans "must take into consideration the communication, transportation, and evacuation needs" of the functional needs population of the schools in the LEA.

- NIMS requirements: LEAs must provide assurance that they have completed, or will complete, all current NIMS requirements in accordance with the DHS/Federal Emergency Management Agency NIMS Implementation Objectives (U.S. Department of Homeland Security, 2009).

Should an award be received, representatives of the LEA are required to attend three meetings at set periods in the life cycle of the grant: an initial meeting to review grant management requirements and introduce grantees to basic emergency management and planning principles; a secondary meeting covering more advanced topics related to school emergency management; and an annual conference held by the OSDFS addressing a broader range of issues related to the school environment, including mental health and physical education, drug and violence prevention, and civic and character education. Grantees are allowed to use a portion of their awarded budget to attend these events.

There is a clear expectation that grantees will increase the capacity of the local school district to develop, implement, and sustain a comprehensive emergency management system, including an internal capability to implement all aspects of emergency management, including conducting vulnerability assessments, developing and updating written procedures, training staff, and conducting exercises and drills. Grantees must outline a plan to evaluate their performance in executing the project and agree to cooperate with evaluation efforts conducted by the DOE to ensure school districts conform to this focus. LEAs are required to:

- Maintain records on the extent to which their program objectives are being met
- Include specific performance measures in their evaluation plans
- Provide a copy of their revised emergency management plan to the Department of Education at the conclusion of the grant period
- Make ongoing project information, findings, and products available upon request (U.S. Department of Education, 2010b)

The evaluation plan is required to include qualitative measures such as improved partnerships with community stakeholders and quantitative measures such as increased response time to drills, as well as process measures such as the number of vulnerability assessments conducted.

Additionally, in an effort to establish a quantifiable measure of success for the program that the DOE can then report under the Government Performance and Results Act of 1993, grantees must document the number of completions by key personnel in any of the training courses offered by the U.S. Department of Security on the NIMS. LEAs must report the average number of NIMS course completions at the beginning of the grant as compared with those completed at the end of the grant. Recommended courses for school staff and administrators focus on graduated training in the Incident Command System, a standardized, on-scene, all-hazards incident management doctrine adopted by the Federal Emergency Management Agency. The training includes the following:

- IS-100.SCa (or IS-100.b): Introduction to the Incident Command System for Schools
- IS-200.b (ICS 200): ICS for Single Resources and Initial Action Incidents
- IS-700.a: National Incident Management System (NIMS), An Introduction
- IS-800.b: National Response Framework, An Introduction
- ICS-300 (G300): Intermediate ICS for Expanding Incidents
- ICS-400 (G400): Advanced ICS Command and General Staff—Complex Incidents

Although this training is invaluable for expanding the knowledge of personnel in the management of an incident, this measurement is certainly not comprehensive enough to demonstrate an increased emergency management capacity among school districts. The OSDFS should consider additional measurements related to the inclusion of certain basic and/or specialized components in individual school plans, such as a food defense or infectious disease plan.

Most of these courses are available in an online, self-paced format and thus do not require a significant portion of a grantee's budget. However, some choose to contract with outside entities to bring the training on-site in a classroom-based environment where table-top exercises and more hands-on interactivity can be included.

CHALLENGES

Although the REMS program has made a significant difference in preparing schools for crisis, challenges still exist. As with all discretionary and competitive grant processes, not all applicants can be awarded funding for their causes. School systems may be required to "layer" funding sources and programs to fully cover the activities necessary to build and maintain their emergency management programs or fund them within their own budgets.

To date, there has been no focused, independent research conducted specifically on REMS recipients to determine program effectiveness beyond the Government Performance and Results Act measurement of the number of NIMS-trained staff. Doing so could help determine best practices and lessons learned, innovation among schools, and gaps in the program where certain requirements should be expanded or altered.

Although a significant number of states surveyed in a 2008 U.S. Government Accountability Office study reported having a requirement for school districts or schools to have emergency management plans (thirty-two states), only eighteen states required the plans to include specific hazards or required review and update of the plans by the school district or other entity. Only nine states required the involvement of parents in the planning process, and only ten states included community partners and other stakeholders (U.S. Government Accountability Office, 2007).

In a 2011 report by the National Center for Education Statistics on crime, violence, discipline, and safety in U.S. public schools, only 41% of schools had a written plan for instances where the National Terrorism Advisory System (which replaced the Homeland Security Advisory System based on color-coded levels) alert indicates an imminent threat. This is compared with 95% of schools that reported having a written plan for natural disasters and 94% of schools that reported having a written plan for bomb threats (Neiman and Hill, 2011). Additionally, the same report notes that a "higher percentage of suburban schools drilled students on a written plan describing procedures to be performed during a school shooting (58%) than did city schools or rural schools (49 and 48%, respectively)" (Neiman and Hill, 2011).

More recent criticism of OSDFS programs in general are focused in three areas:

- *New requirements to include anti-bullying and harassment prevention efforts in programs not intended for this purpose.* Members of Congress and conservative watchdog groups argued the efforts would take limited resources away from the primary purpose of the program and be too narrowly focused on sexual-orientation and gender-identity issues. Changes to OSDFS programs of this nature were credited to the highly controversial OSDFS director, Kevin Jennings, who served in the office from 2009 to 2011, and who was often accused of pushing an agenda specific to lesbian, gay, bi-sexual, and transgender youth, to the detriment of other issues in the school environment (Lott, 2009).

- *Lack of evidence that programs reduce the incidence of violence and drug abuse in schools.* Critics argue that although these are "pressing societal issues, they are problems that rarely occur on school grounds" (Congressional Budget Office, 2009).
- General criticism of discretionary funding programs in tough economic times (Office of Management and Budget, 2011).

INTERVIEW

The interview with Wanda Johnson, Safe Schools Coordinator, Pulaski County, Kentucky School District (2011) included below helps to identify practical challenges in being awarded a REMS grant and implementing an emergency management plan in a school district.

The Pulaski County School District contains:

- Eight thousand students (approximately)
- Eight elementary schools
- Two middle schools
- Two high schools
- One technology center
- An adult education program
- A Golden Age Program for senior community members aged fifty-five and over

How Did the School District Determine Its Needs in Preparation for the Grant Application?

Pulaski County (Kentucky) Schools were recipients of 2005 REMS grant funding (then referred to as the Emergency Readiness Crisis Management Grant). After conducting an annual review of their District Emergency Management Plan, it was determined an update was needed. This formed the basis of their application for the funding.

How Was REMS Funding Used Specifically to Improve School Safety and Preparedness?

The Pulaski County School grant was used in two primary areas:

- To update the district's emergency management plans to include school-level emergency teams and accompanying plans for the teams. The plans will be transferred to an online format with personalized authorization to allow team members to edit the plans as needed. Floor plans and utility cutoff information will also be included in the plans for access by first responders. As designated in the grant application language, the plans will address food defense, special needs populations, and infectious disease.

- To ensure district and school-level teams receive NIMS and emergency management training and subsequently provide necessary education to other staff and students.

WERE THERE ANY DIFFICULTIES IN USING THE FUNDING OR ANY GRANT RESTRICTIONS THAT CREATED BARRIERS TO SUCCESSFUL IMPLEMENTATION (FOR EXAMPLE, LENGTH OF TIME TO SPEND, TYPES OF EQUIPMENT, AND SO FORTH)?

In their application, the Pulaski County School District established a project performance measure related to staff training and certification on critical incident stress management issues. However, the primary provider of training in this area, the International Critical Incident Stress Foundation, has not offered the necessary courses within the certification process. Extra time may be needed to achieve this metric.

WHAT IMPACTS WERE SEEN AS A RESULT OF THE GRANT?

The district has seen an increased awareness among school and district administration of the importance of school and student safety as it relates to emergency preparedness. Additionally, partnerships with first responder agencies have been expanded and reinforced.

WHAT ARE THE SCHOOL'S CURRENT AND FUTURE NEEDS IN THIS AREA?

As the requirements and demands on academic standards remain high, carving out time for staff professional development is even more challenging. The Department of Education should require and fund annual training in school safety and mandate minimum training hours for emergency team members.

CASE STUDY

This case study concerning the Apex, North Carolina/Wake County School District is adapted from one included in the Center for Sustainable Community Design report "Safe Schools: Identifying Environmental Threats to Children Attending Public Schools in North Carolina" (Salvesen et al., 2008).

Across the country, school districts are searching for creative ways to add capacity, such as adaptive reuse of buildings, including industrial facilities. This case study highlights some of the potential hazards of situating a school in an industrial area.

Lufkin Road Middle School (sixth through eighth grades) is located in Apex, North Carolina. About 1,025 students attend the school, of which approximately

30% are members of racial or ethnic minority groups. Rapid population growth in the area has strained the public school system. In response to this rapid growth, the school system has looked for creative ways to increase its capacity, such as adaptive reuse of commercial or industrial buildings. Adaptive reuse can provide useable classroom space on a much shorter time horizon than constructing a new facility from the ground up.

Lufkin Road Middle School was constructed in 1998 as an adaptive reuse of a building that had been the home of the American Sterilizer Company. The 24-acre tract that eventually became Lufkin Road Middle School is located between two facilities that used hazardous chemicals. The vast majority of the one-quarter mile area surrounding the school is currently zoned for light industrial use.

The hazard assessment and buffer/proximity analysis for the school identified both highway and industrial threats in the vicinity of Lufkin Middle School, as well as a CSX-owned railway located approximately one-half mile to the west of the school. The Claude M. Pope Memorial Freeway (US 1), lies only 75 meters from the school. This stretch of the freeway had an average annual daily traffic volume of eighteen to forty-one thousand vehicles.

- What key components should be included in the school's emergency plan?
- Who should the district/school involve in the emergency planning process?
- What challenges does the school face in communication with first responders, students, and parents?

EXERCISES

1. What is REMS designed to accomplish?
2. What topics should be included in an emergency management plan?
3. What financial award might a school expect to receive from an REMS grant?
4. What items must be included in a plan in order to receive REMS funding?
5. What evaluation activity must an LEA engage in?
6. What training modules are recommended for faculty to attend either online or in a classroom setting?
7. What challenges exist when developing an emergency management plan for a school?

REFERENCES

Congressional Budget Office. (2009). *Budget Options* (Vol. 2, p. 115). Congress of the United States, Washington, D.C.
Cune, G. (2011). Joplin tornado death toll revised down to 161. *Reuters News Service.* Accessed November 12, 2011. http://www.reuters.com/article/2011/11/12/us-tornado-joplin-idUSTRE7AB0J820111112.
Elementary and Secondary Education Act (2011). As amended by the No Child Left Behind Act of 2001, Pub. L 107-110, Title IX, Section 9101.
Interview with Wanda Johnson, Safe Schools Coordinator, Pulaski County, Kentucky School District, November 1, 2011.

Joplin Schools. (2011). Accessed November 12, 2011. http://www.joplinschools.org/modules/cms/pages.phtml?pageid=231908&sessionid=dfa6d2a6ed82554a35ad53c82071a242.

Lott, M. (2009). Obama's "safe schools" czar admits he poorly handled underage sex case. *FoxNews.com.* Accessed March 14, 2012. http://www.foxnews.com/politics/2009/09/30/obamas-safe-schools-czar-admits-poorly-handled-underage-sex-case.

McCallion, G. (2008). *Safe and Drug-Free Schools and Communities Act: Program Overview and Reauthorization Issues.* Congressional Research Service, Washington, D.C.

Missouri Department of Elementary and Secondary Education. (2011). Accessed November 2, 2011. http://mcds.dese.mo.gov/guidedinquiry/District%20and%20School%20Information/Missouri%20School%20Directory.aspx?rp:DistrictCode=049148.

National Center for Education Statistics. (2011). Digest of Educational Statistics, 2010. Table 36. Enrollment in public elementary and secondary schools, by state or jurisdiction: Selected years, Fall 1990 through Fall 2010. Accessed March 14, 2012. http://nces.ed.gov/programs/digest/2010menu_tables.asp.

Neiman, S., and Hill, M. R. (2011). *Crime, Violence, Discipline, and Safety in U.S. Public Schools: Findings from the School Survey on Crime and Safety.* National Center for Education Statistics, U.S. Department of Education, Washington, D.C.

Office of Management and Budget. (2011). *Terminations, Reductions, and Savings Budget of the U.S. Government.* Office of Management and Budget, Washington, D.C.

Salvesen, D., Zambito, P., Hamstead, Z., and Wilson, B. (2008). *Safe Schools: Identifying Environmental Threats to Children Attending Public Schools in North Carolina.* The Center for Sustainable Community Design, Institute for the Environment, University of North Carolina at Chapel Hill, Chapel Hill, NC.

Skinner, R. R., and McCallion, G. (2008). *School and Campus Safety Programs and Requirements in the Elementary and Secondary Education Act and Higher Education Act.* Congressional Research Service, Washington, D.C.

Time Select/Families. (1999). How to keep the peace: Adults and students together must guard against school violence. *Time,* 154(11), C7.

U.S. Department of Education. (2006). Food safety and food defense for schools. *ERCM Express,* 2(5), 1–2.

U.S. Department of Education. (2010a). *A Blueprint for Reforming the Reauthorization of the Elementary and Secondary Education Act.* Office of Planning, Evaluation and Policy Development, U.S. Department of Education, Washington, D.C.

U.S. Department of Education. (2010b). *Readiness and Emergency Management for Schools: A Grant Application to Improve and Strengthen School Emergency Management Plans.*

U.S. Department of Education. (2011a). About us: Office of safe and healthy students. Accessed November 1, 2011. http://www2.ed.gov/about/offices/list/oese/oshs/aboutus.html.

U.S. Department of Education. (2011b). U.S. Department of Education principal office functional statements. Accessed November 1, 2011. http://www2.ed.gov/about/offices/list/om/fs_po/osdfs/intro.html.

U.S. Department of Homeland Security, Federal Emergency Management Agency. (2009). FY 2009 NIMS implementation objectives. FEMA. Accessed March 14, 2012. http://www.fema.gov/pdf/emergency/nims/FY2009_NIMS_Implementation_Chart.pdf.

U.S. Government Accountability Office. (2007). *Emergency Management: Most School Districts Have Developed Emergency Management Plans but Would Benefit from Additional Federal Guidance (GAO-07-609).* U.S. Government Accountability Office, Washington, D.C.

26 Academic Continuity

E. Scott Dunlap

CONTENTS

National disasters have helped us to see the need to integrate principles of emergency management into how we operate our schools. Events such as the destruction caused by Hurricane Katrina around the Gulf of Mexico presented the need to not only respond to these events, but also to plan for recovery. This concern has given rise to the discipline of business continuity in industry. The concept is to develop plans that will allow a business to maintain the continuity of operations to the greatest degree possible when experiencing and recovering from a disaster. Although a school is not considered a "business," a best practice in this area exists to develop academic continuity plans that can be used to maintain educational activities when faced with small and large events.

OVERVIEW

An academic continuity plan is a series of steps that allows a school to regain operational capability as soon as possible following an incident. An incident can be any event that impacts a school's operational capacity. On one end of the spectrum, it can be something as small as a simple power outage. On the other end of the spectrum, it can be the total loss of a school because of a natural disaster.

Academic continuity planning will typically include three phases: emergency response, crisis management, and academic resumption. The activity that takes place in the first two phases is similar to what occurs when creating an Occupational Safety and Health Administration (OSHA) emergency action plan. Emergency action plans include an evaluation of what disasters could impact a school and identifying ways for employees and students to escape and seek refuge. Such a plan can easily be integrated into an academic continuity plan.

Emergency response as the first phase of an academic continuity plan may already be addressed if an emergency action plan is in place. In this phase, the type of events that can occur will be identified, and how a school will immediately respond to each one will be determined. For example, on the lower-risk side of the spectrum, the breakdown of a bus transporting students to school might occur. The emergency response to this event might include immediately dispatching another bus to the site of the breakdown so that students can be transferred to the alternate bus and transported to school. On a larger scale, the event of severe weather in the form of a tornado or hurricane might be identified. The emergency response to this event would include gathering at internal sheltering locations.

The second phase of an academic continuity plan is crisis management. Severe weather may impact a school, and all students, faculty, and staff have to be evacuated to shelter. Imagine the devastation that occurred in the Gulf area when Hurricane Katrina struck. On an organizational level, facilities had to manage through the crisis because their employees were instantly without basic resources and may have lost communication with friends and family members. On a community level, violence against individuals, vandalism, and theft quickly arose. Identifying ways to manage through such crisis situations is a critical second element of academic continuity planning.

The third phase of an academic continuity plan is academic resumption. This is the final goal of the complete process. This phase provides the opportunity to identify what resources are necessary to restore operational capacity as soon as possible. Depending on the degree to which these principles are applied, issues ranging from a small property damage incident to a community-wide disaster can be managed.

An example of the benefit of this process was in an organization where the plan actually had to be placed into action. Prior to planning, it was jokingly said that Plan B was to ensure that Plan A never failed. Although that would be nice in a perfect world, incidents do occur. During the planning process, one issue to be addressed was backup power to the facilities. Rather than paying the expense to have backup generators permanently installed, the decision was made to contract with a company that guaranteed delivery of backup generators within a specified period of time following a disaster. Although this service came at a monthly fee, the organization decided it was worth the expense in comparison with the cost of installing and maintaining generators. The worst-case scenario was that the facility would be nonoperational for approximately twenty-four hours, but those who were in a position of leadership agreed that it would be the best avenue through which to manage the risk.

Within the first year of establishing the plan, the entire community was hit by a storm that included several tornadoes touching down. Most of the city lost power. Numerous organizations were impacted by receiving direct damage from high winds and the tornadoes. Fortunately, the facility that had conducted planning was not directly hit, but it was without power, and because of the damage of the storm, there was no clear estimated time of power restoration. The plan was immediately activated. The generators arrived, and the facility was operational within a short window of time, while many of the organizations in the city were still simply responding to what had happened without having done preplanning. The organization was able to proactively manage the situation rather than allow the event to simply happen to them.

METHODOLOGY

Academic continuity planning involves the use of a methodology in order for the results of the process to be effective. One way to visualize this is the four steps in the cycle of business continuity planning illustrated by Hiles (2011).

First, it is important to understand the school. Although this may sound simplistic, it is actually a very complex exercise. In understanding the school, it is important to understand things such as how the school operates, how material and people move through the property, how staffing is managed, what standard operating procedures exist, and what actual facilities exist. This can appear to be an overwhelming expectation, but there should not be an expectation that one individual accomplish this task. Those involved in the academic continuity planning process should feel free to reach out to subject matter experts within the school. No one person can know everything that needs to be understood in the academic continuity process. It must be a team effort.

Second, an academic continuity strategy or approach needs to be determined. Understanding the school will place into perspective all of the activities that must be taken into consideration. A strategy can then be determined that makes sense for the school. For example, based on the needs of the school, one school may choose to take an approach that considers granular issues, such as losing a given piece of equipment in an operational process, whereas another school may chose to use a macrolevel strategy that only considers those issues that would result in complete loss of operational capacity. There are also shades of gray that can be considered between these two extremes.

Once a strategy has been identified, the hard work then begins in the third step, which is to actually develop and implement the academic continuity plan. This will be the most resource-intensive part of the process. This involves such activities as creating written procedures, conducting training, establishing a budget, and identifying individuals who will manage the process. It will be important to establish realistic expectations as to how long this process will take. Imagine that the decision has been made to initiate the academic continuity process for all of its value and a plan is to be created that addresses the loss in a number of tiers, including individual class areas, sections of the school, and the school as a whole. Beginning at the class level, plans will have to be created that address response, alternative areas to hold classes, and recovery for every individual class within the school. Planning will then have to be elevated to include what will happen if a complete area of a school has been taken out of commission. Finally, decisions will have to be made and processes put in place to manage the potential loss of the entire school. The structuring of these details and the associated decision-making processes will require a great deal of time and input from a large team of people.

Once the program has been developed and implemented, ongoing training and review of the program will need to occur. A challenge here is maintaining interest and focus. An academic continuity plan is of value in the event of an incident. Hopefully, loss control efforts occur on an ongoing basis, which in turn should prevent the need for exercising an academic continuity plan. However, the challenge is that when you need such a plan, it is needed in a great way. Constant vigilance and preparation are required so that when the plan is needed, everyone in the school knows how to utilize it. This will require such activity as periodic plan review, training, and drills.

Because of the complex nature of assembling a basic plan, the assistance of those who work in different areas throughout the school will be needed. It is important to know the flow of activity that occurs in every area in addition to how it flows. For example, an incident might occur that takes the school's kitchen out of service at 9 a.m. A plan will have to take into consideration the flow of food preparation and how each of the steps in the process occurs. Alternatives will need to be identified to immediately provide lunches for students in the absence of a functioning kitchen. This can be done by identifying work-around procedures to accomplish the same task in an alternative fashion or to identify a third party source of prepared food that can immediately be engaged.

Although basic, work-around plans will require great attention to detail. For example, such a plan can dramatically increase the amount of personnel needed. Automation is great while it is working, but if it is down, immediate access to additional workers might be needed to manually perform work. This will require the need to work with various departments to determine how many people would be needed and also the need to include the human resources department in the planning process. Human resources personnel will be able to help identify avenues to bring in the people necessary to maintain production. This might include signing agreements with a temporary employment agency to supply workers when needed. It might also involve working with other departments to determine where labor can be shifted to accommodate the immediate need.

Academic continuity planning is no small task. Simply identifying how to maintain academic continuity through the interruption of a simple event can involve a great deal of time and planning. The value in doing this is that in a critical hour when a problem occurs, the resources to maintain academic work will be immediately available.

The strategic decision might be made to not address microlevel incidents and to only address macrolevel loss. Rather than planning for microlevel academic interruption, the decision might be made to accept down time as a random occurrence and simply react by paying for the repairs and restoration as they occur. Macrolevel planning might include the complete loss of a given school. Macrolevel planning will still involve the need to preplan how to bring the school back online. Contractors and internal resources will need to be identified that could be immediately deployed to address the problem. This could include such large-scale activity as identifying an alternative real estate property to be used to conduct school while the original school building is being repaired or rebuilt following a major incident. It might also include the use of a learning management system to convert classroom learning to online learning.

Whether an academic continuity strategy dictates micro- or macrolevel planning, the plan will need to be put in writing. Keep in mind that the most important goal of the academic continuity planning process is functionality. All of the effort that goes into the planning process must work in the end. People must be able to understand what it is that has been created and make it happen. The plan should be so well organized and written that someone who is not closely associated with the process can pick up the plan and manage recovery operations. The plan should be written so that anyone can understand it. It cannot be known who might be around and must be able to make sense of the material once a disaster has occurred. Key personnel that are typically depended upon may not be available. Consider the catastrophic loss that occurred during Hurricane Katrina. Because of personnel needing to take care of their friends and families who have been affected, response efforts might become severely limited in relation to the people that can be drawn upon to restore a school to a stable level of productivity. If something were to happen that caused the cafeteria to go down, an organized and well-written academic continuity plan would allow any trained faculty or staff member, even one who has never worked in the area, to effectively facilitate academic recovery operations. Contact names and numbers should be easily identified, and steps for response and recovery should be clearly outlined. Some plans are very well done but are incredibly complicated and could only be understood by those who put it together. This is not a time to display how much one knows about Microsoft Word, Excel, PowerPoint, or the academic continuity software that has been purchased. It is a time to develop and deliver a product that is well organized and can be understood by all.

PROJECT MANAGEMENT

Each phase of academic continuity plan development must be imbedded in the school culture. The degree to which an academic continuity plan can be imbedded in the school's culture will greatly depend on the level of importance the school gives the plan. This makes the initiation of the plan an important first step.

Hiles (2011) outlines a number of reasons that might cause an organization to focus on continuity planning. Unfortunately, the reason that might drive many schools is that they have experienced an incident and the subsequent consequences. School system administrators might see the development of an academic continuity plan as unimportant until a disaster is experienced and significant cost is incurred. This presents a challenge as to how to accomplish the integration of academic continuity into a school's culture proactively in the absence of having experienced a disaster. Those reacting to a situation may embrace academic continuity strongly but may do so in a dysfunctional fashion in a rush to get something into place. Those who approach the process proactively may face the challenge of how to establish a sense of the urgency and longevity of the process among those who may not sense a great need for it. Depending on the situation that exists, it is important to be prepared for the challenge of integrating academic continuity into a school culture based on the catalyst that has caused the plan to be considered.

A step that can be taken to initiate an academic continuity plan is to help school leadership understand the need for the plan. One way this can be done is through financial modeling that presents the cost and benefit of such a plan in relation to the risk of loss. On the microlevel, an analysis can be conducted on the cost of downtime and loss of productivity if a given area of the school cannot be used because of an incident. The risk of missing time scheduled for academic activity could also be estimated. This concise evaluation could be researched and presented with relative ease. On the macrolevel, the cost to the school if a plan was not in place to manage through the loss of a school could be presented.

After the case has been presented and support has been gained for the process, the task of assembling various teams that will be necessary in the development and implementation phases can begin. Areas of responsibility can include a broad range of disciplines. For example, there might be the need to engage people beyond personnel directly related to the school, such as legal counsel. There might be a need to engage legal counsel and operations personnel to both design services that a third party would provide and create the contracts that would need to be signed. Informal agreements may not be sufficient. Attorneys need to be involved to ensure that properly binding agreements are established where needed.

Team members must be identified who have technical writing skills. As the plan evolves, there will be numerous documents that will be created from standard operating procedures to written programs. Individuals with technical writing skills will be needed to ensure that the documents produced will be readily understood by the respective target audience.

Liaisons will need to be identified to work with numerous outside organizations that will be needed to support the plan. The first phase of plan execution is emergency response. Individuals will need to work with government agencies that will be involved. This will include such activity as preplanning with the local fire department, preparing for OSHA or Environmental Protection Agency response, and working with local emergency medical services to identify areas for equipment staging and triage of those who have suffered injury. Emergency response may also include the immediate need for access to public utility services or other services based on the nature of the event.

The second phase of executing an academic continuity plan is crisis management. Individuals will need to be assigned to identify and develop relationships with each entity that will be required to effectively manage this stage of an event. Crisis management may include ongoing support of local fire and emergency medical services. It will require the need for public relations to engage in providing adequate communications to family members and the public as news media begins to arrive on the scene. Communication will also be critical to identify an internal process by which school administrators receive notifications and to establish and maintain lines of communication among all who are managing the crisis. This can be a challenge given the variety of communication systems that may be in use among all of the parties involved. Communication avenues can include face-to-face discussion, e-mail, cell phones, radios, and written reports. Crisis management may also include the need to establish the accounting of expenditures and insurance coverage considerations. This may require engaging personnel from departments such as finance, accounting, and risk management. Engineering may also be needed in the crisis management phase. Even though a fire has been extinguished, the school building may not be completely safe to enter because of structural damage. A damage assessment must be made to identify the scope and type of damage that has occurred to the structure.

The third phase of academic continuity is recovery. Multiple teams will need to be assembled for this phase as with the previous two phases. Maintenance may be needed to immediately begin managing the flow of contractors that will be needed to repair or replace damaged equipment. Human resources personnel may be needed to manage the return of employees and students and the sourcing of additional labor as needed. The information technology (IT) department may be needed to bring computer systems back online and restore the Internet connection.

Each of the three phases of academic continuity plan execution present the need to identify a system that is comprised of a matrix of teams to accomplish specific tasks as needed. It is the responsibility of each school to determine what makes sense for that school to accomplish the scope of activity within an academic continuity plan.

As material is developed and teams are established within the plan, it is important to communicate the success that is experienced as the plan reaches various milestones. Two things will be accomplished by identifying milestones and communicating the success of achieving them to the organization. First, it will engage the school by demonstrating progress toward the goal of a completed academic continuity plan. This is of critical importance because of the amount of time that will pass between the moment the plan has been approved for development and when it is completed. Celebrating the accomplishment of milestones will help the school to see that activity in the academic continuity plan is occurring and that there is tangible value in the process even though it might not yet be fully complete. Second, celebrating milestones will motivate those working on the plan. It will provide an immediate sense of accomplishment by bringing a given segment of the plan to a close.

Once the plan is fully developed, it will be necessary to ensure ongoing school administration support and employee awareness. An academic continuity plan is not something that is completed and simply placed in a binder or on a website to be left alone. It is a living and breathing process that must have a mechanism for sustainability. This can be accomplished through such activity as periodic communications, training, drills, and new program developments. Although the excitement of program development and implementation has passed, constant vigilance must occur through sustained awareness of the existence and value of the academic continuity plan.

EMERGENCY RESPONSE

A school may already have an existing emergency action plan within the scope of a health and safety management system. Much, if not all, of such a program can be directly integrated into the academic continuity plan in the emergency response and crisis management phases.

Based on the strategy that a school has selected to address academic continuity, the emergency response phase is designed to provide resources to immediately address the situation after it occurs. The first step is to identify emergencies that can occur. On a general level, this could include severe weather or fire affecting a school. On a more granular level, it could be identifying plausible scenarios within the school that could result in the need for emergency response within the scope of an academic continuity plan, such as a water leak that prevents a given classroom from being used. Both ends of this spectrum can result in loss of academic operation, and both include the need for emergency response because of the potential for injury to people and loss of operation.

Once potential emergencies have been determined, the next step is to determine what level of emergency response is needed. Depending on the resources available, emergency response measures can range from dialing 911 for public emergency services to paging first responders internally. A school may operate a first responder team that has been trained at a minimum in first aid and CPR. These individuals can be called upon to respond to medical emergencies within the scope of their training. It is important to understand the limitations of first responder teams so that a decision can be properly made regarding when to engage public emergency services.

A response to emergencies will also involve the consideration and maintenance of physical systems. This could include adequate coverage and operation of a sprinkler system and fire extinguishers. It could also include the location, maintenance, and operation of automated external defibrillators. Systems and equipment need to be identified, and a process must be put in place to ensure that they remain operational.

Emergency response procedures should be written so that steps and resources for emergencies are adequately documented. Documenting such procedures will help to provide guidance to anyone who needs access to the material. Placing things in writing will provide an authoritative document for activity and will extract things from the minds of those who simply "know" what to do. Such procedures are also beneficial when training faculty and staff on what must happen within this first phase of the academic continuity plan. Written procedures may already exist in a school emergency action plan. If this is the case, the procedures should be reviewed and possibly expanded upon to meet the intention of a comprehensive academic continuity plan.

Once emergency response procedures have been documented, individuals within the school must be trained as to their respective responsibilities. Training should take place initially when the academic continuity plan is created, whenever changes are made to procedures, when an individual enters a new position of responsibility, and on an ongoing basis. One cannot assume that individuals will properly respond based on common sense. They must be sufficiently trained to meet the expectations of the plan.

CRISIS MANAGEMENT

The crisis management phase of an academic continuity plan is a second area in which an existing emergency action plan can be integrated, depending on how much information is included. The intent of this phase is to provide specific guidance on how to address the various areas of crisis that have been caused by the incident. For example, if a tornado strikes a school, a number of issues could arise. Emergency response would include warning and getting everyone to an area of shelter and then possibly responding to a fire and the search and rescue of trapped employees. This phase will take place in a very short period of time. Crisis management evolves immediately from emergency response, and problematic issues will be managed, such as how to triage injured employees, shutting off gas lines that may be damaged and leaking, responding to the media who heard of the event and have started to arrive, and interacting with family members of faculty, staff, and students. An existing emergency action plan might have a number of these elements already present. Media relations could be one of these items. Schools may go to the extent of including a media relations section in their emergency action plan. This information can be easily incorporated into the crisis management material of an academic continuity plan. However, there may be areas that a typical emergency action plan may not address, such as identifying resources to provide counseling for those who experience critical incident stress as a result of experiencing the event. These elements will need to be created and added to the crisis management section of an academic continuity plan.

The crisis management phase may also involve a shift in responsibilities from those in the emergency response phase. For example, the fire department may immediately take control of the situation during the emergency response phase by responding to a fire. However, leaders within the school may begin to take control of the situation as the crisis management phase comes into effect. This presents the need to have open communication between the school experiencing the incident and public emergency services. It is an issue of working together as a team to accomplish the goals of the program, not a matter of who has how much power. Emergency response and school personnel each bring unique strengths to the emergency response and crisis management phases.

The flow of communication in both the emergency response and crisis management phases is of critical importance. Decisions will need to be made in the heat of the battle. A school may have an individual designated as an emergency response coordinator. During the emergency response and crisis management phases, a great deal of information will be generated, and it is important to determine how this information will flow and the select few individuals with whom the emergency response coordinator will have direct contact. Defining this flow of communication will assist the emergency response coordinator in not becoming overwhelmed with information and will also facilitate more effective decision making based on the control of the information.

Work will need to be done to identify the most efficient way to incorporate emergency response plan information into the emergency response and crisis management sections of an academic continuity plan. On the surface, it might appear to be most efficient to simply fold an emergency action plan into an academic continuity plan. This could be a viable option, but it will be important to ensure that the approach to creating the academic continuity plan will accommodate all of the components of the emergency action plan. This may be easily accomplished if an academic continuity plan addresses detailed issues, but it may not be effective if an academic continuity plan addresses only macrolevel issues.

RECOVERY

The recovery phase is the final step in the process that will bring the school back to some form of working order. This might be manifested in a number of different ways. Alternate sites may have been identified to conduct class while operations are underway at the school to repair or rebuild from damage that occurred. Alternate sites can include such things as areas within the school that are undamaged, trailers temporarily brought onto school property, or facilities that are apart from the primary school property.

Online education is a tool that can be integrated into the recovery phase. Students can utilize a course management system to continue their coursework while repair and rebuilding efforts are underway at the school building. Technology has opened a new frontier to continue academic work while enduring time away from school. Technology can be utilized to continue work rather than accumulating the cost and logistical challenges of scheduling additional school days in the summer because of school cancellation days in the winter from snowfall.

Recovery to normal operations will include the management of a wide variety of functions. Salvage companies may be involved through the collection and removal of debris. Various contractors will be involved to repair or rebuild from damage that has occurred. Public utilities may be involved through the restoration of power, water, or gas to the school. This dynamic activity will require the skills of someone who is adept at project management to ensure that all activity is occurring as needed to bring the school back to full operational capacity as quickly as possible.

PLAN TESTING

Once all of the documents necessary to establish an academic continuity plan have been created, third party agreements have been secured, and training has been conducted, the remaining item is to conduct ongoing testing of the plan. There are no limits to the testing that can be done, with the possible exception of money. Testing takes time, and time costs money. This is why testing should be included in the budget for academic continuity plan management.

The decision might be made to start with the least invasive type of testing, which is to conduct a table-top exercise to test the plan. In this type of test, an intricate scenario will be designed to determine whether the plan has taken into consideration all of the variables that may unfold. This exercise will require the presence of all of the stakeholders in a room to walk through the exercise. These individuals might include the school principal, certain teachers, maintenance manager, security manager, safety manager, and finance manager. The team present needs to represent the scope of activity that will occur as the plan is executed. From the scenario that has been created, information on cards can be recorded that could be provided to each team member at the table. These cards can represent the chronology of events that are used to test the plan. For example, the first card can be read that provides background information that defines the day of the week, time of day, and other details that will directly impact the event that is described as having occurred, such as a tornado striking the school causing complete power loss and specific damage to the building. It is important to provide as many details as possible regarding the scenario that the director of the exercise can use to develop the scenario as it unfolds. After the card has been read, the group can be asked to walk through the plan to determine whether sufficient contingencies are in place for the emergency response phase of the academic continuity plan.

The next person can then read their card that represents a challenge and falls in chronological order with the scenario. For example, a challenge may be presented that not only has the tornado struck the school, it has also caused city-wide destruction. In this situation, city infrastructure and other entities are given priority by public emergency services, such as hospitals, so it may be a period of time before professional assistance arrives. Once this challenge card has been read, the group can review the plan and draw on their training to determine whether appropriate contingencies are in place to address the problem that has been presented. The group can continue to work its way through the cards chronologically as each team member at the table presents a new challenge to the group. The exercise should walk through the three phases of emergency response, crisis management, and academic recovery.

The limitation of table-top drills is that the group is not able to physically see what is occurring. Although this level of reality cannot be achieved, table-top exercises do present a great opportunity to utilize imagination to create a detailed scenario against which the academic continuity plan can be tested. Table-top exercises can be used to test all or part of an academic continuity plan. A scenario may be created that will be addressed by the complete management team or a scenario may be chosen to conduct a focused table-top drill with those who work in a given department or area of the school. Departmental table-top drills may provide the opportunity to also engage employees in the process rather than only involving members of management.

A second form of testing is to actually test a component of the academic continuity plan. An example of this that may be very familiar is a fire drill. A fire drill will allow a school to test the evacuation of the building as a part of the emergency response phase of the academic continuity plan. A pull station can be activated that sounds an audible alarm and strobe lights, which should in turn result in everyone exiting the building and being accounted for in assigned assembly areas. Although many schools will notify employees on the day of a drill that a drill will be occurring, it is preferable to only inform those who truly need to know, such as the people responsible for gaining permission from the principal for the drill and communicating with other individuals in high ranking positions. A more typical response to an alarm can be witnessed by not informing faculty, staff, and students that a drill will be occurring.

Another way that a component of the plan can be tested is to work with the IT department. The integrity of computer infrastructure, including all hardware and software applications, is becoming a more sensitive component of academic operations. Therefore IT may need to become a focal point of the academic continuity plan. Although the person charged with creating the academic continuity plan may not be an expert in computer information systems, the IT manager can get involved by walking through all of the possible failures that could occur and what the plan currently accommodates, from backup generators to transferring the work to another school or a third party.

Specific components of the crisis management phase of the plan can also be tested. This is important because of the infrequent exposure to many of the things that might occur during this phase of the plan. For example, a portion of crisis management might be to bring counselors on site to help faculty, staff, and students deal with what has occurred. This part of the plan can be tested simply by contacting the organization that was included when the plan was created to ensure they still have the capability to provide services that were originally discussed. It may be found that they cannot because of budgetary cutbacks or because they are no longer in business. This could similarly be done with a number of the third party arrangements that may have been made for both the crisis management and academic recovery phases of the plan.

A third way in which the plan can be tested is to conduct a full-scale exercise. Among plan testing options that are available, this will probably be the most time- and cost-intensive exercise. A great deal of work will need to be done in planning the exercise, including developing the scenario, coordinating with all of the stakeholders, and conducting the exercise at a convenient time for school operations. Such an exercise will involve the physical presence and activity of everyone engaged in the plan. In a table-top exercise, activity is only discussed. In a live exercise, personnel will actually be executing responsibilities. A number of groups can be engaged, including the fire department, school board offices, emergency medical services, police department, utility companies, contractors, and everyone in the school.

The benefit of a live full-scale exercise is that it allows the opportunity to actually see whether the plan can function. Although a focused fire drill will verify whether the building can be effectively evacuated, a live full-scale drill will verify whether there is enough room to stage fire apparatus and supply water for long-term fire ground operations. It will verify whether the triage area is adequate for mass casualties. It will also verify whether communications are effective between the school, public emergency services, and the school board offices.

Conducting any form of testing, whether it is table-top, component, or full scale, will require planning as to how to evaluate the exercise. Plan testing is of limited use unless there is adequate design to record information, make an evaluation, and implement plan changes. For example, you may choose to conduct a fire drill. How would you evaluate it? Would you simply click a stopwatch when the alarm is activated and click it again once the last person has exited the building or been accounted for? Although the use of a stopwatch will provide basic information regarding the evacuation of the building, there is a great deal of information that will be missed. An evaluation sheet can be designed that includes a broad range of information. The evaluation sheet can include the need to capture information such as how long it took to evacuate an area of the building, how well the alarm could be heard, the proper shutdown of equipment, and how faculty, staff, and students behaved during the evacuation. Monitors can be assigned an area of the building moments before the drill and an evaluation can be conducted from the moment the alarm sounds until the last person has left their area. In addition to reviewing these evaluations with the team following the drill, the security system can be reviewed to generate a report to determine which exit doors were activated during the alarm to ensure people actually left the building through their nearest exit. The detail of this evaluation is very valuable. It can provide much more information regarding the execution of the plan than simply how long it took to clear the building. Capturing valuable data will need to be a priority in the testing process. This will allow the team to make a substantive evaluation of the process and determine what changes may needed to improve the effectiveness of the plan.

CASE STUDY

Flooding has not occurred in Townville since the 1930s. However, a particularly harsh winter has resulted in flooding that has impacted the property where the area high school stands. Flood waters have reached the level of the locker handles and are slowly rising. It is clear that the building will not be habitable until the fall because of the amount of time that it will take to conduct salvage operations, and repair or rebuild the school. Ten weeks remain before the end of the school year.

- In what ways can an academic continuity plan be used to address this situation?
- What specific contingency plans will assist school administration, faculty, and staff maintain school operations immediately following the event and continuing through the end of the school year?

EXERCISES

For the following questions, identify a single school environment in which you would like to situate your responses and answer each question accordingly.

1. What are the three phases of academic continuity?
2. What methodological steps can be used to create an academic continuity plan?
3. What issues are involved in initiating an academic continuity plan?
4. What activity must occur in the emergency response phase?
5. What activity must occur in the crisis management phase?
6. What activity must occur in the academic recovery phase?
7. What options are available to you to test an academic continuity plan?

ACKNOWLEDGMENTS

The majority of the material for this chapter was taken from a course developed for an online master's degree program in safety, security, and emergency management at Eastern Kentucky University (SSE 890: Business Continuity).

REFERENCES

Hiles, A. (2011). *The Definitive Handbook of Business Continuity Management* (3rd ed.). Wiley, West Sussex, UK.

Section IV

---

*Program Development
and Execution*

27 Creating Written Programs

E. Scott Dunlap

CONTENTS

The phrase "safety program" can mean two different things. First, a school "safety program" is an overarching program that addresses all activity that is directed to protecting students, faculty, staff, contractors, and visitors. Within this larger effort, numerous individual safety programs and activities will exist. It is this more specific area of "safety programs" that will be discussed here. For example, within your comprehensive safety program you will have specific safety programs such as:

- Hazard communication program
- Fire prevention program
- Fall protection program
- Playground safety program
- Access control program

Two elements are necessary in the beginning stage of developing written safety programs. The Occupational Safety and Health Administration (OSHA, n.d.) emphasizes the need for leadership within an organization to support the development and implementation of a safety program and the need for employees to be engaged in the process. Friend and Kohn (2010) echo the need for management support and employee engagement. Management support will help to set the tone of the need for the program within the organization. It will then be deemed as important by faculty, staff, and students. Petersen (2001) further defines activities that leaders at various levels can perform to demonstrate support for the implementation of safety programs, enforcing the use of procedures delineated in a program, and training employees on responsibilities that are assigned through a safety program. Employee engagement will help everyone to take ownership of the program rather than it being developed in an isolated office and then thrust upon everyone (Geller, 1996).

A familiar saying within the safety profession is, "If it isn't documented, it didn't happen." Whether there is a concern regarding regulatory compliance or establishing a high degree of internal performance, written safety programs provide the action plan to achieve a safe work environment. It also establishes the documentation that is needed to demonstrate that the program is functioning. A model will be presented in this chapter that can be applied to the development of any individual safety program. The sections of a written safety program include:

- Purpose of the program
- Scope of the program
- Delineation of responsibilities among those impacted by the program
- Equipment needed to implement the program
- Procedures used to carry out the program in the workplace
- Employee training content and requirements
- Corrective action that is provided when individuals are found to be non-compliant with the program
- A revision history of changes made to the program
- Appendix material

Populating each of these sections will help to establish a comprehensive written program for each of the specific areas contained within the scope of a comprehensive school safety program.

PURPOSE

The purpose section of a written program is a concise statement of why the program exists. Two areas of focus can be addressed in this section. The first is legal compliance. A primary reason for having most written safety programs in place is to comply with a piece of legislation. For example, within OSHA standards, there is a law that focuses on the requirement to inform employees about the hazards of chemicals that are used in their work activity. This regulation is cited as 29 CFR 1910.1200 and is commonly referred to as the hazard communication standard. This standard could be applicable to school janitorial or maintenance staff that use hazardous chemicals in the performance of their job. The purpose of the school hazard communication program will be to comply with 29 CFR 1910.1200, OSHA's hazard communication standard.

A second purpose of any written safety program goes beyond regulatory compliance to the moral and ethical responsibility of an employer to provide a safe and healthful work environment. In addition to focusing on regulatory compliance as a purpose of a written program, a second purpose is to prevent injury and illness of students, faculty, staff, contractors, and visitors. The purpose section might be as brief as the following statements:

> The purpose of this program is to comply with the OSHA Hazard Communication Standard 29 CFR 1910.1200. The purpose of this program is also to ensure that all individuals who are affected by this program are provided with the information necessary to protect themselves and others from injury and illness when working with hazardous chemicals.

SCOPE

The scope section of a written safety program will be equally as brief and focused as the purpose section. Two issues can be addressed within the scope section of a written safety program. First, scope can be addressed in light of defining the environments where the program applies. One example of this might be to view the application of the written program from the perspective of individual schools. If the program is written from a state, county, or district perspective, then the program would apply to all schools within that entity. If the program is written from an individual school perspective, then the scope of application would be only that school.

The second aspect to be addressed within the scope section is to clearly define the people to whom the program applies. Remaining with the example of creating a hazard communication program, the people to whom the program applies could include faculty, janitorial staff, maintenance staff, and contractors. People within each of these three groups could become engaged in activity that requires the use of hazardous chemicals in a school or on school property. The scope section might read as follows:

> This program applies to all schools operated within the county school system. This program applies to all individuals who work with hazardous chemicals, including faculty, janitorial staff, maintenance staff, and contractors.

DELINEATION OF RESPONSIBILITIES

Where the purpose and scope sections were very short and focused, the section that presents a delineation of responsibilities will begin to take the written program to a different level of complexity. Here, it is important to specify who will perform various tasks and be accountable for various responsibilities within the program. As an issue of ongoing written program maintenance, it will be helpful to establish responsibilities by job title where feasible. This will eliminate the need to continuously revise the program as individuals change positions or when turnover occurs among faculty and staff.

Responsibilities should be delineated to cover the spectrum of levels within the school or school system depending on the scope of the program. This might include faculty, administrative staff, custodians, maintenance personnel, vice principal, principal, and superintendent. The delineation of responsibilities within a hazard communication program might read as follows:

- Principal: Provide support for the program and allocate funds as needed within the school operating budget to address the financial needs of the program
- Faculty: Be aware of the requirements of the program; ensure chemical labels are provided as needed and that material safety data sheets are available for hazardous chemicals that are used in the classroom
- Custodians: Receive training and comply with all requirements of the program
- Maintenance staff: Receive training and comply with all requirements of the program
- Administrative staff: Procure chemical labels and signs as needed to clearly communicate the hazards of chemicals that are used in the school
- Contractors: Provide copies of material safety data sheets for all hazardous chemicals to be used on school property prior to their use; must execute job activity that meets or exceeds the requirements of this program

The goal is to evaluate all activity that must occur to implement the program and assign responsibilities to individuals who are involved. The school or school system must be examined to determine who these individuals are and what activity they must engage in to effectively implement the program.

EQUIPMENT

A list of equipment needed to manage a given safety program is helpful in that it provides a dedicated area of the program where employees can reference what physical items are needed to implement the program. Delineating this in the written program will also help to ensure that the awareness of these items can be more easily carried into practice. This section of a written program can be brief and simply contain a list of the equipment that is needed to support the program. For example, a hazard communication program might include the following information:

Equipment needed to implement and sustain this program includes:

- Material safety data sheet stations to be placed in areas where hazardous chemicals are used and stored
- Labels to be placed on secondary containers in which hazardous chemicals are stored from the primary container, such as a spray bottle being filled from a drum of hazardous chemical
- Personal protective equipment, such as gloves, respirators, safety glasses, and goggles that must be worn when hazardous chemicals are being used
- Safety signs posted in areas where hazardous chemicals are used and stored

In addition to providing a list of equipment that is utilized within the scope of the program, it might also be useful to provide a list of suppliers that accompanies the list of equipment. Sharing this information could make it easier for individuals to know where to obtain equipment in the event that it needs to be procured. Providing this information in a written safety program will need to be done in consideration with ensuring that school system procurement processes are also being followed.

PROCEDURES

Written procedures will occupy the majority of space in a written safety program. Procedures will be presented that address how each component of the safety program must be executed. It is important to differentiate operational information from job safety information when writing safety procedures. A written safety program is typically focused on only the safety aspects of the topic that is being addressed. For example, OSHA has promulgated a regulation that deals with welding in the workplace. Welding is a spark-producing activity that is used to repair damage that has occurred to a metal object, such as a storage rack that might be damaged. When welding occurs, there are certain safety procedures that must be executed prior to, during, and following the activity of welding to ensure that a fire does not occur in the workplace. A "hot work program" will address these safety procedures, but it will not address the operational activity of how to weld, such as selecting the type of welder and welding rods. Operational training such as this is typically addressed in separate skills training and written procedures.

A resource that can be used to begin the process of writing safety procedures in a written program is the applicable OSHA regulation. Each OSHA regulation will stipulate requirements that must be met in relation to the topic that is being addressed. The OSHA regulation for hazard communication is contained in Chapter 29 appendix. Although the regulation is extensive, it addresses many issues that are not applicable to a given school or school system, such as creating a material safety data sheet. Schools are consumers of such materials and not the creators of them, so the applicable information for a school would be requirements that address the use of hazardous chemicals. This would include such things as:

- Procuring material safety data sheets
- Making material safety data sheets available in the workplace
- Chemical container labeling
- Employee training
- Providing personal protective equipment for employees who must use a hazardous chemical
- Maintaining a list of hazardous chemicals

The regulation must be explored to determine all issues that must be complied with, and then these items can be converted into procedures in a written program. Procedures will clearly articulate how each of these items will be physically manifested in the school environment. In addition to issues identified in the regulation, procedures must also be considered in light of what things will need to occur when managing the program. For example, a best practice is to have a clearly communicated method by which new chemicals can be introduced into the school environment. Information can be included in the procedures section of a written hazard communication program that delineates who must approve chemicals and the process by which these chemicals can be introduced into the school environment, including things such as procuring a material safety data sheet, ensuring proper container labeling, and employee training. For example, such a section in a written program might be presented as follows:

INTRODUCTION OF NEW CHEMICALS

Employees and students are prohibited from independently purchasing chemicals to be used in the school environment. All new chemicals to be purchased must be reviewed by the director of facilities for approval. This approval will include a review of the hazardous chemical's material safety data sheet to determine whether it is acceptable for the intended application. In situations where the chemical hazards are considered to be too hazardous, a safer alternative product must be identified and procured. Contractors working on school property must provide copies of material safety data sheets for all hazardous chemicals to be used when performing work. The director of facilities must review the material safety data sheets and provide approval or denial of the use of the chemicals based on the hazards that the chemicals present. If a chemical is denied, the contractor must identify a safe alternative product.

Greater detail can be provided based on the unique situation and the needs of a given school or school system. The benefit of providing detailed information in the form of written procedures is that the program will outline specific expectations of performance. This information can be accessed by everyone so that safe work practices can be executed.

EMPLOYEE TRAINING

Information regarding employee training can be included only from the perspective of general content, methodology, frequency, and documentation. This section of the written program will not include actual training material, such as PowerPoint slides and tests. That material can be included as an appendix to the program. This section will specifically address how training occurs.

Although a great deal of training occurs only in the classroom, it is considered a best practice to also integrate practical skills training to engage participants in the event rather than them being passive. This can be accomplished through exercises in the classroom as well as hands-on activities in the school environment. The general process that is used for training can be presented here in the written program.

The frequency with which training will be conducted can be addressed. Typically, this will include stating training requirements upon initial hire and periodic refresher training to be provided on an ongoing basis. Regulations will often stipulate when training must occur. For example, the hazard communication standard includes information and training requirements in 29 CFR 1910.1200(h). This section indicates that training must occur at the time of assignment to a job where hazardous chemicals are used and whenever a new hazard is introduced to which employees have not been previously trained. These two intervals must be included in the frequency at which training occurs in a hazard communication program. In addition, freedom exists to go beyond these minimal requirements to include annual refresher training on the core components of the program.

Documentation to be used in the training can also be delineated. A standard component of training documentation is the training log. This is a document that contains a chart listing printed participant names, participant signatures, name of the trainer, topic of the training session, and the date on which the training occurred. A second common piece of documentation is a test that is administered at the close of the training session as a form of comprehension verification. A third critical piece of training documentation, although not often used, is the certificate of completion. This document outlines the core components that were reviewed in the training. Participants will sign and date the document, indicating that they agree that the material was covered in the training session. The trainer will also sign and date the document. The certificate of completion is important in being able to hold individuals accountable for the training they received. For example, an employee might attend hazard communication training and later be found to be using a hazardous chemical without wearing the appropriate personal protective equipment. When someone calls the employee's attention to the problem and attempts to implement a form of discipline resulting from to the infraction, the employee might respond, "That was not covered in the training session I attended." It can be difficult to prove that the information was covered in a training session without an employee's signed acknowledgment that the material was included. Although a PowerPoint presentation can be presented that is used in such sessions, proving that every slide was reviewed can be difficult. This challenge can be avoided by utilizing a certificate of completion that contains the core elements of the training session.

CORRECTIVE ACTION

Accountability can be integrated into written safety programs similar to all other school programs and policies with which individuals must comply. It is unfortunate that even the best efforts in implementing a program will still result in some who choose not to comply with the requirements of the program. Accountability systems typically include progressive discipline where employees are given the opportunity to improve once the violation is brought to their attention. This might include:

- Verbal warning
- First written warning
- Second written warning
- Final warning (resulting in termination)

Although the system is progressive, the human resources department may deem a certain situation to be worthy of advancing to higher levels depending on the severity of the program violation. This section of the program must be in harmony with human resources policies on discipline and corrective action. A general statement that can be included in written safety programs might be:

> All school system employees and contractors must comply with the requirements of this program. Violations of this program will result in corrective action up to and including termination (employees) or expulsion from school property (contractors).

This statement clearly indicates that there will be consequences for violating the program and provides flexibility when determining the level at which corrective action will be administered based on any given infraction.

The actual document to be used when administering corrective action can be included as an appendix to the written safety program. This document can outline:

- What program was violated
- What specifically was done to violate the program
- What the employee or contractor needs to do in order to achieve performance improvement
- What will occur if performance improvement is not achieved

This information will clearly communicate to the individual receiving the corrective action what behavior occurred that was unacceptable and what must be done to improve. Although this communication might be perceived as being "written up," it can be perceived in a positive light in that open communication has occurred in a way that seeks to rehabilitate the relationship with the employee or contractor rather than simply attempting to terminate it. It provides the individual with the ability to take control of the situation and alter behavior that can result in negative consequences.

REVISION HISTORY

A revision history will be the last primary section of a written safety program. This can be presented in the form of a simple chart that records the evolution of the program. Changes can be recorded to ensure that everyone is abreast of the most recent events in the life of the written program. It can also serve as a form of document control to ensure that current forms or procedures are being used in program implementation activity. The chart could include an item number that chronicles each event of the program document, the page number that was changed, a brief description of the change that occurred, the name of the individual who made the change, and the date when the change was made. The chart could appear as shown in Table 27.1.

APPENDICES

Each written safety program can contain appendix material that is comprised of documents or resources that are used to implement the program. Each appendix can serve to easily assist employees in identifying tools that are critical to the implementation of the written program. Each document or resource can be determined based

TABLE 27.1

Revision History

Item Number	Date	Page Number	Change	Person Responsible
01	2-20-12	All	Program created	John Doe
02	10-18-12	5	Changed "facility manager" to "risk manager"	John Doe
03	11-11-12	Appendix	Included copy of the school corrective action form	John Doe
04	12-10-12	8	Inserted list of where material safety data sheet stations are located	John Doe

on the unique nature of the written program. A hazard communication program might have limited information in an appendix, such as a chemical inventory and a corrective action document. An emergency action plan might have a much larger appendix section including things such as school maps, emergency contact lists, a bomb-threat checklist, and a facility inspection form. The appendix section is an opportunity to identify all existing resources that are available to implement certain safety programs in addition to exploring the opportunity to create new resources that will facilitate the implementation of the program.

It is important to consider the degree of access to such tools compared with simply referencing where the tools are located. For example, a copy of OSHA's hazard communication program might be included as an appendix to a hazard communication program because of the value of having quick reference to the information while reading the program compared with only providing information on where this information can be obtained. The value of including the OSHA regulation for quick reference as an appendix supersedes the value of simply providing directions as to where this regulation can be found on the Internet.

OUR MOTIVATION

There are at least two primary motivating factors for having written safety programs. The first might be seen as the "stick": regulatory compliance. Fundamentally, the consequence of not maintaining written safety programs could be some form of penalty. This penalty could be in the form of citations that are given by OSHA when written plans are not present or are found to be inadequate. This penalty could also be poor publicity in the media if an event occurs and it is found that written safety programs were not present or were lacking and were not properly implemented.

The second motivating factor might be considered the "carrot": the positive incentive of having written programs in place because of the positive impact they have on protecting human life. Rather than seeing only penalty as a motivating factor, the positive result of protecting people should be seen as a motivating factor for having written safety programs. Beyond simple compliance, the objective of workplace safety is to safeguard individuals from injury and illness. Adjusting to this program can aid in broadening the spectrum of value of written safety programs from compliance to a level that provides the greatest protection to individuals.

AREAS FOR PROGRAM DEVELOPMENT

Written safety programs can be utilized to develop school safety efforts in two areas. The first area includes all topics that are regulated. This might include the need to provide an OSHA-compliant written safety program that addresses personal protective equipment to be worn by custodial staff, maintenance workers, or those who work in a laboratory. It might also include the need to provide a written hazard communication program for employees who work with hazardous chemicals. Each of these programs will be dictated by the requirements of OSHA or some other government agency.

A second area of opportunity to develop written safety programs is in the area of nonregulated issues. This might include the development of a written ergonomics program that addresses repetitive motion risks, such as working at a computer workstation. Ergonomics is the science of how the human body interacts with the environment. Poorly designed workstations can result in back or wrist pain caused by unnecessary force being placed on these parts of the body. There is no regulation that addresses ergonomics, yet we know that ergonomics is the cause of a great number of work-related injuries and illnesses. Another example is fleet safety. A school system might own standard vehicles that are operated on public roads, such as pickup trucks and sedans. From a safety perspective, these vehicles are not regulated by the Department of Transportation (DOT) because these vehicles do not meet the definition of a commercial motor vehicle. OSHA is concerned about the place of work, so these vehicles are not covered by OSHA regulations because they are operated on the road away from school property (the place of work). However, a fleet safety program can easily be developed by identifying transferable elements from DOT regulations and other accepted best practices, such as:

- Obtain a motor vehicle report for all fleet vehicle drivers
- Ensure that each fleet vehicle driver has a valid driver's license
- Implement a documented preventive maintenance program for fleet vehicles
- Conduct preuse inspections on fleet vehicles
- Require fleet vehicle drivers to wear seatbelts
- Implement a distracted driver policy that addresses the prohibition of behavior such as texting while driving
- Provide driver safety training that addresses safe driving issues such as inclement weather and adverse road conditions

There are a number of areas where written safety program development can occur that will be beneficial to school system employees. These areas include both regulated and nonregulated areas. The objective will be to evaluate all of the risks that impact a given school and develop the appropriate written programs to protect students, faculty, staff, and visitors.

PROGRAM SUCCESS

Some form of program evaluation will need to be in place in order to determine whether the written programs are successfully implemented. Prior to such an evaluation, one element that is critical to the success of programs is to involve those who will be impacted by programs in the development phase. This will help everyone to see the programs as something that they had a hand in creating, which will also aid in achieving a high degree of acceptance of the programs within the school system. If people are not involved in the creation process, the programs might be viewed simply as requirements that are being thrust upon them. Encouraging participation in program development will help to establish a sense of ownership among all who are affected by the programs.

Once programs are in place, it will be necessary to periodically audit them to determine how well they have been implemented and are performing. This can be done by auditing the process in three areas:

- Documentation review: Written safety programs and supporting documentation should be evaluated to determine whether the appropriate information is present and that documents show evidence of program implementation. For example, a fire safety program might include the use of a weekly school inspection to confirm that exits are not blocked, fire hazards are properly managed, and fire system sprinklers are not obstructed. In this situation, both the written program and a file of the documented inspections that have been conducted would be audited to determine whether the program is adequate in scope and is functioning well.
- Facility inspection: A facility inspection can be conducted to determine whether what is recorded in the written program and is presented in evidence of documentation is verified to be true in the actual environment. Although the written fire safety program may appear to be in order, it may be found that exits are blocked routinely throughout the school when a facility inspection is conducted.
- Employee interviews: A third area in which auditing can occur is to conduct employee interviews. Arrange times when faculty, staff, and contractors can be asked questions regarding the program. For example, faculty can be asked what fire hazards need to be managed within their classrooms. Their ability to effectively respond and converse on safety program topics will serve as an indicator of the degree to which the programs are being fully implemented.

A goal of auditing is to measure the degree to which programs are being implemented. A danger exists in that written safety programs can be incredibly well crafted but not implemented. This can be manifested by a situation where the materials are very well written on paper, but no training occurs, and materials are not purchased to support the implementation of the program. Auditing can be performed to determine how well-written safety programs are being implemented by evaluating all written documentation, conducting a facility inspection, and conducting employee interviews.

CASE STUDY

Sarah has worked as a custodian for Phillips Middle School for three years. She routinely works with a number of chemicals to perform housekeeping throughout the school. While mixing two chemicals together, Sarah is overcome with fumes that are generated and loses consciousness. A nearby teacher witnesses what happened and calls for an ambulance. Upon investigating the incident, it is found that Sarah's behavior was routine. She even reported that she was trained to mix the chemicals when she was initially hired for the position. It is also found that the chemical container labels expressly state that the chemicals are not to be mixed with any other chemical. When asked why the chemicals were mixed together, Sarah reported that the combined solution resulted in a much stronger cleaning solution. The investigation further revealed that no formal training or communication regarding chemical hazards occurred, and no written program was in place or implemented.

- Would the presence of a written program have helped to prevent this incident? Why or why not?
- What options are available to prevent this incident from occurring again in the future?

EXERCISES

For the following questions, identify a single school environment in which you would like to situate your responses and answer each question accordingly.

1. What general sections should be included in a written safety program?
2. Why should written safety programs be created and implemented?
3. What role does auditing occupy in the life of a written safety program?
4. Should only regulated areas of safety be considered for written program development? Please explain your answer.

REFERENCES

Friend, M., and Kohn, J. (2010). *Fundamentals of Occupational Safety & Health*. Government Institutes, Lanham, MD.
Geller, S. (1996). *Working Safe: How to Help People Actively Care for Health and Safety*. Chilton, Radnor, PA.
Occupational Safety and Health Administration. (n.d.). Safety & Health Management Systems eTool. Accessed November 17, 2011. http://www.osha.gov/SLTC/etools/safetyhealth/index.html.
Petersen, D. (2001). *Safety Management: A Human Approach*. American Society of Safety Engineers, Des Plaines, IL.

28 Organizational Training

E. Scott Dunlap

CONTENTS

Those who work in different capacities within a school system need to be trained on safety issues that affect how their job is performed. Training can be manifested in a wide range of formats and content. There are a number of best practices that can be implemented in the training process that will make the time and expense dedicated to training worth the investment.

LEVELS OF TRAINING

Successful organizational safety training must resist the one-size-fits-all approach. Although it is efficient to design one training presentation and expose everyone to it, there is a need to consider the informational and skill needs of various levels that exist within the school system. Specific training strategies can then be implemented that address each of these levels. A needs assessment can be conducted to identify what training should be presented to various individuals in the school system (Cekada, 2010). This assessment can be the first step to establishing a training strategy to address all members of the faculty and staff. Groups to consider might include:

- Staff: Staff members in a school will need safety training that is function specific. Janitorial staff will need detailed training in hazard communication if hazardous chemicals are used. Maintenance technicians will need detailed training on lockout/tagout. Each staff member will need to be evaluated to determine what specific safety training is needed.
- Faculty: Faculty will need a mixture of function-specific training and general awareness training. They will need function-specific training in

areas that impact their safety and the safety of their students. Function-specific topics can include emergency response, bullying prevention, and fire safety. Faculty may also need to receive general awareness training in areas that do not directly impact them but with which they may come into contact throughout the work day. General awareness topics might include lockout/tagout because of maintenance technicians working on electrical equipment that affects activity in the classroom or hazard communication because of the chemicals that janitors might use in the classroom.

- School administration: Those who work in school administration, such as principals and vice-principals, may need to receive some function-specific training with a larger volume of general awareness training. These individuals will need to be aware of all safety programs that exist within the school, but their direct engagement might be limited to such topics as emergency response, where they will be responsible for leading efforts in responding to emergencies.

- School system administration: Similar to administrators at the school level, administrators at the system level, such as board of education members, may need function-specific training in areas such as emergency response where they will be directly engaged. They will also need to receive general awareness training in other topics so that they are familiar with all safety programs that exist within the school system.

A training matrix can be established that can be used to monitor training that each individual has received based on the role that they occupy within the school system. This can be done with something as basic as a word-processing-program chart or as complex as a database. The system must simply function to meet the needs of those who manage the school safety program so that training sessions can be scheduled and executed that consider all of the groups that exist in the school system.

TRAINING FORMATS

The training format selected is an area of significant challenge when designing a training strategy. An inherent issue is that what may work well for the trainer might not be the best format for the participants. Consideration must be given to the spectrum of formats available to make the experience meaningful and memorable for those who attend the training sessions.

A variety of training formats exist. Rather than simply relying on one-way communication in a classroom, various formats can be integrated to make the training session a meaningful experience for participants. An evaluation of the target audience and resources that are available will help to establish a training strategy that is most effective for a given school.

CLASSROOM

Classroom training is a commonly used format to conduct training. The classroom must be effectively utilized to ensure that a positive experience is had by those attending the sessions. The problem is typically not the classroom itself but how it is used. There can be environmental issues that will affect the training session, such as lighting, cleanliness, and comfort of chairs and tables. These factors should be considered before selecting the room in which training will occur. A tendency might be to conduct one-way communication in the training session by simply walking participants through a PowerPoint presentation. Such a format is easy for the trainer because the topic can be easily captured in an outline form and presented across a number of slides. Preparation and delivery are expedited. However, participants need to be engaged in the process in order to enhance learning. This can be accomplished through the addition of breakout group discussions, problem-posing scenarios, and role playing. The audience must be understood to determine the exercises that will be most effective in a given training session. Fanning (2011) identified role-playing as a successful training tool to engage participants, but it must be well-planned and implemented in order to be effective.

PRACTICAL SKILLS

Practical skills exercises can be provided as an opportunity for participants to engage in activities related to the topic that is covered in the training session. For example, rather than showing a diagram of emergency response exit routes and gathering and sheltering locations, participants can be taken into the school and walked through these areas. Similarly, rather than showing a copy of a material safety data sheet (MSDS) to participants in a hazard communication training session, participants can be taken out into the school were MSDS stations are located and shown how to read an MSDS and demonstrate safe work procedures that are related to various hazardous chemicals.

COMPUTER OR WEB-BASED TRAINING

Computer or web-based training modules can be purchased or developed internally. An evaluation will need to be made to determine the target audience for such a delivery system. Those who must navigate such training will obviously need access to a computer, so it might be difficult to utilize this format for certain employees based on their job activity. PowerPoint provides a great deal of functionality in developing point-and-click computer or web-based training modules through the use of timed slide transitions, word or photo hyperlinks, narration, and customized content. Canned training modules can also be purchased, but care must be taken that the content is appropriate for the participants and the work environment.

SCHEDULING TRAINING

The time at which training occurs is another variable that must be considered when creating a safety training system. One option is non- or low productive time that may occur while employees are working their normal schedule. An evaluation can be conducted to determine when such times occur for various positions. For example, maintenance staff may have periods of the year or week when there are fewer demands on them. These times can be considered as opportunities to conduct safety training.

Teaching schedules present a unique challenge in that time spent with students cannot be utilized as an opportunity for training, so teacher training may need to occur during dedicated time throughout the year when they are at school, but not in a class. In situations where training is conducted via computer or web-based modules, it would not be advisable for teachers to be expected to navigate the training during the course of a teaching activity. For example, a teacher might be administering a test that requires students to be independently occupied for a period of time. It might be assumed that teachers could navigate the training while monitoring their students. This division of concentration could serve as a significant limitation of the teacher being able to absorb all of the safety information, as well as being available to students for assistance.

Training could be scheduled at a time that is in addition to an employee's normal work schedule. Training could be scheduled for a weekend or at a time that is prior to the start or following the completion of a normal work schedule. This brings up the financial consideration of overtime pay for those who are paid at an hourly rate. Budgetary concerns will need to be included if this option is utilized to ensure that appropriate funds are available to pay for the training time. Employee morale should be considered when evaluating work schedules and training times. Hourly employees might be excited at the opportunity to report to work on a Saturday morning and earn overtime pay during safety training. However, those who are salaried may see it as time taken away from family activities with no additional pay.

The specific time of training during a given day should be considered in relation to the attention span of participants. Scheduling training at the end of a workday could cause employees to be distracted by thinking about the activities that they have planned immediately following the training. Scheduling training at the beginning of the day could help participants to remain focused in light of it simply being the first of many activities that will occur during the workday.

A number of variables exist that should be considered when planning training sessions for faculty and staff. Evaluating each variable within the context of each job function can help to identify training times that will result in the most positive outcome in participant learning.

TRAINING SESSION

Once the training format and timing have been identified, consideration can then be given to the actual delivery of the training session. It is important to plan for the right amount of content. If a training session is scheduled to last one hour, an appropriate amount of content must be planned that will fit within the hour. Participant frustration can easily rise when forty-five minutes have passed and only one-third of the slides in a PowerPoint presentation have been covered. Going beyond the allotted time can generate additional payroll cost and frustration among participants. Simply stopping the session at the one-hour mark will not be acceptable because of the need to eliminate training content or establish another time to complete the training at a later date. Although circumstances might arise periodically where this cannot be avoided, planning for the right amount of content should avoid this risk as a normal matter of course.

Delivering training sessions should allow for participant interaction. One-way communication from a trainer to participants will result in the lowest form of learning. Opportunities for engagement should be integrated in the delivery of the training. This can be fundamentally achieved by frequently asking the participants questions regarding the topic matter. Some participants might feel uncomfortable speaking in the larger group. Smaller group discussions of problem-posing scenarios can serve as a more structured opportunity for all participants to interact.

Creative thinking can be utilized to determine what training materials will be useful. A typical handout might be a hard copy of PowerPoint slides that are to be covered throughout the training session. In addition to this, other items can be utilized that will allow participants to see information. For example, handing out a copy of a material safety data sheet during hazard communication training will allow participants to actually see what one looks like and the information that is contained on the document.

The background of those in a training session should be considered to establish a level of communication that is effective. The trainer should be a subject matter expert in safety, so there could be a tendency to use jargon or vocabulary that is not understood by the audience. A level of communication must be established that accommodates the participants. Cullen (2008) found that in identifying with those in the training session, story telling can be an effective tool. Rather than the trainer being a professionally designated person, the trainer could simply be a veteran peer of those attending the training. The peer can use stories of things that have occurred in the work environment to focus on why safety is an important consideration while at work.

Visual aids can be used to assist participants in understanding material that is being presented in the training session. Technology has greatly increased our ability to execute training in this area. YouTube provides a wealth of educational videos that can be used to provide examples of issues presented in a training session. Although YouTube has been stereotyped as a tool for entertainment, a large volume of videos are available that have a great deal of educational value. These videos can be hyperlinked in a PowerPoint slide to provide immediate access during a training presentation. Personalized photos or videos serve as a visual aid that can have a great impact on participants. Rather than using canned clip art to enhance PowerPoint slides, photos or videos of school employees performing various tasks safely can be used to customize the presentation to the audience.

Technology has also become a tool that can be used to conduct training beyond the basic use of PowerPoint. LaRose (2009) presents the value of integrating audience response systems into training sessions. Audience response systems utilize a handheld device commonly referred to as a "clicker" to gauge audience response to various questions. For example, a multiple choice or true/false question can be presented on a PowerPoint slide. Each trainee in the room will have a clicker that can be used to press a button that selects the response that they believe to be correct. In addition to clickers, smartphone applications can be used to convert the phone into a clicker. The trainer can then immediately present the response to the question. This tool can be used innovatively in training to include such applications as anecdotal questions and questions that reference significant issues associated with the training session.

Melnik (2008) encourages the use of an approach that considers the rational, emotional, and physical needs of participants. From a rational perspective, the training must have some type of application to what the participants experience on a daily basis. This can present the risk of a one-size-fits-all approach in training. For example, there might be a need to train faculty and staff on hazard communication. The application of chemical use among different jobs will have different meaning, so a need exists to ensure that material presented in class applies to the daily needs of those who are in the class. The emotional perspective can be addressed by helping participants to see the personal benefit that can be gained as a result of implementing the material that is covered in the training session. This can be achieved by evaluating how personal meaning can be made from the material being presented. The physical perspective can be addressed by integrating the realities of the physical environment of the workplace into the training presentation by addressing challenges and opportunities that exist to improve safety. The goal is to address these three components in a holistic training presentation that touches on the various needs of the participants.

It can be easy for a trainer to simply prepare a training session where PowerPoint is used as an enhanced outline to cover applicable material. However, implementing a variety of tools throughout the training session can greatly enhance the experience of participants as well as their knowledge retention following the training session.

TRAINING DOCUMENTATION

Documentation of training is necessary to be able to prove who has been trained in various topics. There are three forms of documentation that can be generated during a training session:

- Training log: A training log is the most common form of training documentation. It is typically comprised of five pieces of information. The title of the training is the first piece of information and is typically recorded at the top of the training log. The remaining pieces of information are recorded in columns presented in a table on the training log. One column represents the date on which the training was conducted. A second column will include the name of the trainer who conducted the training session. A third column will include the signature of each employee who attended the training. A fourth column will include the printed name and employee identification number or other identifier that can be used to identify the employee in the event that the signature is illegible. Employees will typically sign the training log at the beginning of each training session.
- Comprehension verification: A test may be given at the end of each training session to document that each employee comprehended the information that was presented. Tests may be in the form of multiple choice, true/false, fill in the blank, matching, or short answer.
- Certificate of completion: A certificate of completion will identify four pieces of information. The name of the training session will typically

appear at the top of the page followed by a statement of agreement that certain information was covered in the training session. Each core item covered in the session will be presented in a bulleted list. A section will be provided at the bottom of the page for both the attendee and the trainer to sign and date. This document can serve as a type of contract between the employee and the school that the employee agrees that the information presented in the bulleted list was covered in the training session. This establishes a document that can be used to hold employees accountable to execute on material that they agree was covered in the training session.

These forms of documentation can be evaluated to determine what type of training documentation needs to be maintained for a given school or school system. Training logs can be maintained in binders that can easily be transported to training sessions, while tests and certificates of completion can be maintained in employee training files. A spreadsheet or database can also be created and used to easily track training that has been completed. Such a tool can be used to easily identify where training gaps occur that can be converted into a strategy to ensure that each employee receives the needed safety training.

Merli (2011) addressed the need to create training documentation in such a way that it can be understood by the participant. She recommended writing the material at a sixth-grade level, which will allow for the reading skills of most adult learners.

CASE STUDY

John is the risk manager for Fairfield County Schools and has recently conducted a safety audit of each of the schools in the county. A significant deficiency that he identified was in the area of safety training. Each school in the county had independent ways of doing training and different schedules used to conduct the training. He determined that approximately 30% of new hires in the county had not yet received any initial safety training and also found that approximately 50% of the remaining employees did not have training that is up to date. Training files were lacking in general as to documentation that was being maintained among the schools.

- What system design might John consider implementing among the county schools to ensure that safety training is being conducted?
- What format of training would be most appropriate for the various employee groups within the school system?

EXERCISES

For the following questions, identify a single school environment in which you would like to situate your responses and answer each question accordingly.

1. How might training content and delivery differ based on the various levels of employees in a school or school system?

2. What training format would be most appropriate to provide training to teachers at a given school?
3. How would you design a training schedule that takes into consideration each employee group at a given school?
4. What variables should you consider when determining how to deliver a training session?
5. What documentation would you maintain as a record of training that has occurred? Why?

REFERENCES

Cekada, T. (2010). Training needs assessment: Understanding what employees need to know. *Professional Safety*, 55(3), 28–33.

Cullen, E. (2008). Tell me a story: Using stories to improve occupational safety training. *Professional Safety*, 53(7), 20–27.

Fanning, F. (2011). Engaging learners: Techniques to make training stick. *Professional Safety*, 56(8), 42–48.

LaRose, J. (2009). Engage your audience: Using audience response systems in SH&E training. *Professional Safety*, 54(6), 58–62.

Melnik, M. (2008). The rational, emotional and physical approach to training. *Professional Safety*, 53(1), 49–51.

Merli, C. (2011). Effective training for adult learners. *Professional Safety*, 56(7), 49.

29 Program Auditing

E. Scott Dunlap

CONTENTS

Once a school safety program is in place, it will then be important to evaluate its degree of implementation by utilizing a program audit. Auditing provides an opportunity to measure the performance of the program against established standards of performance. Much like a financial audit, a school safety audit will explore each aspect of the school safety program to determine what standards are being met and where opportunities for improvement exist. Safety auditing can be accomplished through three areas of concentration: documentation review, physical inspection, and employee interviews (Dunlap, 2011).

DOCUMENTATION REVIEW

A school safety audit can begin with a review of all documentation associated with the program. Although this step does not have to occur first in the audit process, a benefit of performing a documentation review as the first step in a school safety audit is that it provides the auditor with an understanding of what is occurring in the given school. Evaluating the written documentation first will help to prepare the auditor to more effectively engage in the school inspection and interviews by knowing what level of written program development has occurred. This program development can be confirmed by investigating the success of implementation activity by performing the school inspection and interviews.

An auditor reviewing documents will need to consider a spectrum of items. First, all written safety programs should be reviewed. The content of these documents can be measured against regulatory requirements as well as school system policies.

Second, supporting documents for each program can be reviewed. This will include such things as:

- Fire system acceptance test documentation from when the system was initially installed
- Alarm test documentation
- Fire system testing documentation (sprinkler system and smoke detectors)
- Fire extinguisher inspection and testing forms
- Fire hose test/reracking documentation
- Weekly and/or monthly safety inspection forms
- Documentation of follow-up performed on identified safety deficiencies
- Hand tool and equipment inspections
- Powered motor vehicle inspection forms (forklifts)
- Aerial lift inspections
- Lockout/tagout procedure observation forms
- Monthly eyewash station inspection forms
- Preventive maintenance documentation
- New equipment inspection/testing records to ensure safety of operation
- State inspection records on any applicable pressurized vessels
- Thermographic test results on equipment and electrical panels
- Employee training documents
- Hot work permits
- Injury and illness recordkeeping
- Occupational Safety and Health Administration 300 logs
- Medical records
- Safety committee meeting minutes

This list is not intended to be comprehensive but provides ideas on documentation that should be reviewed within the scope of a school safety audit. Reviewing such documents will provide an initial indication of how much safety program implementation activity has occurred. For example, if it is found that there are no documented safety inspections, then the inspection aspect of the school safety program is lacking in implementation.

SCHOOL INSPECTION

After reviewing written program documentation, the school inspection provides an opportunity to verify that what was recorded in the programs is actually being implemented. The school inspection will include a physical walk through of every part of the school. It will be important to explore every room, closet, storage area, and maintenance workstation. The inspection might include an escort, someone who is intimately familiar with the property, such as a maintenance technician or custodian. It is important for the auditor to maintain a degree of control of the inspection and avoid the temptation of simply going where the escort leads. The auditor will need to open every door, climb every ladder, and walk every passageway to ensure that the facility as a whole has been observed.

The inspection phase is designed to observe physical conditions. This will include the need to observe such things as:

- Emergency exit routes being free of obstruction
- Lack of chains and locks on emergency exit doors and the presence of panic hardware
- The presence and level of charge of fire extinguishers
- Storage of material so as to not block sprinkler heads
- Condition of tools used by maintenance
- Condition of fleet vehicles
- Presence of material safety data sheets
- Functioning of emergency lighting

The inspection phase is also designed to observe human behavior. Procedures outlined in written safety programs should be witnessed where possible in practice as the inspection occurs. This indicates that the inspection will need to extend to all shifts worked by school system employees. It will require the auditor to start early, as soon as the school is opened for the arrival of faculty, staff, and the first bus delivering students. Daytime activity can be observed as the school day unfolds. The audit must continue into the evening in order to evaluate the activities carried out by custodial staff that might perform their work after the end of the school day. Special events may also need to be taken into consideration, such as football games, to observe issues related to event safety. The scope of the audit will present unique challenges in scheduling the inspection in a way that allows the auditor to observe as much behavior related to school safety as possible. Behavior to observe during the inspection might include:

- Maintenance employee locking out the energy source of a piece of electrical equipment prior to working on it
- Access control of students, faculty, staff, and visitors entering the property and the school building
- Personal protective equipment being properly worn
- Faculty using ladders or step stools to access items at a height rather than climbing on desks and chairs
- The occurrence of a fire or severe weather drill
- Chemical disposal
- Ergonomic condition of staff who work at computer workstations throughout the day

The goal will be to observe as much behavior as possible throughout the inspection. An obvious concern is that not all behavior will naturally occur when the inspection is underway. It is acceptable in some situations to manufacture the behavior that needs to be observed. For example, although maintenance employees may not be working on equipment that requires the use of lockout while the inspection is occurring, it is acceptable to ask a maintenance employee to demonstrate the behavior. Although this is not ideal, it will confirm at a basic level if maintenance employees can properly execute lockout procedures that are recorded in a written safety program. Another option is to perform the audit across all hours that will account for the scope of activity that needs to be observed (Cahill, 2001).

INTERVIEWS

Interviews provide an additional level of verification of the implementation of written safety programs beyond the school inspection. Interviews are opportunities to measure the degree to which faculty and staff can converse on safety-related topics. Open-ended questions are the desired format for interview questions because they prompt comprehensive responses from the individuals being interviewed. Conversely, closed-ended questions should be avoided in that they only require the interviewee to respond by saying "yes" or "no." For example:

- "Have you received emergency response training?" This closed-ended question will only result in a "yes" or "no" response. This information can actually be confirmed through training documentation review and yields no substantive value in an interview.
- "What should you do if the fire alarm is activated?" This open-ended question solicits a much more robust response in that the interviewee can openly explain what procedures will be executed.

The environment in which interviews occur can vary. It will be important to consider the operational needs of the school when conducting interviews. For example, it would be inappropriate to interrupt a class in order to ask a teacher interview questions. The schedule of school operations can be reviewed to determine when teachers are on a break or in a planning period when it might be more appropriate for the interview to occur. The comfort of the interviewee should also be considered. Interviews should be conducted in an environment in which the interviewee feels comfortable. A custodian might feel stress if called into a formal office for an interview but might feel comfortable being interviewed in a quiet hallway while classes are in session.

Interview questions should be scripted in advance as a component of the audit program. This will help to ensure a degree of consistency among different schools that are audited in that each school is presented with the same battery of questions during the interview phase. However, it is appropriate to supplement scripted questions with probing questions in situations where the scripted question does not yield a robust response. Probing questions can be used to more specifically approach an issue in a way that directs the interviewee toward the information that needs to be obtained during the interview.

AUDIT SCORING

Scoring audits begins with the format in which question are phrased. An audit is simply a series of questions that are asked regarding various issues of compliance or program management. It is important to ensure that each question is phrased in a way that elicits a positive or "yes" response indicating that compliance is achieved. Below are examples of how this can be a challenge:

- "Are emergency exits blocked?" The anticipated answer to this question would be "no" in that for a safe environment to exist, exits would not be blocked and would be clear for faculty, staff, students, and visitors to exit.
- "Are emergency exits clear for easy access?" The anticipated answer to this question would be "yes" in that faculty, staff, students, and visitors should be able to easily leave the building.

Establishing questions that result in a positive response when compliance is achieved will assist in creating a proper scoring methodology by ensuring that all correct responses are uniform. Phrasing questions so that a correct response is in the negative (no) with some questions and in the positive (yes) with other questions can result in confusion when evaluating an audit. It can also create a problem in properly scoring the audit. For example, an audit document can be created by using spreadsheet software. The document can be created by designing one column for "yes" responses, one column for "no" responses, one column for "not applicable" responses, and one column for the audit questions. In order for the spreadsheet to properly calculate the score, it is necessary to have responses that are uniform in relation to compliance or noncompliance:

- "Yes" response: The school is in compliance on a given question.
- "No" response: The school is not in compliance on a given question.
- "Not Applicable" response: The question does not apply to the school.

An X can be placed in the appropriate "yes," "no," or "not applicable" box next to a given question. When the audit document has been fully populated, the score that has been calculated by the spreadsheet can be used to communicate the level of the school's performance. All X markings in the "yes" column can be totaled to arrive at the number of areas in which the school was in compliance, and all of the X markings in the "no" column can be totaled to arrive at the number of areas in which the school was not in compliance.

Principles of quality management have taught us that the things that are measured receive the attention of an organization. It is for this reason that audits should be scored. A score of an audit provides a metric that can be used to determine whether a given school is improving from year to year, as well as to determine how one school is performing when compared with other schools within a system that have received the same audit. Scoring can occur using a number of methodologies:

- Yes/No: Scoring can be conducted by applying one point for a "yes" response where the school was in compliance on a question and zero points for a "no" response where the school was not in compliance on a question. This methodology provides little ambiguity on how an auditor is to assess any given situation. However, it might also fail to adequately represent what is occurring at a school. For example, a question on an audit document might address the need for all fire extinguishers to be properly charged. After checking one hundred fire extinguishers throughout the school inspection, it might be found that ninety were properly charged while ten were not fully charged. Although the school performed at 90% as a whole, the findings would result in a "no" response and zero points awarded because all of extinguishers were not fully charged.

- Yes/Average/No: Similar to the first example, scoring can be conducted by applying two points for a "yes" response where the school was in compliance on a question, zero points for a "no" response where the school was not in compliance on a question, and introducing a middle option that would score one point for an "average" category where some work had been done but the school was not yet fully in compliance. This methodology would reflect some point values in the above example of fire extinguishers in that the school would at least be awarded one point for having 90% of its fire extinguishers fully charged, although the school could not yet achieve two points for a "yes" response. A potential behavioral challenge in this scoring methodology is that the auditor might frequently select "average" as an option rather than fully committing to a "yes" or "no" response in questionable situations.

- 0-2-8-10: Scoring can occur on a more detailed level by providing four categories in which a given question could be scored. Similar to the "Yes/No" methodology, this methodology would allow for zero points to be awarded in a situation where the school is not compliant on a given question and ten points in a situation where the school is fully compliant. However, it also introduces an option for the auditor to award two points for a situation where a school has performed some work on a given question and major issues still exist and eight points for a situation where a school has performed some work on a given question and minor issues still exist. This scoring methodology provides a dynamic representation of activity that is occurring at the school while also eliminating a middle ground selection for the auditor to select. The scoring for each question will clearly reside on the positive or negative side of activity being performed.

These scoring methodologies provide three basic formats to choose from in establishing a scoring system for an audit. They can be evaluated and modified for what will be best received by the school system. The goal will be to identify a methodology that when communicated will have the greatest impact on improving school safety.

The audit can be divided into sections that provide detailed levels of scoring in certain areas while also providing a total score for the audit. The audit can be divided into dedicated areas of documentation review, school inspection, and interviews. This will clarify where work needs to occur to improve school safety. For example, a school might achieve a 95% in the documentation review section of the audit. However, the score might drop to 85% in the school inspection section and 65% in the interview section. This will indicate that the school has done great work in creating written program material but has not made exceptional progress in implementing the material in the physical environment of the school and has made even less progress in effectively training faculty and staff. Similarly, the audit can be divided into sections based on the safety topic, such as creating individual sections of the audit for such topics as hazard communication, ergonomics, and emergency response. Organizing the audit in this fashion will help to identify specific programs that have been implemented by including documentation review, facility inspection, and interview audit questions within each topical section. The design of the audit is flexible and will need to be determined based on what measurement is most useful for the school system.

AUDIT REPORT

A report will need to be generated following the audit that provides a high-level summary of the performance of the school. Although the audit document might be included with the audit report, the report is an independent document containing general information from the audit. The audit report should be a one- or two-page summary of the audit. This will provide high-ranking individuals within the school system a brief summary of the audit that can be quickly read and understood. The audit report can include:

- Audit information: Primary individuals who were involved in conducting the audit can be identified by name and position title. This will provide those reading the audit report with a list of individuals who can be contacted if there are questions regarding the audit. The date the audit was conducted can be recorded. The name and general information of the school that was audited can be included to specifically identify the school that is the topic of the report.
- Scoring: The total score and section scores can be provided as a metric of the school's performance on the audit.
- Opportunities for improvement: Although a number of deficiencies might have been found during the audit, the audit report can contain a focus on the most important areas of risk that were revealed by the audit.
- Best practices: A few best practices that are occurring at the school with regard to safety program management can be identified. This can help to provide a balanced view of what is occurring at the school instead of only presenting opportunities for improvement. Best practices that have been identified can also be communicated throughout the school system to assist others to improve their school safety management program.
- Follow-up: A summary can be provided that outlines how follow-up will occur. This will include a general description of the categories of risk in which deficiencies fall and the time frame in which they must be brought to closure. Delineation will also be provided of who will be responsible for following up on audit deficiencies.

The audit report can be distributed based on the needs of the school system. Recipients might include board of education members, the school system superintendent, the school system risk manager and/or safety manager, the principal of the school receiving the audit, and individuals at the school with safety responsibility. The distribution list will be based on the individual needs of the school system. A decision must also be made in conjunction with school system legal counsel as to whether the audit report will be distributed as attorney client privileged. This will need to be evaluated based on the risk of the audit report being considered as discoverable material in the event that a lawsuit might occur that is in some way related to school safety.

FOLLOW-UP

Once an audit has been conducted, it is critical that a process be in place to monitor the correction of deficiencies that were identified in the audit. This process of follow-up will help to ensure that all deficiencies are brought to closure. The auditor will identify deficiencies that will typically fall into three categories of risk:

- High: High risks are those that can have an immediate impact on the safety of faculty, staff, students, and visitors. These risks will generally need to be fixed in a time frame that can range from immediate action to thirty days.
- Medium: Medium risks are those that can have a moderate impact on school safety. These risks will generally need to be addressed within ninety days.
- Low: Low risks are those that represent minimal impact on school safety. These risks will generally need to be addressed within six months.

In addition to providing a categorized list of deficiencies that must be addressed according to the level of risk, the system will also need to delineate the process through which follow-up will occur and the person at the school that is primarily given the responsibility to ensure that audit deficiencies are brought to closure. This can include a schedule of meetings to be held between the auditor and the individual at the school to review progress. It can also include the need for periodic reports to be sent to the auditor. A defined system for follow-up assigns accountability for correcting deficiencies to systematically bring audit deficiencies to closure.

AUDITOR SELECTION AND TRAINING

A challenge in developing a school safety auditing program will be to define who will conduct the audits. Auditors need to be qualified to engage in this activity through formal education, work experience, specialized training, or a combination of each of these things. Auditor candidates may exist within a given school or school system. Someone who is already vested with school safety responsibility might be a logical choice to become an auditor. In such situations, it will be important to determine whether the person can be objective in properly assessing the safety performance of a school. A conflict of interest could occur if the success of a school on an audit reflects on the person conducting the audit. Although this may often be the case, the auditor can focus on objectively rating the school in each section of the audit.

A decision might be made to utilize a third party to conduct the school safety audit, such as the insurance company that insures the school's property risks. Most insurance companies provide loss control services that can be incorporated, including auditing. The benefit to this option is that the auditor might be less biased than an internal auditor. A potential problem is that the auditor might not be as familiar with the school's operational procedures, which can affect how certain areas in the audit are evaluated. Cost is also a concern. It might be less expensive to train an internal auditor compared with the cost of having a third party perform the audit. A benefit to utilizing a third party is that subject matter expertise in the area of safety auditing can simply be contracted rather than determining how to develop the skill with internal personnel. Careful consideration will need to be given to determine whether an internal auditor or a third party auditor will be most appropriate for a given situation.

Auditor selection can be affected by the basic professional skill requirements to be an auditor. A number of inherent and learned skills must also be considered when selecting an auditor. These skills can include:

- Communication: An auditor will need to be able to communicate constructively with individuals at all levels within the school system, including faculty, staff, contractors, and school executive leadership. This will require a high level of interpersonal communication skills. An auditor who appears condescending or weak in communication can have an adverse impact on the success of a school safety auditing program.
- Endurance: Auditing is a physically demanding process. It will require walking for hours to cover all areas of the school property. It will require climbing stairs and ladders to access each part of a building.
- Knowledge: The auditor will need to be knowledgeable about both school operations and safety implications in order to effectively assess safety performance and the unique challenges that might exist at a school. The auditor must be a subject matter expert on all topics covered within the scope of the audit. Certain situations may arise that will challenge the knowledge of the auditor that can be researched by the auditor for a definitive answer. The auditor must be able to clearly articulate why each question on the audit is in compliance or not in compliance.
- Computer skills: An efficient audit system will be constructed in an electronic format in order to calculate the audit score and share information throughout the school or school system. At a basic level, this will require the auditor to be able to use the software in which the audit is created and the medium used to communicate information regarding the audit electronically.
- Time management: The auditor will need to be sensitive to the schedule of the school and conduct the audit accordingly. The auditor will need to accomplish a review of documentation, school inspection, and interviews while taking into consideration faculty and staff work schedules, class schedules, and hours of school operation. This will require the use of time management skills to address all of the school scheduling variables and issues that need to be addressed within the audit.

Thorough planning and preparation can result in an effective audit program being developed to evaluate a school safety program. Although much work and effort is required to create and implement a school safety program, an audit program is equally as necessary to ensure that safety programs are sustainable. Ongoing measurement of program effectiveness is required in order to maintain the multiple areas of ongoing work necessary to maintain a school safety program.

Auditors may be internal to the school system or external through a third party. A common benefit of third party audits as identified by Rains (2011) is that a greater degree of objectivity can be achieved. Rather than the auditor being internal with a potential vested interest in the outcome of the audit, a third party audit places the auditor apart from the organization, with the intended result being greater objectivity while conducting the audit. Each school system will need to determine whether internal safety auditing or an audit performed by a third party would be most effective for the intended goals of the audit program.

CASE STUDY

Jan is in the process of conducting her third school safety audit. In previous months, she has conducted an audit at Madison Elementary and Sherman Middle School, and she is now at East High School. She is conducting the physical inspection while being escorted by Sue, the head custodian. Jan notices a school maintenance employee working on an air conditioning unit, so she decides to conduct an interview. Upon interviewing the maintenance worker, she finds that he is a Hispanic employee who can speak English moderately. He appears to understand the safety aspects of the job, but communication is difficult.

- Would this event be considered a deficiency in the safety audit? Why or why not?
- What concerns might Jan have in working through her assessment of the situation?

EXERCISES

For the following questions, identify a single school environment in which you would like to situate your responses and answer each question accordingly.

1. What types of documentation should be reviewed during a school safety audit?
2. What physical condition items should be explored during a school inspection?
3. Should open-ended or closed-ended questions be used to conduct interviews? Why?
4. Should audits be scored? Why or why not?
5. Who should receive a copy of the audit report?
6. What process would you put into place to ensure that deficiencies are brought to closure following a school safety audit?

7. What qualities do you consider most important when selecting a school safety auditor?

APPENDIX
Sample Hazard Communication Audit

Documentation Review

Yes	No	N/A	Questions
			Is a comprehensive written hazard communication program present?
			Has a chemical register been created that lists all hazardous chemicals?
			Are material safety data sheets present for each hazardous chemical?
			Has documented training been conducted?
			Has documented refresher training been conducted?
			Is there evidence of a process that is used to approve the use of new chemicals?

School Inspection

Yes	No	N/A	Questions
			Are material safety data sheets available to employees?
			Are employees using hazardous chemicals properly?
			Are chemicals properly stored?
			Are chemical containers properly labeled?
			Is appropriate personal protective equipment available?

Interviews

Questions
What personal protective equipment should you wear when using a chemical?
Where do you go to gain information on the hazards of a specific chemical?
What things did you learn during hazard communication training?
What information is important when reading a chemical container label?
Where do you store chemicals?

REFERENCES

Cahill, L. (2001). *Environmental Health and Safety Audits.* Government Institutes, Rockville, MD.

Dunlap, S. (2011). *Loss Control Auditing: A Guide for Conducting Fire, Safety, and Security Audits.* CRC Press, Boca Raton, FL.

Rains, B. (2011). Process safety management: Finding the right audit. *Professional Safety,* 56(10), 81–84.

30 Injury and Illness Recordkeeping

E. Scott Dunlap

CONTENTS

The Occupational Safety and Health Administration (OSHA) injury and illness recordkeeping requirements are designed to quantify the types of injuries and illnesses that are experienced in various categories of work environments. The data collected from this process can also be helpful to the employer by examining the types of injuries and illnesses that are occurring within the organization so that appropriate measures can be taken to prevent the recurrence of work-related injuries and illnesses.

The OSHA recordkeeping requirements are found in 29 CFR 1904. An interesting issue with the recordkeeping standard is the format in which it is written. Rather than simply presenting factual information as occurs in other standards, OSHA revised this standard to be presented in a question-and-answer format in an effort to make the information more clearly understood.

EXEMPTIONS

Employers who have ten or fewer employees at all times throughout a given year are exempt from recordkeeping requirements according to 1904.1. In addition, Appendix A (shown in Table 30.1) of the regulation delineates specific organizations that are exempt from the recordkeeping regulation. The second column of organizations indicates that "schools" are exempt from recordkeeping requirements. Because schools appear in Appendix A of the regulation, there is no regulatory requirement to maintain injury and illness records except in the following situations:

- All employers must comply with the requirement to report within twenty-four hours fatalities and incidents that require three or more employees to be hospitalized.

TABLE 30.1

Nonmandatory Appendix A to Subpart B: Partially Exempt Industries

SIC Code	Industry Description	SIC Code	Industry Description
525	Hardware stores	725	Shoe repair and shoeshine parlors
542	Meat and fish markets	726	Funeral service and crematories
544	Candy, nut, and confectionery stores	729	Miscellaneous personal services
545	Dairy products stores	731	Advertising services
546	Retail bakeries	732	Credit reporting and collection services
549	Miscellaneous food stores	733	Mailing, reproduction, and stenographic services
551	New and used car dealers	737	Computer and data processing services
552	Used car dealers	738	Miscellaneous business services
554	Gasoline service stations	764	Reupholstery and furniture repair
557	Motorcycle dealers	78	Motion picture
56	Apparel and accessory stores	791	Dance studios, schools, and halls
573	Radio, television, and computer stores	792	Producers, orchestras, entertainers
58	Eating and drinking places	793	Bowling centers
591	Drug stores and proprietary stores	801	Offices and clinics of medical doctors
592	Liquor stores	802	Offices and clinics of dentists
594	Miscellaneous shopping goods stores	803	Offices of osteopathic
599	Retail stores, not elsewhere classified	804	Offices of other health practitioners
60	Depository institutions (banks and savings institutions)	807	Medical and dental laboratories
61	Nondepository	809	Health and allied services, not elsewhere classified
62	Security and commodity brokers	81	Legal services
63	Insurance carriers	82	Educational services (schools, colleges, universities, and libraries)
64	Insurance agents, brokers, and services	832	Individual and family services
653	Real estate agents and managers	835	Child day-care services
654	Title abstract offices	839	Social services, not elsewhere classified
67	Holding and other investment offices	841	Museums and art galleries
723	Beauty shops	87	Engineering, accounting, research, management, and related services
724	Barber shops	899	Services, not elsewhere classified

- If asked to do so in writing by OSHA or the Bureau of Labor Statistics (BLS).
- If asked to do so in writing by a state agency operating under the authority of OSHA or the BLS.

Although there is no legal requirement for schools to maintain injury and illness recordkeeping beyond these few items as stated in the regulation, having an understanding of what OSHA requires for a comprehensive reporting system can assist school administrators to establish a form of internal records so that analyses can be conducted on the occurrence of incidents at a given school or within a school district. This information can help to establish the direction of workplace safety initiatives.

RECORDKEEPING

OSHA has established recordkeeping regulations in 29 CFR 1904 that delineate how various injuries must be categorized and recorded by the employer on a form known as the OSHA 300 log. Fundamentally, an injury is considered recordable on the log if the employee had to be taken to a physician for treatment. However, there are exceptions to this, such as incidents where only first aid was needed to treat the employee. OSHA generally is not concerned with the large volume of first aid injuries, such as applying a band aid to a small cut. OSHA is interested in knowing about and measuring the volume of injuries that require treatment beyond basic first aid, such as stitches for a much larger cut. Therefore, the term "recordable" means that the employer must record the injury on the OSHA 300 log for that given year.

The OSHA 300 log requires specific information to be recorded for each injury. The first piece of information is a case number. This numbering system is determined by the employer. In the calendar year 2012, the first injury might be recorded as 12-001, where "12" indicates the calendar year, and "001" indicates the first recordable injury for that year. The second piece of information is the employee's name. The employee's proper name will be recorded as it appears on employment records.

The third piece of information is the job title of the employee. This job title should be consistent with that which appears in employment records. Although a "maintenance technician" may informally be called a "mechanic," the title "maintenance technician" should be used if that is what the human resources department indicates as the employee's job title.

The fourth piece of information is the date the injury occurred. This date might be different from the date the injury was reported. A key component in an injury management program is to require that employees immediately report injuries, regardless of how mild or severe the employee perceives the injury to be. For example, an injured employee may feel an odd sensation in his back while lifting a box. He feels like he can continue working and assumes the pain will go away. However, he wakes up the next morning and finds that it is very difficult to stand up and reports the injury upon arriving at work. Immediate reporting of these injuries can assist in lowering the severity of the injury through early intervention of treatment. In this case, although the employee reported the injury on the second day, the date of the previous day when the injury occurred is what must be recorded on the OSHA 300 log.

The fifth piece of information is where the injury occurred. This can be recorded using descriptive language that is common to the facility such as in the "chemistry lab."

The sixth piece of information is a very brief description of the injury. The description must include the type of injury or illness, the parts of the body affected, and what specific event or action caused the injury. For example, an employee may have received a cut that required stitches while opening a box with a box cutter. The description could be recorded as "cut on the left palm from the blade of a box cutter." This short description specifically identifies what occurred.

The seventh piece of information is to categorize the injury based on the worst type of injury. Categories on the OSHA 300 log are identified as:

- Death
- Days away from work
- Job transfer or restriction
- Other recordable case

Categorizing a case as a "death" obviously means that the employee died as a result of a work-related incident. Categorizing the case as "days away from work" means that the treating physician believed the injury to be severe enough that the employee could not return to work and had to remain at home at least one full day other than the day that the injury occurred in order for proper healing to occur. Categorizing an injury as "job transfer or restriction" indicates an injury where the treating physician believed that the employee could return to work but would be limited in performing their full duties, such as not being able to lift a certain amount of weight because of a back injury. The final category of "other recordable case" refers to an injury that was treated by a physician, but the employee returned to work without limitations.

The eighth piece of information on the log is only required if the injury was categorized as either "days away from work" or "job transfer or restriction." In this section, the number of days away from work or the number of days of restricted duty will be recorded.

The ninth, and final, piece of information is to check the box that indicates the nature of the injury. Options provided are:

- Injury
- Skin disorder
- Respiratory condition
- Poisoning
- Hearing loss
- All other illnesses

There are a number of unique issues to consider when determining how to record a work-related injury or illness. The first is related to first aid. If an injury occurs and the employee is sent to a doctor, he may actually only receive first aid as treatment by the physician. In this case, the injury will not be recorded on the OSHA 300 log. 29 CFR 1904 defines first aid as:

- Using nonprescription medication at nonprescription strengths
- Administering a tetanus shot
- Cleaning, flushing, or soaking wounds on the surface of the skin
- Using wound coverings such as bandages, Band-Aids™, gauze pads, etc.; or using butterfly bandages or Steri-Strips
- Using hot or cold therapy

- Using any nonrigid means of support, such as elastic bandages, wraps, or nonrigid back belts
- Using temporary immobilization devices while transporting an accident victim, such as splints, slings, neck collars, and back boards
- Drilling of a fingernail or toenail to relieve pressure or draining fluid from a blister
- Using eye patches
- Removing foreign bodies from the eye using only irrigation or a cotton swab
- Removing splinters or foreign material from areas other than the eye by irrigation, tweezers, cotton swabs, or other simple means
- Using finger guards
- Using massages; however, physical therapy or chiropractic treatment are considered medical treatment for recordkeeping purposes
- Drinking fluids for relief of heat stress

If any of these treatments are administered either at the facility where the injury occurred or at a clinic, the injury is not considered as being recordable.

Another issue to consider is changes that may need to be made to the log as the claim advances. An employee might receive a cut requiring stitches and return to work. The injury might be initially recorded on the OSHA 300 log and categorized as "other recordable case." Two days later, the employee reports that the injury feels very sore. An appointment is arranged with the treating physician for the employee to have the injury examined, and it is found that an infection has set in. The treating physician now restricts the employee from using the arm that was injured. The OSHA 300 log must then be updated to reflect this information. The check under "other recordable case" is erased and is now checked as "job transfer or restriction," which reflects the change in the injury.

A final issue to consider is the restriction that has been placed on the employee. The presence of a restriction issued by a treating physician does not necessarily require the injury to be recorded as restricted work. The restriction must actually affect the employee's job. For example, if a treating physician issues the restriction of not being able to lift more than twenty pounds, this will directly impact a worker in maintenance or a janitorial staff person who routinely lifts products. Therefore, the injury would be recorded as restricted work. However, this same restriction placed on an office employee who only works at a computer for the entire shift has no impact on the ability of the employee to perform normal duties. Therefore, the injury would remain as a standard recordable injury and would not be elevated to being recorded as restricted work. Restrictions must be interpreted based on the job of the employee and recorded appropriately on the OSHA 300 log.

In addition to the OSHA 300 log, OSHA has two other documents that must be completed within the recordkeeping requirements. For each injury recorded on the OSHA 300 log, the employer is responsible for completing the OSHA 301 form. This document is commonly referred to as an incident investigation form. Here the employer will record specific information regarding the incident, such as what work the employee was performing, what occurred, and what caused the injury. OSHA allows employers to use an in-house version of the 301 form as long as the information required by the 301 form is captured on the employer's form. The final document is the OSHA 300A summary form. This document is a high-level overview of all of the information contained on the OSHA 300 log for a given calendar year. OSHA is concerned about engaging employees and informing them of workplace health and safety information. Toward that end, OSHA requires that the 300A summary be posted in a prominent location for employees to have access to the information. A summary containing the previous calendar year's information must be posted from February 1 through April 30.

A challenge in OSHA recordkeeping is that the requirements are based on the calendar year. A school may operate on a fiscal year that differs from the calendar year. For example, the school year may begin on August 1 and conclude on July 31 to include fall, spring, and summer activity. Similar to other operational data, it might be necessary to maintain injury data on this cycle, which differs from the calendar year. It is important to maintain records via the calendar year for OSHA and via the fiscal year for the school. Although cumbersome, this system may be necessary to both maintain compliance with OSHA and provide school administrators with meaningful data based on the fiscal year. A school may need to maintain records internally on a cycle that is appropriate for other school operational data while also maintaining data on a calendar year for OSHA.

INJURY DATA

Maintaining injury and illness data within a school system can prove to be informative in that it provides a clear picture of where workplace safety issues exist that need to be addressed. Although schools are exempt from most OSHA recordkeeping requirements, the information contained in the OSHA recordkeeping regulation provides a framework that a school can use to establish an internal injury and illness recordkeeping system that provides an opportunity to track incidents that occur. It is difficult to target workplace safety opportunities for improvement without gathering and analyzing this information.

The first step that can be taken is to establish an injury and illness reporting process by which employees must immediately report work-related injuries and illnesses to a supervisor. The supervisor can then forward the information to the person responsible for workplace safety so that an investigation can be completed. The goal of the investigation is to identify all of the causes of the injury or illness. Rather than targeting who to blame, the investigation should seek to uncover useful information that can be used to improve workplace safety. Corrective action can then be taken in the work environment or through employee training to ensure that the incident does not recur.

The second step will be to gather the information from completed investigation reports so that trends can be identified. This can be accomplished by categorizing incidents by type and severity. Tracking incidents by type can help to identify trends of common exposures that result in a number of employee injuries or illnesses. For example, it might be found that 60% of all injuries suffered in a given period of time were the result of falling from chairs while using the chairs as a stepladder. This can lead to implementing the use of actual stepladders in the classrooms to provide a greater degree of safety for school employees. Injuries can be tracked by severity to identify where high-risk issues might exist. This can be accomplished by tracking the days that employees are away from work on restricted or lost days that are required for employees to heal from injuries or illnesses. For example, it might be found that there were a total of seventy-two lost work days caused by injuries and illnesses in a given period of time. It might be found that sixty of these lost days were related to back injuries. This can lead to the implementation of an ergonomics program that addresses proper lifting technique and other body mechanics issues.

Injury and illness data must be generated, collected, and analyzed in order to identify the best strategy to reduce the incidents that are experienced within a school. Although not required to by OSHA, a school can proactively gather these data to identify things that can be addressed to provide a safer work environment for faculty and staff.

CASE STUDY

Sarah has been the principal of East Elementary School for only one year. She has noticed a high rate of teachers being injured in the classroom in light of frequent workers' compensation checks that cross her desk. She is struggling to identify what to do to address the problem. She has searched through school records from past years and has not located any data related to the injuries. The current state of performance is not acceptable, and Sarah needs to identify a course of action.

- What information would help Sarah to address the situation?
- What system could she implement that will provide ongoing injury and illness data?

EXERCISES

For the following questions, identify a single school environment in which you would like to situate your responses and answer each question accordingly.

1. What is the applicability of the OSHA recordkeeping regulation to schools?
2. Regardless of applicability status of the full regulation, what incidents must schools immediately report to OSHA?
3. What categories of injuries can exist based on severity?
4. What categories of injuries can exist based on the nature of the injury?
5. Why is it important to collect and analyze injury data?

REFERENCES

Occupational Safety and Health Administration. (2003). Recordkeeping. Accessed November 17, 2011. http://www.osha.gov/pls/oshaweb/owasrch.search_form?p_doc_type=STANDARDS&p_toc_level=1&p_keyvalue=1904.

31 Injury Management

E. Scott Dunlap

CONTENTS

A strategic plan must be in place to effectively manage injuries when they occur. This will require a general understanding of workers' compensation and Occupational Safety and Health Administration (OSHA) recordkeeping, as well as of best practices in injury management. Combining these issues into an injury management system will create an environment where employees can be returned to work as soon as possible following an injury while also controlling the costs associated with an injury.

COMPENSABILITY

Case law and workers' compensation law indicate that there are dozens of variables that must be considered when determining the compensability of an injury that occurs in the workplace. Larson and Larson (2000) have presented a comprehensive overview of workers' compensation compensability issues. There are two fundamental principles that affect compensability. The first of these two issues is whether the injury arose out of and in the course of employment. "Arising out of" refers to the injury being the result of on-the-job activity. For example, a teacher falls from standing on a chair while hanging decorations in her class. The job she had required her to perform this function. Therefore, the injury arose out of her job.

The second issue involves an injury that is suffered in the course of work. Using the previous example, a teacher falls while standing on a chair. She is at work performing her duties during a time period that she is required to be there. Therefore, the injury has also occurred in the course of her employment.

In general, if an injury occurs as a result of an employee's job and the injury is suffered while that job is being performed at the time it is supposed to be performed, the injury is considered as compensable. This is a very simplistic way of looking at compensability, but it presents a picture of general circumstances that must be present in order for an injury to be considered as compensable.

It is important to understand who determines compensability. Workers' compensation adjusters will make the decision as to whether an injury is compensable. I have had the opportunity to work for four of our nation's largest organizations. In two of these organizations, the decision was made to operate a self-insured workers' compensation program, which meant that we maintained money in reserve to pay injury claims from within our company rather than purchasing workers' compensation insurance. We hired an outside organization, known as a third-party administrator, to manage our program. Workers' compensation claims adjusters were assigned within the third-party administrator to manage our various facilities. Workers' compensation law can vary greatly from state to state, so these adjusters were experts in the law as it applied to our different plants based on where they were located. It was these adjusters who determined compensability.

In both of these cases, the benefit is that the decision of compensability was made by an unbiased individual. Imagine a situation where you assume that an injury that is being reported is fraudulent and you believe the claim should be denied. Such a decision, based on emotions, assumptions, and lack of considering the full scope of workers' compensation law, can result in liability for your school if the claim is denied without foundation. A school system risk manager or other similar professional should be consulted to determine how compensability is determined for injuries that occur within a given school.

OSHA RECORDKEEPING

Although discussed more fully in another chapter, a brief discussion of OSHA recordkeeping is necessary within the context of injury management. In addition to complying with workers' compensation law, employers must comply with OSHA recordkeeping requirements when an injury occurs. Workers' compensation and OSHA recordkeeping laws must be understood independent of each other because they are not always in harmony, meaning that the generation of a workers' compensation claim does not necessarily mean that it is an OSHA recordable injury.

In basic terms, an injury is considered to be "recordable" if the employee requires the expertise of a physician to be properly treated. A workers' compensation case will be generated the moment the employee sees the physician because costs will begin to immediately accrue. However, the physician might only need to provide basic first aid, which is exempted by OSHA as a recordable injury. The injury will be considered as a workers' compensation claim because it incurred minor medical cost but would not be considered an OSHA recordable injury. In the event that the physician had to treat the injury beyond first aid measures, such as using stitches to care for a cut, then the injury would be considered both a workers' compensation claim because of cost incurring and an OSHA recordable injury because of the level of care that was provided to treat the injury. Employers must manage injuries both from a workers' compensation claim perspective and considering variables that impact OSHA recordkeeping.

WORKERS' COMPENSATION

Workers' compensation can be a sensitive issue. This simple phrase might elicit skepticism because of the ambulance-chasing attorney commercials on television and the stories of fraud that make the headlines. Nothing good or productive can come from experiencing a workers' compensation claim. Claims can be incredibly expensive. They can drain the time of staff, particularly if the injured employee is represented by an attorney. A school's injury rate and its insurance experience modification rate increase. Administration might be skeptical of the legitimacy of a claim each time a subsequent injury is reported. The injured employee can suffer physically, emotionally, and financially. Simply, it is bad for everyone involved.

However, a more positive outlook on managing workers' compensation claims can be developed. Although there is a great deal of workers' compensation fraud, the great majority of injuries that are reported are legitimate. The employee truly is hurt. A few isolated cases of fraud should not tarnish the reputation of a system that is designed to help employees in their time of need.

Many of us may have had experience directly or indirectly with a work-related injury. As a child, I remember my dad coming home from the coal mine with his fingers wrapped in black electrical tape. When I asked him what happened, he explained that he had hurt one of his fingers in the mine and was using what I saw as a homemade brace to protect the injured finger. Throughout my career I have interviewed numerous employees who have experienced very minor to very serious injuries. These employees truly were hurt and needed help. As Pennachio (2008, p. 65) indicates, "There's no evidence that competent, honest and loyal employees abuse the system; rather, it is the system itself that induces needless disability and high costs." He goes on to say that the key issue is to have a system in place that moves the injured employee through an efficient process toward recovery.

In general, workers' compensation is designed as a no-fault system to pay for expenses generated from a work-related injury. The phrase "no fault" means that payment is provided regardless of who was found to be at fault for the injury as was found during an incident investigation. For example, an investigation may reveal that an employee was in a hurry to finish a job and used a short cut to complete a task prior to going to lunch with his friends. Being in a hurry and using a short cut caused the injury to occur. Because workers' compensation is a no-fault system, the employee would still receive benefits even though the injury was his fault through volitionally using a short cut that was outside established operating procedures. One might look at that situation and say that it is unfair for employers to pay for an injury that they did not cause. I look at this situation from a different perspective.

Although the injury in the example would be considered the fault of the employee, he did not do it on purpose. He was driven by the desire to go to lunch with his friends and thought he could complete his work and still make it to lunch on time. He did not go about his work with the intent of injuring himself. An employer inherits a lifetime of habits and assumptions held by the people that are hired. In many cases, these habits and assumptions can easily lead to a work-related injury.

Consider yourself. When was the first time that someone taught you how to lift properly? When did someone teach you that you needed to stand close to the load, lift with your legs, and avoid twisting? I was in my middle to late twenties. Up until that point, I inflicted a great deal of abuse on my back. Through work and play, I injured my back over a period of time. I took that behavior into every job I had. Fortunately, I never suffered a work-related back injury, but many employees do. They bring habits that place them and their employer at risk for an injury to occur. The employee assumes that the way they have been doing work is okay because they have not yet been injured. There lies the opportunity for the employer to intervene and educate employees as to how they can work safely. Rather than looking at fault directly related to the incident that caused the workers' compensation claim, look at the whole picture of life experiences that culminated in the injury occurring.

An extreme challenge with workers' compensation claim management is to understand what laws apply because laws are managed state by state. I began managing workers' compensation claims in the state of Ohio. I was able to work with state adjusters and experts within my company to learn how to comply with workers' compensation laws in that state. I then moved to a corporate role where I was responsible for managing claims across dozens of states. Because my primary role was leading health and safety, there was no expectation that I could become proficient in all of the workers' compensation laws in every state. In that situation, I became familiar with the key issues among the states and relied heavily on our adjusters to assist us in making decisions on managing claims. Although there are a number of differences between states, Wertz and Bryant (2001) point out similarities among all states:

- First, employees must be compensated regardless of fault. It does not matter that an incident investigation reveals that they are at fault, compensation must still be provided.
- Second, compensation is limited to specific types of costs. Medical bills must be paid that are generated from treating the injury. Disability payments must be paid if the injury results in short-term or long-term loss of use of a body part. Rehabilitation must be made in cases such as when an employee must receive physical therapy to recover from a back injury. In the case of a fatality, death benefits are delineated.
- Third, although popularized in the media among other types of injury claims, workers' compensation does not permit payment for pain and suffering and punitive damages. In most cases, it also prohibits the employee from suing the employer for additional money. Workers' compensation is provided as the sole remedy for responding to work-related injuries.

Typically, there are two avenues that an employer can use to maintain the appropriate coverage. One option is workers' compensation insurance. The system identifies an insurance company that provides such insurance policies and pays the designated premium. A second option is to become self-insured. Large organizations may select this option because it enables the company to absorb the entire cost of workers' compensation. Smaller organizations may select insurance because they can manage paying an insurance premium without the risk of a single claim having a devastating impact on their ability to remain in business.

When a workers' compensation claim is reported, there are a number of roles that must be considered and included in the management process (Wertz and Bryant, 2001). The first is the injured employee. The individual who suffered an injury will report it to the second role, which is the school. The school includes all members of administration, such as the human resources manager, the safety manager, and the supervisor of the employee who was injured. Once the employer (the school) is aware of the injury, the third role is engaged, which is the claims adjuster. This individual might be someone who works for the insurance company that provides workers' compensation coverage, or it might be someone who works for a third-party program administrator if the organization is self-insured.

Parallel to engaging the adjuster will also be activity to engage the fifth role, which is the healthcare provider. The employee will need to be seen immediately by a treating physician who can provide care for the injury. A sixth role that follows up on the healthcare provider is that that of rehabilitation services that might be needed to provide follow-up care in the event that the injury was severe.

A final role is the engagement of legal services. This can be seen from two perspectives. First, the injury may involve uniquely sensitive issues, so it might be necessary to immediately contact in-house council for guidance on certain decisions. Second, the employee might retain council, which will in turn cause school administration to contact in-house council. The complete dynamic of claims management changes once attorneys become involved in workers' compensation claims. Prior to being represented by an attorney, the workers' compensation claim manager typically has open communication with the employee. However, this changes once an employee retains an attorney. At that point, the primary avenue of communication regarding the claim will take place through attorneys, and normal communication with the employee ceases.

MINIMIZING CLAIMS

One way to proactively manage injuries in a school is to establish an effective safety and health management system that lessens the volume of injuries that occur (Wertz and Bryant, 2001). Workers' compensation is a reactive system designed to compensate an employee after an injury occurs. Lessening the volume and severity of these injuries requires the presence of a comprehensive safety and health management system.

The first step in developing a safety and health management system is to perform a risk assessment of a school to determine what issues are present that can result in employee injury. The goal is to identify potential risks that exist for employee injury. Risks may include things that are very obvious, such as the potential for a fall of a maintenance worker replacing a light in a gymnasium ceiling. Risks may include less obvious things, such as carpal tunnel syndrome that can develop over a period of repetitive motion. The challenge in identifying risks is to establish a team of people from various disciplines who can look at a job being performed from different perspectives. A supervisor will know the details of the job and risks that are typically encountered while performing a job. However, this same person may fail to see a risk because that is how the job has always been performed. Janitorial staff, maintenance staff, faculty, and others who can evaluate the job from a fresh perspective can be engaged in the process.

Once risks have been identified, it will be necessary to create programs that address each risk. For example, if it is found that an employee uses a chemical while performing her job, this will lead to considering the hazards that the chemical might present. OSHA regulations will need to be investigated to determine whether there is a law that mandates requirements for this work activity. It might be found that OSHA does indeed have a regulation that is referred to as the hazard communication standard. As the regulation is read it is realized that there are requirements for things such as material safety data sheets (MSDS), personal protective equipment, and employee training. All of this information will be gathered and translated it into a written program that delineates requirements for what must occur in the school to manage the use of chemicals.

With the written program in hand, the phase of training employees will occur. A freedom of training is that there are typically no designated requirements for how the training is conducted but simply a list of what must be covered in the training. As with other freedoms, this freedom can be abused. Training can look like a lot of things. An easy option in abusing the process is to simply gather everyone in a room and give a PowerPoint presentation with only one-way communication from the trainer. The employees sign documentation, and technically training occurred. But will the employees remember any of the information, let alone apply it in their work? Conducting training in this fashion has a number of benefits for school administration. It is cost effective in that it can be conducted in a short time frame. It is easy in that someone only needs to create a PowerPoint presentation that outlines the requirements of the program. It is efficient in that large volumes of employees can be rapidly taken through the process in scheduled sessions. But is it effective? Probably not.

Training needs to occur in a way that engages employees and not in a way that allows them to be passive. Although there might be the need to cover basic information in a classroom using PowerPoint, this process can be improved by asking the employees questions throughout the session and providing exercises. Rather than putting an MSDS on a PowerPoint slide, copies of an MSDS from a chemical that employees use can be distributed. The trainees can then be asked to read through it and discuss the information that they see. Following this classroom experience, employees can be taken into the school where practical skills training can be conducted. Employees can put on personal protective equipment, such as gloves and safety glasses that might be needed to protect them while using a chemical. They can be given an opportunity to demonstrate how they will safely use the chemical while protecting themselves and those around them.

Employee involvement is a general principle to keep in mind when developing a school safety and health management system. The more employees that are involved in the process, the more they will take ownership of the process. Rather than creating these requirements and handing them down as simple rules that must be followed, administrators can engage employees by asking for their feedback on risks and how to properly control them. Employees do the work every day, and they are the people who are best positioned to identify the potential for injury and to create solutions to protect them from getting hurt.

A more formal way to involve employees in a safety and health management system is to create a safety committee. Safety committees can be used to include employees from various departments and shifts to engage in safety activities. This can include things such as performing facility inspections, being part of incident investigations, offering safety recommendations, and working on follow-up activities to ensure that safety interventions are being properly implemented. Membership on a safety committee should not be taken lightly. High-performing and enthusiastic employees need to be identified who will add value to the work that the committee will do. One word of caution is to beware of "donut eaters." These are people who join groups simply for the benefits that being a member of the group provides, such as eating donuts during meetings. They become part of the group for the recognition, treats, or other benefits that the group affords and not because they have a sincere desire to contribute to the team. The members of the team will have a great impact on the perception that others in the school have of the team.

One thing that has commonly become part of a number of safety and health management systems is an incentive program. The challenge is to design and implement the program in such a way that truly results in desired behavior without having undesired effects. Geller (1996) outlined very useful information to consider when creating a safety incentive program. There are a number of mistakes that are commonly made. One mistake is to create a system that is based on not having injuries. For example, an incentive might be earned if no injuries occur in a given month. Although it is our goal to have fewer injuries, having such an incentive program could encourage bad behavior, which is to not report injuries. This type of incentive program can cause employees to drive injuries underground. Although school administration may believe the program to be a great success when injuries are reduced, the reality could be that employees have simply stopped reporting them.

Another common mistake is to establish an incentive that is based on group performance. An example might be everyone in a certain group receiving an award if the group as a whole works through a given period of time without an injury. Again, this might cause employees to not report an injury because of the stress of knowing that their injury will result in the entire department losing the award.

A final common mistake is to establish an incentive program that rewards only a few people. For example, everyone who has not experienced an injury during a certain period is entered into a drawing where three employees will be selected to each win a new television. The problem is that though everyone involved may have met the expectation, only a small number were actually rewarded.

Incentive programs can be designed that encourage positive activity to occur rather than simply avoiding injury. An incentive program can be established that rewards employees for becoming engaged in safety. A number of legitimate safety recommendations could be set to encourage employees to identify issues and recommend safety solutions that can prevent injury. Opportunities can be defined for employees to volunteer and participate, such as being on the safety committee or first responder team. This is also an area where employees can become engaged by asking them what will motivate employees to focus on safety and then establish an incentive program that rewards employees accordingly.

RETURN TO WORK

Waiting until an injury occurs is not the best way to manage safety. A school system can be proactive with injury management by creating a transitional return-to-work program. This might also be referred to as a light-duty program. Although the difference in these two program titles might appear to only be semantics, the phrase "transitional return-to-work" sets a much more positive tone for what the program is designed to achieve. There are two ideas communicated in this program title. First, the word transitional communicates that the employee will be in a process that is not permanent. Each job that the injured employee will be placed in is designed to accommodate physical restriction needs at the time. As the employee heals, he or she will be graduated to a new job. The phrase "return-to-work" communicates that the goal of the program is to return the employee to his or her normal job. This stands in contrast to the phrase "light-duty" which is ambiguous. By titling a program "transitional return-to-work," it clearly communicates that the goal is to walk injured employees through a process that brings them back to their normal job.

Planning a transitional return-to-work program requires the assistance of virtually every member of administration in a school. Involving people from various areas can assist in managing risk that can result in injury and identify ways to transition employees back to work (Gonzales, 2010). The first thing administration can assist with is identifying every task that can be performed by someone who is on work restrictions because of an injury. This might involve tasks that are not physically demanding, such as processing paperwork or conducting inventory audits. The goal will be to accumulate a comprehensive list of tasks that an injured employee can be transitioned into and through as the injury is managed. The challenge with generating this list is to think outside the box. It is easy to identify routine and recurring tasks, but it is also important to identify those tasks that might occur infrequently. For example, an office manager might identify processing paperwork as a fairly routine task where an injured employee can be assigned. However, in the one week that the program is needed to utilize that option, the office manager reports that there is no paperwork to process, and they are caught up on their work. Fortunately, the inventory control manager contributed tasks associated with conducting the annual inventory, and the injury has occurred during the time of the inventory. The employee can be assigned to conducting work within the inventory.

It is important to not only identify routine tasks but also those that occur infrequently throughout the year. As the list of tasks is generated, points throughout the year that each task is conducted can be identified. This list will provide a resource to immediately bring an injured employee back to work once the treating physician has assessed work restrictions. Establishing this process is important in bringing the injured employee back to work as soon as possible following an incident. This process is also important in most effectively managing the category of injuries within your organization. Rather than resulting in a lot of time away from work because of an injury, this process will help to immediately bring an employee back to work. This will limit it to being a restricted work injury within OSHA recordkeeping requirements.

A second area in which administration can be engaged is assisting with management of the injured employee. First, the manager of an area in which the injured employee has been placed must be aware of the presence of the employee and the needs of the employee related to work restrictions. The manager can engage with the employee and the workers' compensation claim coordinator to ensure that restrictions are complied with and that the employee is transitioned to new tasks as the injury heals. Second, the normal manager of the injured employee needs to be engaged with what is happening with their teammate. The employee's manager needs to be aware of when the employee is brought back to work and where he or she is in the transitional return-to-work process. Not only is this a responsibility in helping manage the injured employee, but this also allows the manager to replace the injured employee on a temporary basis and maintain work that needs to be accomplished. The manager may need to request a temporary employee or may need to divide work responsibilities among existing staff to meet work demands. By being engaged, the manager will be able to effectively plan work until the employee returns to full capacity.

A risk in this process is for managers to abdicate responsibility for managing the employee to the workers' compensation claim coordinator. This can create a number of communication problems. The employee may get lost in the system, causing him or her to remain in a restricted work position long after healing has occurred. Managers need to be engaged to ensure that the employee is transitioning through the process to full duty.

It will be important to inform the treating physician of the transitional return-to-work program. The school may have an established relationship with an occupational medicine clinic that specializes in treating work-related injuries. One proactive step that can be taken is to ensure that a treating physician is aware that a robust transitional return-to-work program exists that allows the flexibility to immediately bring an employee back to work based on restrictions that the physician deems necessary to heal the employee.

A second step that can be taken is to actually have the physician or a representative from the clinic come to the school for a tour. This will provide an opportunity to clearly explain the physical requirements for each primary job, as well as to review the degree to which transitional return-to-work tasks have been identified that can be used to immediately return an employee to work, regardless of an employee's primary job. Developing this level of relationship with treating physicians will allow them to make more informed decisions on how to communicate the status of an employee following the treatment of an injury. For example, an uninformed physician may immediately place an employee off of work for three days to heal from an injury. However, if that same physician was aware that a mature transitional return-to-work program existed, the employee might be returned to a role that accommodates work restrictions that have been placed on the employee.

Work restrictions issued by a treating physician are not recommendations. They are requirements that schools must comply with in managing an injured employee. Following the initial treatment of the injury, the treating physician will complete a physical capability form that outlines what, if any, restrictions that the employee must abide by throughout the treatment process. These restrictions will be very specific, such as the employee not being able to stand for more than four hours per day. It is the responsibility of the employer to immediately become aware of these restrictions and ensure that they are followed when assigning the employee to transitional return-to-work tasks. Similarly, the employee must follow the instructions of the treating physician. Failure of the employee to follow the treatment plan could result in the claim being denied and workers' compensation benefits being revoked.

In addition to workers' compensation regulations, a second legal issue to consider when creating a transitional return-to-work program is the Americans with Disabilities Act (ADA). This act is designed to require employers to make accommodations for employees who have qualifying disabilities. The ADA may come into the picture if an employee reaches maximum medical improvement that still results in a disability that is within the scope of the act. For example, an employee might initially suffer a crush injury to an arm. Over a period of months of treatment, the medical decision is to amputate the arm. The employee later recovers from the procedure and is now faced with a life of a having a new disability. The employer must then evaluate the employee's job to see if a reasonable accommodation can be made that will allow the employee to perform his or her job. Another issue within the scope of the ADA that could apply to managing an injury in this type of situation is the threat of safety that the disabled employee presents to themselves or others. When determining whether or not to return an employee to his or her normal job, an evaluation will need to be performed to determine whether the disability could result in a risk to the employee or others in the work environment.

There are many benefits to maintaining a transitional return-to-work program. The employee will be more meaningfully engaged in work and the healing process by remaining as a part of the organization rather than sitting at home. The cost of the claim can be minimized by allowing the employee to engage in productive work and achieving maximum medical improvement in the shortest time possible by working to ensure that the treatment plan is followed. It also demonstrates the level of commitment to employees. Rather than discarding injured employees, they are brought back to contribute valuable work toward accomplishing the goals of the organization. This also creates a culture where employees feel they are valued and taken care of in their time of need (Bose, 2008).

INVESTIGATIONS

Once an injury occurs, an incident investigation becomes an opportunity to gather all of the facts surrounding the incident. Incident investigations should be performed immediately following the incident. A goal is to complete the initial investigation within twenty-four hours of the incident so that as much relevant information as possible can be gained. Although this is feasible for basic incidents, such as an employee slipping and falling and fracturing a wrist, it is probably not feasible for an injury that occurs during a large-scale disaster. Complex investigations will require days, weeks, and even months to complete depending on the incident that occurred.

An incident investigation should answer the questions who, what, where, when, and why. It should identify who was involved in the incident. This would include employees, contractors, and visitors. The investigation should also identify who saw the incident occur or was a witness to subsequent events. It should identify what occurred, including that the injured employee was engaged in prior to and during the incident as well as related activity that others may have been involved in that touched on the occurrence of the injury. The investigation should identify exactly where in the facility the injury occurred. Rather than generally stating "in the school," the information should be specific, for example, stating that the injury occurred in the library near the computer workstations. It should identify specifically when the incident occurred. Rather than stating "in the morning," the information could include "at approximately 9:30 a.m." Finally, the investigation should identify why the injury occurred. The first four questions can typically be answered within close proximity to the time of the incident, but the answer to why the injury occurred might take a day or much longer to determine based on the complexity of the incident. The goal of this last step is to identify root causes.

In order to most effectively answer all of the questions that will be raised during an investigation, it will be important to involve a number of people in the process. Rather than relegating the incident investigation solely to one person, such as a safety manager, the investigation should be conducted by a number of people who have knowledge about the processes that were involved. This will help to ensure that information from every perspective is explored to properly arrive at the causes of the injury. An employee who does the same or a similar job can add insight into the day-to-day activities of the job that was being done when the injury occurred. A maintenance technician can add insight regarding certain pieces of equipment and safeguards that are provided. The administrator of the area where the injury occurred can add insight into the general processes that occur and the specific responsibilities of the employee within that context. The involvement of each of these people will help to ensure that nothing is missed as the investigation is conducted.

Corrective action will need to occur once the investigation is completed. Corrective action will occur on two levels. First, corrective action must occur to prevent the incident from recurring. This might include providing training to employees where gaps in knowledge or skill exist, and it might include making changes to the physical environment to improve general conditions.

A second and much more sensitive area in which corrective action will need to occur is with the employee who was injured. This might also be perceived as discipline. The investigation might reveal that the employee was injured because of a volitional unsafe act that was performed with disregard to previous training that he received and established workplace safety policies. This type of corrective action will typically follow a series of four levels. The first level is a record of coaching, which is followed by a first written warning, second written warning, and lastly, a serious warning. A serious warning is when termination of employment might occur. The idea behind this process is that if an employee is found to be in violation of a safety requirement, then a record of coaching will be given to address the need to comply with training that was provided. If the behavior does not change and an offense occurs again, then a first written warning is given. If the behavior does not change and an offense occurs again, then a second written warning is given. If the behavior does not change and an offense occurs again, then a serious warning is given, which will typically result in termination. Steps in the process may also be skipped depending on how egregious the offense is found to be. For example, if an employee willingly disobeys a safety requirement and puts the lives of others at stake, the decision might be made to move directly to termination because of the severity of the incident. A significant part of this process is to establish a consistent way of implementing corrective action across various incidents as they occur. It is inconsistent to terminate an employee for a simple error, while issuing a record of coaching to an employee who was guilty of a severe infraction. Such a process will be executed in conjunction with the corrective action policy established by the human resources department.

When issuing corrective action, it will be important to address four things in written documentation.

- The company policy or program violated by the employee will need to be clearly stated. For example, if an employee volitionally failed to lockout a piece of equipment, then the school's lockout/tagout program would be identified as the program that was violated.
- What the employee did to violate the program will be specifically stated. For example, it might be stated that the employee simply turned the power switch to the "off" position without locking the equipment out at the power source.
- What the employee needs to do in order to demonstrate acceptable behavior will need to be stated. It can simply be stated that the employee needs to comply with all aspects of the lockout/tagout program.
- Future corrective action will be delineated if the employee fails to execute acceptable behavior. This might be a general statement communicating that future violations will result in corrective action up to and including termination.

All documentation that results from an investigation should be maintained confidentiality. This will include the investigation form, witness statements, and corrective action forms. However, general information might be shared across the school system to assist in preventing such injuries.

Conducting quality and extensive investigations of all injuries is a critical component of managing workplace injuries. It is instrumental in preventing future injuries. An investigation will also help to provide support in accepting or denying the claim. For example, an employee might report that he injured his back shortly after beginning his shift. While conducting the investigation you are able to access footage from a security camera that happened to film the employee doing his job from the moment he began work the day in question until he went to lunch. The evidence supports that an injury did not happen on the job and can be used to potentially deny the claim. In the same respect, an investigation may reveal facts that support the occurrence of the injury while on the job and the claim can be accepted. Although time consuming, a thorough investigation is incredibly valuable in the injury management process.

FRAUD

Employees actually experience injuries that are work related and have caused them to be in pain. However, there are claims that are reported that are fraudulent. Although we should not always suspect fraud, we must exercise due diligence while investigating claims to ensure that fraud is not occurring. Wertz and Bryant (2001) provide information that is helpful in addressing the potential for fraud as follows.

In general, fraud is lying in order to gain financially. Fundamentally, this might be a reported back injury that results in the employee being off duty for a number of days. Although the employee did not actually suffer an injury, it is reported, and the benefit is staying at home for a period of time while collecting a portion of their wage without actually working.

Such incidents are typically defined as fraud, but workers' compensation fraud can extend beyond acts committed by an employee. A physician may commit fraud by wrongly reporting and treating an injury as being more severe than it actually is in order to submit bills and collect compensation. It can be financially beneficial to the physician to extend the life of the claim in order to generate more revenue through billing for treatment that was provided.

An attorney representing an injured employee can commit fraud by manipulating the case to increase the cost of the claim unnecessarily in order to maximize the amount of money garnered by a client or the law firm itself. We have all seen the commercials declaring, "If you have been injured on the job, we will fight for you." Do not necessarily think that all of these firms are tainted by fraud. They really do provide a service to employees who have been injured and are being treated unjustly by their employer. However, attorneys may manipulate the system for personal gain beyond that which should be considered as part of the actual injury.

Employers can commit workers' compensation fraud. A lot of money is at stake, whether it is through workers' compensation insurance premiums or the payment of claims. An employer may be tempted to falsify facts in order to receive lower exposure to losses associated with workers' compensation claims and insurance coverage.

Fraud is tempting. It is tempting for the employee because workers' compensation legislation is typically written in favor of the employee, which might make it easy for an employee to take advantage of the system. Therefore, it can be tempting for the employer to develop a combative mentality regarding workers' compensation and work diligently to refute claims that are actually legitimate. This can lead to shading information rather than simply presenting facts.

One simple way to combat fraud is to build relationships with employees. Supervisors should frequently engage with employees to demonstrate a presence in the workplace. Rather than being isolated in an office, supervisors should be visible throughout the workplace. This indicates support of the employees and provides first-hand knowledge for the supervisor of the activities in which employees are engaged. This lowers the opportunity for fraud to occur in that employees are aware of the visibility of the supervisor.

Managers and employees can be educated on the system used to deeply investigate and manage all reported injuries. This level of communicated scrutiny will help everyone in the school to understand that legitimate cases will be compensated, whereas those that are fraudulent will be identified. A designated member of administration who is responsible for managing workers' compensation can help to establish a productive atmosphere by diligently responding to all injuries and gaining truthful information regarding the claim.

Benefits can be evaluated to determine whether the school system is providing a sufficient package for employees. For example, if adequate healthcare insurance is not provided, the stage could be set for an employee to report an injury that occurred at home as actually occurring at work in order for it to be compensated through workers' compensation. Vacation time could also be an issue. If sufficient vacation time is not provided, an employee might fraudulently report a back injury in order to receive time away from work.

School administration can take advantage of the assigned workers' compensation insurance claim adjuster. This person will have an objective perspective of a claim and will be able to identify issues with a claim that might point to the potential for fraud.

Physicians can be identified who will work to ensure that proper treatment is provided for all reported injuries. Physicians can be invited to the school, and a relationship can be developed that fosters the treatment of employees.

Open communication can be maintained among all parties associated with a claim, including the injured employee, the treating physician, the manager of the employee, the manager of an area where an employee might be placed on restricted duty, the workers' compensation insurance claim adjuster, and attorneys. Communicating with each of these individuals will help to ensure that accurate information is obtained regarding the claim. It will also help to identify the potential for fraud where information from one party is not consistent with that from other sources.

CASE STUDY

Sally is a tenth grade math teacher and is decorating her class in preparation for Math Field Day. Just prior to the end of the work day, she stands on a chair in order to attach a streamer to the ceiling. As she reaches for the ceiling, she loses her balance and falls to the floor. She immediately feels a sharp pain in her right ankle. She limps to the office and reports the incident to the principal and is embarrassed because all of the teachers were recently trained on how to safely work from heights. They were told of the new prohibition of climbing on chairs, but Sally's new stepladder has not yet arrived. She is taken to the doctor and an X-ray reveals a minor fracture. Treatment is provided, and Sally is placed in a cast. The treating physician requires Sally to stay at home for three days to rest with her leg elevated prior to returning to work.

- What OSHA recordkeeping and workers' compensation issues are involved?
- How would you transition Sally back to work?
- What corrective action should occur?

EXERCISES

For the following questions, identify a single school environment in which you would like to situate your responses and answer each question accordingly.

1. Can a claim be considered under workers' compensation and not be an OSHA recordable? Provide examples.
2. What tools are available that can help you to avoid experiencing workers' compensation claims?
3. How would you design a safety incentive program? Explain why you would include each component.
4. Why is it important to have a transitional return-to-work program? How does a transitional return-to-work program differ from "light duty"?
5. Who would you include in the process of investigating an injury? Why?
6. What tool would you consider to be the most effective in addressing fraud? Why?

ACKNOWLEDGMENTS

The majority of the material for this chapter was taken from a course developed for an online master's degree program in safety, security, and emergency management at Eastern Kentucky University (SSE 890: Business Continuity).

REFERENCES

Bose, H. (2008). Returning injured employees to work: A review of current strategies and concerns. *Professional Safety*, 53(6), 63–65; 67–68.

Geller, S. (1996). *Working Safe: How to Help People Actively Care for Health and Safety.* Chilton Book Company, Radnor, PA.

Gonzales, D. (2010). Integrated approach to safety: Fewer lost time incidents and greater productivity. *Professional Safety*, 55(2), 50–52.

Larson, L., and Larson, A. (2000). *Workers' Compensation Law: Cases, Materials and Text.* Lexis Nexis, Newark, NJ.

Pennachio, F. (2008). The myth of the bad employee. *Professional Safety*, 53(6), 66.

Wertz, K., and Bryant J. (2001). *Managing Workers' Compensation: A Guide to Injury Reduction and Effective Claim Management.* CRC Press, Boca Raton, FL.

32 Metrics

E. Scott Dunlap

CONTENTS

The previous chapter regarding school safety auditing established the need to measure school safety performance in order to know how well a safety management system is functioning. There are numerous metrics in addition to audit scores that should be evaluated to determine the degree to which a school safety program is functioning. A great deal of discussion has occurred in the safety community regarding the use of leading and lagging metrics when measuring workplace safety performance. Blair and O'Toole (2010) present the need to evaluate the types of metrics available and select those that will be effective in influencing the development of a safety culture. Although lagging metrics, such as injury rates and volume of workers' compensation dollars spent, are often used, opportunity exists to proactively measure things in the organization that can influence future positive performance. These leading metrics can include such items as behavioral observations to identify the degree to which employees are working safely or the volume of safety inspections that are conducted and acted upon. Manuele (2009) presents challenges in using leading metrics to predict positive safety performance in light of the large volume of variables that can impact the occurrence of a given accident. The opportunity for school administrators is to gain an understanding of safety metrics that are available and utilize those that will have a positive impact on the organization.

INJURY RATE

A common metric used to evaluate safety performance is the injury rate. An injury rate is derived from an equation that takes into consideration the number of hours worked within the school being evaluated. An injury rate is calculated by multiplying the number of injuries that occur in a certain year by 200,000 and then dividing that number by the actual number of hours worked in the year. The number 200,000 is a constant in the formula that equates to one hundred people working forty hours per week for fifty weeks. The rate communicates the number of injuries experienced per one hundred people in that year. Below are two examples:

- School A: Faculty and staff worked 75,000 hours in the past year and experienced three injuries. This equates to an injury rate of 8.0.
- School B: Faculty and staff worked 140,000 hours in the past year and experienced five injuries. This equates to an injury rate of 7.1.

These examples illustrate the fallacy of simply evaluating the number of injuries as a metric of safety performance. Although School A had two fewer injuries than School B during the year, School B actually had better safety performance in that the injury rate is lower than School A. Using an injury rate as a metric allows the comparison of two differing schools by converting the number of injuries experienced to a rate based on the number of hours worked by each school.

NATURE OF INJURY

The type of injuries that occur within a school or school system can be evaluated to determine where certain risks exist. Injuries can be categorized by type, such as:

- Cut
- Strain/sprain
- Broken bone
- Particle in an eye
- Burn

A second way in which injuries can be categorized is by injury cause, such as:

- Slip/trip/fall (on a walking surface)
- Fall (from a height)
- Lifting
- Impacted by an object

A third way in which injuries can be categorized is by the severity of the injury. The Occupation Safety and Health Administration (OSHA) utilizes the following categories for this application:

- Recordable
- Restricted work
- Lost time
- Fatality

A recordable injury, in general, is one that required the attention of a physician to treat. Although there are a number of exceptions to this, as discussed in the chapter on OSHA recordkeeping, an injury is typically considered to be recordable if an employee must be taken to a clinic or hospital for treatment. A basic recordable injury can be elevated to restricted work if the treating physician restricts the use of a body part that affects the performance of an employee's job, but feels that the employee can return to work under such restrictions. An injury is considered to be lost time if a treating physician believes that the employee needs to stay at home for one or more days apart from the day of the injury in order to properly heal. An incident is deemed as a fatality if the death occurred because of work-related circumstances.

A thorough incident investigation will be necessary to arrive at the information needed to evaluate the nature of each injury. An incident investigation will need to be done that captures information required by OSHA on the form 301. Employers are permitted to use this form or an in-house form as long as the information captured reflects what is identified on the OSHA 301 form. The following information should be recorded on the incident investigation form:

- Employee information
  - Full name
  - Address
  - Date of birth
  - Date hired
  - Gender
- Treating physician information
  - Name of treating physician
  - Name and address of healthcare facility
  - Whether or not the employee was treated in an emergency room
  - Whether or not the employee was hospitalized overnight
- Information about the incident
  - Case number (assigned by the employer, such as 12-001, where "12" represents the calendar year 2012, and "001" represents the first incident of that year)
  - Date of the injury
  - Time of the injury
  - Time the employee began work
  - Time of the incident
  - What the employee was doing immediately prior to the incident
  - What occurred
  - The nature of the injury or illness
  - Object or substance that harmed the employee
  - If a death, the date death occurred

Additional pieces of information that are helpful to record on the form include:

- Witnesses: Record the names and position titles of anyone who saw what occurred. It is also helpful to gain written statements from these individuals in order to fully verify what occurred.

- Corrective action: Once the cause of the incident has been substantiated, it is helpful to record what corrective action has occurred to prevent the recurrence of the incident. For example, if a teacher falls from a chair while hanging decorations in a classroom, corrective action might include implementing the use of stepladders and training in their proper use and storage.

The objective an investigation is to identify the causes of the incident. Although it is tempting to arrive at the most obvious conclusion, such as the unsafe behavior that immediately contributed to the incident, a deep exploration of all of the causes might reveal additional causes that are very important. For example, in keeping with the example of the teacher falling from a chair while hanging decorations, a fundamental investigative tool is to ask the question "why" five times to arrive at a spectrum of information that needs to be considered.

- Why did the teacher fall? She was climbing on a chair.
- Why was she climbing on a chair? A stepladder was not available.
- Why was a stepladder not available? One had not been placed in her class.
- Why had one not been placed in her class? Budget money was not available to provide the stepladder.
- Why was budget money not provided? Funds were allocated to other projects.

Although basic in nature, this example illustrates how it might be easy to immediately blame the teacher for unsafe behavior when in reality the primary issue is that money was not budgeted to provide her with the tools to safely perform her job. It is important to move beyond surface causes of incidents to determine the root cause so that appropriate corrective action can be taken to prevent the incident from recurring.

NEAR MISSES

Similar to tracking the types of injuries that are occurring, near misses can be measured as a proactive tool. A near miss is simply an incident where the environment contained all of the elements for an injury to occur, but nothing actually happened. For example, the teacher might mention that earlier she was on a chair hanging decorations and slipped but caught herself before falling to the floor. An injury did not actually occur, but the environment could very easily have fostered the occurrence of an injury had she not caught herself.

The occurrence of near misses serves as a warning that a flaw exists in the system. It also provides an opportunity to act on information gained from investigating the near miss and take corrective action prior to an actual injury occurring. This indicates that there is a need for faculty and staff to report near misses when they occur so that they can be fully investigated and corrective action can be implemented prior to someone being injured. A volume of near-miss reports can then be evaluated to determine whether certain trends exist of unsafe acts or conditions that need to be addressed through employee training, safety communications, or improving the work environment.

BEHAVIORAL OBSERVATIONS

Behavior-based safety arose in the early 1990s as a tool to be even more proactive beyond investigating near misses. Rather than waiting on an injury or near miss to occur and then reacting to the information provided in an incident investigation, behavior-based safety utilizes workplace observations to proactively observe behavior. The use of such audits and reaction to information gained will strongly communicate to employees that safety is of concern within the organization (O'Toole and Nalbone, 2011). Behavior is quantified on an observation sheet as to whether it is considered to be safe or at-risk. Table 32.1 shows an example of a form that might be used to perform such an observation on a custodian.

This information can be printed on a card that can be easily carried and utilized. It is important to note that in the general information provided at the top of the card, a place to record a name is missing. This is due to the philosophy that anonymity needs to be achieved in order for employees to feel comfortable being observed. They need to feel that they will not get into trouble if something negative is found during the course of the observation. The date, activity, location, and job title provide areas in which the data can be categorized and evaluated to determine trends.

Behavior is measured by counting the times that safe or at-risk behavior occurs and placing an X or a tick mark in the appropriate box each time it occurs. In this example, a custodian is using a special cleaning solution to remove graffiti from lockers. During this observation, the custodian had an opportunity to communicate what was occurring to three bystanders that came into the work area at different times. He communicated clearly with two of the bystanders but did not say anything to the third person. This resulted in two safe behaviors being observed and one at-risk behavior being observed. A portable safety sign was the only sign required to be placed in the work area, and he properly placed it, resulting in one safe behavior being observed. The task required the use of safety glasses and gloves, but he only used gloves. This resulted in one safe behavior being noted for the use of the gloves with one at-risk behavior being noted for not wearing safety glasses. Two safety

TABLE 32.1
Sample Behavioral Observation Form

Date: _____ Activity: _____

Location: _____ Job Title: _____

Activity	Safe	At-Risk
Communication	X X	X
Use of safety signs	X	
Personal protective equipment	X	X
Followed safety procedures	X X	
Total	6	2
Percent Safe	75%	

procedures were needed to accomplish the task. The first was to use a stepladder when reaching heights and the second was to use a fan to increase ventilation in the work area. He utilized both procedures, which resulted in two safe observations.

Once the observations have been recorded, the data will be added to arrive at a total number of behaviors for each category. In this example, there were a total of six safe behaviors and two at-risk behaviors observed. Of the eight total behaviors that occurred, it was found that six were considered to be safe, so this task measured as 75% safe. Rather than focusing on failures, such as are measured by injury rates, this metric provides a proactive positive measurement of safety performance. Observations can be accumulated in a database or spreadsheet and measured over time with the goal of increasing the level of safe performance of jobs that occur.

Behaviors that are measured can vary among different jobs. An evaluation will need to be conducted for each job within the school system to determine what behaviors are critical in order to work safely. These behaviors can then be recorded on a card that is specific to that job and then utilized to measure safe work behavior across time.

SAFETY COMMITTEE ACTIVITY

Safety committee activity can be measured to determine the level of activity and engagement that is occurring by the committee and its members. Safety committees are designed to provide an opportunity for individuals who perform work in an organization to become actively involved in safety rather than safety being relegated only to a safety manager or risk manager. Because of the various groups involved in schools, ratios can be created to determine the volume of membership. For example, if 70% of the employees represent faculty, then roughly 70% of the safety committee should be comprised of faculty members. A safety committee should be limited to approximately twelve members in order for it to function smoothly. A typical school safety committee could be comprised of:

- Faculty (6)
- Administration (2)
- Custodial (2)
- Maintenance (2)

This membership can be changed to meet the needs of the school and groups that exist within the organizational structure.

The activities of the committee can be measured to ensure that it is productive. Activity that can be measured within the safety committee could include:

- Number of meetings held
- Number of safety recommendations made
- Number of safety recommendations implemented
- Number of incidents investigated
- Number of corrective actions identified to prevent recurrence
- Number of corrective actions implemented

Safety committee membership should be designed so that it is seen as an honor to be part of the organization. Credibility must be achieved through who is permitted to be on the committee and the resulting activity that occurs and is communicated to others.

INJURY COST

Injury cost is a metric that can be used as one way to evaluate the severity of incidents that have occurred as well as general financial loss caused by incidents. It is important to realize that there may not be a correlation between lowering an injury rate and lowering the financial cost of injuries. One way in which this can be seen is a single occurrence of carpal tunnel syndrome by a member of the administrative staff. Carpal tunnel syndrome falls under the general category of cumulative trauma disorders where repetitive work at a computer workstation can result in the inflammation of the carpal tunnel, which is in the wrist. Such an event can result in surgery, lost work days, and physical therapy to properly heal. Although this event may only represent one injury when calculating an injury rate, the cost in dollars can be very expensive.

The cost of injuries directly affects a school's operating budget. This will occur whether the school is self-insured or insured through a workers' compensation insurance policy. Self-insured workers' compensation programs are designed to have money set aside that is dedicated to covering the cost of work-related injuries. An injury that occurs will deduct money from this account, thus causing the school to suffer a direct loss whenever an injury occurs. Workers' compensation insurance policies can easily be increased when a history of injury and payouts in claims occur. Regardless of how the cost is managed, it is in the best interest of the school to monitor the financial cost of work-related injuries and identify ways in which injuries can be prevented that will in turn reduce cost.

AUDIT SCORES

The metric of audit scores is a proactive measurement of safety performance. Audits are used to measure the safety performance of a school against established organizational standards or government regulations. Questions are asked related to each aspect of a safety program being audited. The responses to these questions on the audit are then translated into a score that can serve as the metric that is used to measure performance. Types of safety audits can include:

- Annual: Annual audits are typically comprehensive in that they measure each component of the school safety management system. This will include auditing in the areas of documentation review to determine whether written programs and supporting documentation are appropriate, school physical inspection to determine whether written programs are being implemented in the work environment, and interviews to determine the degree to which employees can converse on safety topics on which they have received training. Annual safety audits may require in excess of one day to complete.
- Housekeeping: Housekeeping audits are focused solely on how well the school environment is maintained. Such an audit is designed to identify

the presence of any hazards that could cause a slip/trip/fall injury, such as a spilled liquid that has not been cleaned in a timely fashion, and fire hazards, such as the presence of an excessive amount of combustible material. Housekeeping audits can take place on a weekly or monthly basis in a period of a few hours.

- Fire: Similar to housekeeping audits, fire audits are focused on life safety issues that are applicable in the event of a fire. This can include the verification of fire pump operation, clearance of sprinkler heads from obstruction, and clear paths of egress in the event that an evacuation must occur. Also similar to housekeeping audits, fire audits can take place on a weekly or monthly basis in the span of a few hours.

This list provides three examples of how audits can be implemented and used to measure safety performance. Other audits can be developed and implemented based on the needs of a school, such as access control audits or behavior-based audits to monitor for the presence of bullying among students. An audit is a flexible tool that can be used to accommodate the needs of a school. The quantitative findings can be measured across time to monitor performance and communicate the performance to faculty and staff.

GOALS AND STRATEGY

Evaluating progress on safety goals and strategy can be measured to determine how well a school is advancing toward safety goal attainment. Safety goals can be established that are unique to each school based on safety performance challenges. Each goal should contain strategy points that can be used to facilitate achieving the goal. Strategy points serve to provide direction in activity for personnel associated with the school to achieve the goal. Below is an example of safety goals and strategy points that can be used to achieve the goals:

- Goal 1: Reduce the school injury rate by 10%
  - Strategy 1: Conduct safety training among all faculty and staff within the school year
  - Strategy 2: Conduct behavior-based observation to determine where safe behavior is occurring to share best practices and to identify where at-risk behavior is occurring to identify opportunities for improvement
  - Strategy 3: Implement an ergonomics program to address repetitive motion injuries that account for the largest percentage of the school's total injuries
- Goal 2: Increase the level of safety communication
  - Strategy 1: Create a monthly electronic safety newsletter to be sent to faculty and staff
  - Strategy 2: Include safety on the agenda of all relevant faculty and staff meetings throughout the year
  - Strategy 3: Create a school safety website that can be accessed by faculty, staff, contractors, and parents

The two goals listed in this example are simple, achievable, and memorable. They do not include complex words or concepts but speak directly to the challenge that is present. Faculty and staff can look to the strategy points to determine ways in which they can engage to assist the school in accomplishing the goal. An administrative assistant could volunteer to create a school safety website with other staff and faculty members volunteering to provide content. A faculty member could volunteer to attend extensive ergonomics training in order to create an ergonomics program that is applicable to the school. This process of engagement will assist in everyone taking ownership of the school safety program and working toward the improvement of school safety performance.

EMPLOYEE WELLNESS

School safety needs to be seen as a twenty-four hours a day, seven days a week proposition. It is difficult to expect someone to perform safely at school when that behavior is not congruent with behavior outside of school. Wellness is one area where safety can expand beyond the confines of the school property to affect individual behavior.

Although employee medical records are confidential, healthcare insurance carriers can provide general data on medical claims that have been made among employees. This information can be used to establish initiatives and programs to target the causes of the claims. For example, it might be found that an exceptional amount of healthcare cost has been related to pregnancy complications. This can give rise to the opportunity to implement a healthy pregnancy program that provides information and counseling for expectant mothers. It might be found that a great deal of healthcare cost is associated with problems that stem from obesity. This information could result in the implementation of a program that offers free or supplemented gym memberships along with an incentive program to encourage faculty and staff to get needed exercise.

Embracing a wellness program can help a school system in two areas. First, employees live healthier and safer lives. Problems associated with a range of health issues can be addressed that can result in faculty and staff experiencing things such as lowering weight, cholesterol, and blood pressure while increasing the knowledge of how to live a healthier lifestyle. The school system benefits by experiencing less absenteeism and lowered medical costs.

CASE STUDY

Principal Ellis is struggling with the safety performance at her school. Bullying is on the rise, and employee injuries remain at an undesirable level. She has no full-time safety resources at her school and needs to find a way to begin addressing her school's safety performance. The faculty and staff have proven to be effective in helping to address other areas of school performance.

- Can Principal Ellis move forward with addressing her safety concerns even though she does not have a full-time safety resource to contact for assistance? Why or why not?
- What metrics might prove useful in helping her to make progress toward improving school safety performance?

EXERCISES

For the following questions, identify a single school environment in which you would like to situate your responses and answer each question accordingly.

1. How can the injury rate for a school be calculated?
2. What would be the best way to categorize the nature of injuries to measure issues that need to be addressed?
3. Name a specific task that occurs at a school and describe how you would design a behavioral observation card that could be used to evaluate work performance.
4. What types of audits would be useful in measuring safety performance at a school?
5. List five safety goals that would be appropriate for a school and provide strategy points for each goal that will assist in goal attainment.

REFERENCES

Blair, E., and O'Toole, M. (2010). Leading measures: Enhancing safety climate and driving safety performance. *Professional Safety*, 55(8), 29–34.

Manuele, F. (2009). Leading and lagging indicators: Do they add value to the practice of safety? *Professional Safety*, 54(12), 28–33.

O'Toole, M., and Nalbone, D. (2011). Safety perception surveys: What to ask, how to analyze. *Professional Safety*, 56(6), 58–62.

33 Communication and Program Development

Rebbecca Schramm

CONTENTS

When an unsafe or unhealthy societal problem plagues a city, state or nation, a large-scale educational movement that starts in the K–12 educational setting may be the best long-term solution for change. Classrooms are full of free handouts reminding kids to be heart healthy by following the food pyramid and to avoiding stepping on or going near downed electrical power lines after a destructive storm. Going "green," antibullying, cyberbullying, and avoiding texting while driving (or possibly texting while walking) are all emerging topics of interest that are discussed from elementary school through high school in lesson plans to teach young people how to avoid harm and keep themselves and their environment safe and healthy.

Many teachers are excellent at creating student-centered activities that are hands-on or somehow visually and mentally stimulating to their students and are perfectly suited to be part of a widespread educational movement. There are many different theories that explain ways to help students understand concepts, and one such theory is Howard Gardner's theory of multiple intelligence. Nolen (2003, pp. 115–118) states that there are eight kinds of intelligences including linguistic or verbal, musical, mathematical-logical, spatial, bodily-kinesthetic, interpersonal, intrapersonal, and environmental or naturalist intelligence, which requires teachers to adjust their instruction strategies in order to meet students' individual needs. Therefore, teachers often find themselves constantly thinking about creating great lesson plans. Whether it is the perfect discovery of a prop to purchase at a garage sale or leftover marketing "freebies" from a local fundraiser, many teachers are great at creating activities for their classrooms. When the community offers pre-created activities, curriculum plans, handouts, games, prizes, and sometimes even ready-to-serve volunteers to aid in teaching the objectives, a methodology for linking teachers to the dissemination of safety education is created.

CURRENT SAFETY-RELATED PROGRAMS

An example of a program that exemplifies the usage of safety education curriculums in classrooms was the 1980s educational program titled D.A.R.E. (Drug Abuse Resistance Education). According to Aniskiewicz and Wysong (1990), the combination of new media reports of trafficking of drugs along with a rise in drug-related violence helped create a "drug crisis" in the United States (p. 727). D.A.R.E was aimed at school-age children and became a hot topic at approximately the same time that the simple 1980s commercial from the Partnership for a Drug Free America became popular (Partnership for a Drug Free America, n.d.). It depicted a close-up of a hot frying pan of butter with a voiceover stating, "This is drugs." Then an egg is dropped into the frying pan and is immediately fried to hard white, and the voiceover says, "This is your brain on drugs . . . any questions?" This was meant to compel young and old to stop frying their brain with drugs.

In fact, Trebach (1987) stated "The variety of intervention plans created to respond to the drug crisis were collectively characterized as the *war on drugs*" (as cited in Aniskiewicz and Wysong, 1990, p. 727). Aniskiewicz and Wysong (1990) report that the D.A.R.E. curriculum consisted of seventeen weekly sessions that were conducted by uniformed police officers, minimized use of lectures, and it explored topics about "saying no" to peer pressure (p. 728). D.A.R.E allowed teachers a chance to have heroes in their classroom. Police officers were dispatched to classrooms all over America to offer their services to motivate kids to avoid using drugs. Guest speakers in uniform came to schools with ready-made lesson plans, making it easy for teachers to talk to their students about safety even after the motivational speakers left.

Another excellent example of connections between students, teachers, and safety education opportunities in K–12 classes include firefighters handing out red plastic firefighter helmets to young children. Teachers have also been known to pass out the bright green Mr. Yuk stickers to elementary classrooms: "self-adhesive labels depicting a green-faced grimacing man with a protruding tongue which is intended to allow parents to label all common household poisons to teach their children that labeled items are not to be tasted, ingested, or inhaled" (Fergusson et al., 1982, p. 515). Poison control flashcards, games, pamphlets, and other lesson plan activity builders are also available to teachers. Organizations such as the Centers for Disease Control also offer teachers free tips and ideas for starting antismoking campaigns in their classrooms. Emerging issues like texting while driving are also being responded to by organizations so teachers have the chance to make a difference through the use of "canned" curriculums. What better way to motivate teachers that are constantly looking for ways to inspire students than to offer them easy ways to incorporate lessons about keeping their students safe?

The benefit to this style of communication of integrating safety and security into daily curriculums does not stop there. Students that learn at school also teach their parents, grandparents, and day-care providers about the safety education after they are exposed. Safety topics that teachers enlighten their students about influence the culture through such areas as seatbelt campaigns, heart-healthy 5Ks, gun safety, and preventing the spread of germs.

Mothers Against Drunk Driving (MADD) is another example of how teachers have been provided with readily available motivational speakers. Family members that have been affected by drunk driving are sometimes willing to discuss how they have lost their loved ones to drunk-driving accidents. Therefore, if the country needs to be safer, it may require incremental change, but it can occur through our education system with predesigned lesson plans, curriculum assistance, and premade handouts and craft ideas or games provided on the Internet, through the government, or non-profit organizations.

D.A.R.E. and MADD could actually be categorized as *passive curriculum development* methodologies for teachers or passive safety communication strategies in the classroom. Although teachers can usually determine whether or not to include a certain topic in their unit plans during a semester, many of these programs are often supported as a school-wide "push" or supported yearly by parent organizations or community volunteers. For example, MADD volunteers might be best used to discuss the tragedies of drunk driving annually the week before prom at a high school. Antismoking or antidrug programs might be facilitated before summer break in seventh or eighth grade because students start experimenting when they have extended periods of time during the summer and, for the first time, may not have childcare professionals watching over them. Firefighters might be asked to be regular guest speakers for the entire incoming student population during the beginning of the kindergarten year for a school-wide shared unit called "Heroes" to enhance the fire safety education of students.

Sometimes the use of these large-scale programs are more time or event based rather than personal teacher choice based. They could also be used as a substitute as "important to cover" but not "required on the mid-term or final." Consequently, another category of communication regarding safety in schools is mandatory safety communication strategies in the classroom. These types of safety rule presentations are the same ones that travelers are accustomed to hearing after they board an airplane or start a new job when they sit through the "safety training" about evacuation maps or how to call for assistance in the event of a medical emergency.

DRILLS

Fire or severe weather drills are often practiced in schools each year to ensure that students and teachers know how to get out of the building in the event of a fire or what to do in the case of a hurricane/tornado. Teachers may think that the administration (i.e., the principal) is running the drill, but in all actuality the most important part of the entire activity is to protect the lives of students and staff. Teachers are the key to a successful drill and their attitude about the training they are providing is extremely important.

Whether it is a drill or a true emergency that affects the classroom, all eyes are immediately on the leader of the classroom. When a drill sounds, screams or shots are heard, police are on the scene, or even a fight breaks out, the teacher is the commander of the scene. He may have practiced for years how to teach creative writing, algebraic equations, or Spanish, but how much practical effort was placed into more than just basic discipline practices? In an article about designing classroom management plans, Capizzi (2009) states:

> The amount of autonomous control the teacher holds over the classroom should be individually suited to each classroom situation and group of students and is often dictated by the appropriate level of structure warranted by various classroom factors such as grade level, subject matter being taught, and academic skill level (p. 3).

For those who have run a classroom, this is often easier said than done. However, when it comes to safety, it is very important to know what to do when the "alarm" sounds. How a teacher communicates and coordinates during a real or perceived safety situation will either help or hinder the process toward effective and efficient closure of the incident. These "incident management skills" can be practiced, discussed, and written down before the situation occurs to greatly increase the chances of successful outcomes.

So what alarms are the most important? Should beeping on a panel be ignored or should a teacher respond? When do we evacuate and when do we take shelter inside the building? Do students take a book to put over their heads or do teachers close windows or shades before leaving the classroom? Severe weather events cause significant fears in young and older children alike. Quite honestly, any teacher that has been in a classroom can attest that a common bee or wasp flying around in a classroom can cause mass hysteria, so there are many types of issues that cause students false and real alarm in the classroom.

TEACHER BEHAVIOR

Alarming situations occur when the sky gets dark, and students may become unsure if they are going to be safe. If their teacher is in control and looks calm, then the rest will usually follow. According to Mendler and Mendler (2010), when teachers and administrators realize the importance of remaining as calm as possible, especially during moments of disruption and chaos, students behave best (p. 29). However, older students might use this as a chance for horseplay. For example, high school-aged students may make a joke out of a fire drill, even though a real fire is very serious. Being trapped on the fourth floor of a building with a raging fire outside with thirty high school students will turn grave very quickly. Teachers that have their "game face" on and are able to be serious, calm, and lead with determination are going to be much more successful than ones that do not create boundaries between themselves as the commander and their students as the followers. "Students behave best when they are around a confident teacher not easily rattled by the ups and downs of life" (Mendler and Mendler, 2010, p. 29). In fact, if you are not leading your students quietly and the other teachers are doing their part, then you are contributing to lackadaisical safety support (and in the process probably frustrating the teachers that are doing their best).

In addition to running drills, there are rules and policies that must be shared each year regarding safety and security. Within many educational systems, policies exist regarding "drug-free" and "weapon-free" zones and strict Internet policies for students and staff alike. Each school year, new enrollment packets often include documents that must be shared with students. Pentz (2000) found "local policy change efforts include direct approaches like the development and monitoring of drug-free zones and community policing for drug use and restricting youth access to tobacco and alcohol" (p. 260). A teacher of an individual classroom is the gateway to the entire school safety program. At the beginning of each semester or term all educators need to communicate the importance of drug-free and weapon-free zones or the importance of appropriate Internet usage according to school policies. These constant reminders will help students make better decisions when it comes to disobeying the policy and in turn will keep the campus more safe and secure.

Policy review and curriculum development are important, yet students need to know that the teacher will lead in a safety situation and the students must follow, regardless of their age or the topic of the lesson at hand (whether it be art, gym, math, music, second grade, or physics). In a study about discipline in the classroom, Mendler and Mendler (2010) found "by modeling the skills students should use when they are faced with difficult moments, educators show real ways that real people who are upset can solve problems without anybody getting hurt" (p. 30). In the case of safety communication, the stakes are terribly high. Not following procedures or polices could lead to the serious injury or death of a student. Teachers of any subject, any grade, and any age literally hold the key to using effective safety communication to save a student or to possibly reduce the impact of a situation by the way they handle an incident.

COMMUNICATION MEDIA

Training students is not merely about standing in front of the classroom. Although most schools have replaced chalkboards with whiteboards and overhead projectors with TV screens, there is one method of communication that never seems to go out of style even in our modern day setting: the bulletin board. Bulletin boards plastered with artwork, fundraiser flyers, upcoming event calendars, and assignments are usually commonplace in the K–12 education setting. Various colors of paperboard also litter the hallways of schools to sell everything from candy bars to buckets of frozen cookie dough or to invite students to an upcoming musical or book fair. PowerPoint displays on screens throughout the building are also available to share information and communicate upcoming events. Murgatroyd (2010) states, "thinking, making, and doing are critical components of teaching. Students need to be challenged to produce tangible products, services or activities which are valued by others" (p. 266). Having students use teamwork and their expertise to share safety topics visually is a good component of any safety communication program. Having students put their ideas into posters and on PowerPoint flyers will be useful classroom exercises.

Teachers do not only display exemplary work on boards, they also hang up additional unit material depending on the season, the curriculum, and the special visitors touring the school for events like open houses, carnivals, spaghetti dinners, or parent-teacher conferences. A curriculum guide almost always includes additional learning resources such as available pictures or graphics aligned with the chapter and various hands-on activities that increase learning to engage the mind. Teachers have access to fifteen to two hundred or more students per day, as well as their parents, depending on schedules and their extracurricular activity involvement. Teachers have a constant platform for sharing new ideas each day that can be used to make positive change regarding safety and security in the community.

If a teacher wants to reinforce an everyday disciplinary (i.e., safety) policy, they can do so in a multitude of ways in their classroom. When a principal wants to change a school, they can start with helping the teachers be successful. School districts that want to increase awareness of a particular safety topic can start with campaigns in schools. Pentz (2000) found that "community organizations and programs that promote policy change can also be used to sustain community prevention efforts, particularly policies that mandate setting aside funding, requiring program implementation, and engaging in enforcement and monitoring" (p. 267). *Mandatory safety communication strategies* regarding security are vitally important to emergency planning in a school and can help reduce the risk to health and protect the lives of all students each day in the event of a crisis.

STUDENT PRESENTATIONS

Large-scale pushes by a school district are important, but smaller programs can also develop within schools that become a communication strategy for safety. Teachers might make an assignment that requires a group/student presentation, and the list of topics can include safety topics that are applicable to a certain geographic region or national current events. Current war-time activities or environmental disasters like oil spills can be included on the topic choice list. In this case the teacher is actually creating a *mixed safety communication strategy*. Instead of actually working toward a unit structure or a specific topic, the lesson plan or the curriculum is geared toward teaching students how to work in a group or how to practice good communication skills.

Public speaking skills and small-group interaction skills learned through presentations in the classroom engage students in learning. One saying that teachers often hear is "those that can't *do*, teach." It is usually said jokingly, but teaching is a process of learning what to *do* each and every day. What can a teacher *do* to make ideas sink in and end up creating that "aha" moment? Safety communication strategies revolve around what teachers do to find what works. The tough part of teaching is that each student is an individual and brings different emotional, spiritual, physical, and relational abilities to the table each day. According to Nolen (2003), "When teachers center lessons on the students' needs, it optimizes learning for the whole class and it's part of a teacher's job to nurture and help the children develop their own intelligences" (p. 119). When handing out a group presentation topic choice list that includes "protecting the water supply" or "Harry Potter books," the group of students will be able to choose their plan of action. If they want to dig deep into a topic that could lead to making the world a safer place, the seed that was planted by the teacher included a challenge and an opportunity for safety awareness.

COMMUNICATION STRATEGIES

Finally, teachers may deliberately set up opportunities for growth for students to become more aware of safety topics through *targeted safety communication strategies* in the classroom. Teachers with a personal interest in a specific sports team or children's animated character may set up their entire classrooms with a thematic decoration because it is part of their personality. If teachers care deeply about sustainable practices like recycling then they will probably have a recycling center and teach their students about recycling. In this way, teachers may also become role models for safety and security in the classroom environment they create.

Additionally, world events like wars or presidential campaigns that unfold around teachers also affect classrooms. Many educators were able to turn on televisions in their classrooms on September 11, 2001, to watch the World Trade Center terrorist attacks with their students in real time. According to Felix et al. (2010), "Both the Oklahoma City bombing and the September 11, 2001, terrorist attacks occurred on a weekday during school hours within the academic year; consequently, most school-age children and adolescents in communities near attack sites were in school at the time of the events" (p. 592). Events such as these affect how the teacher teaches. In the years after 9/11, some teachers may have written a targeted unit about terrorism in the United States in social studies or English classes. The study by Felix et al. (2010) illustrates that "overwhelmingly teachers wanted to know how to identify and deal with the psychological and emotional aftermath of a traumatic event on students" (p. 602). Being a good educator includes finding ways to connect with students, and the aftermath of a security or safety event presents an opportunity to connect through lesson activities and advance programs about safety and security.

In our educational system, teachers have a wide range of "control" over what they teach in their classrooms. Extremely organized school systems have tight control over the curriculum development process as demonstrated by multiple classroom teachers using similar strategies (group jobs, portfolios, journals, etc.) and teaching concurrent objectives at about the same base time (i.e., all seventh grade English classes using the same novel during the first two weeks of October for a creative writing class). Less organized school systems have limited control over their curriculums and standardized training, text book usage, and testing requirements. Therefore, the discussion of safety and security program development and the involvement of teachers within that overall program and structure may be dependent on the organization of the school that they work within.

It would seem that the more organized the school, the more organized the safety and security program is apt to be. However, according to Murgatroyd (2010):

> The narrative about schools and change is that they are at the forefront of change. The reality is they look, feel and are almost the same as 25 years ago even though some technology has crept in. New targets have been set and rarely met and new assessment regimens have mushroomed but are distractions from learning (p. 260).

If schools are not as advanced as we hope they are for the youth of our country (regardless of how strong the curriculum development is for the school), that does not mean teachers cannot be relatively autonomous in their own classrooms. Murgatroyd (2010) proposes, "Rather than simply producing essays, book reports, research posters for science—all of which have value—students should be required to work with community organizations on projects that will make a difference for the community" (p. 266), and safety and security topics are perfect options for teachers to use. One of the great joys of being a teacher is that each one brings his own strengths to his classroom and his autonomy to how to teach regardless of what type of system they are working within.

Unfortunately, in many situations safety and security quickly become integral right after a significant security or safety event occurs as a reaction instead of as a proactive push. As explained in an overview and summary of U.S. school health policies by Kann et al. (2007), "the majority of schools did include crisis preparedness, response, and recovery plan elements consistent with current recommendations such as evacuation plans, lock down plans, periodic review of plans, and mechanisms for communicating with school personnel" (p. 394). Yet one tenant of education is teaching a person *before* they are required to carry out a task unless it is a pretest or baseline. In safety-related events like an active shooter situation or building losses based on fire or severe weather damage, the situation seems very backward to an educator. The first thoughts of a teacher involved in a classroom safety event that suddenly gets a significant amount of attention might be "Why didn't I get this warning ahead of time?" or "Why didn't I pay more attention to that training when it occurred?" There will usually be an increased amount of discussions surrounding the event immediately after those things that affect teachers, students, and the education setting. Parents, communities, police, firefighters, and even news stations may all be involved immediately. When the dust settles, and the programs are being quickly reviewed and developed in the aftermath, the teacher will still be responsible for daily operations and a new host of safety communication topics. Integrating these safety and security topics into the curriculum in one way or another may become a reactive requirement instead of implementing them proactively to prevent such incidents.

It is important for educators to know that a good safety program can take years to develop, and the aftermath of an event is often charged with emotions. There is no simple one-size-fits-all plan for facing a student with a gun. After the decision is made by the school to become safer and more informed about serious safety topics, the teacher has to first take the efforts seriously. As reported by Papastylianou et al. (2009), there are certain causes of stress in teaching:

> Work pressure, job conditions, the ambiguities and conflicts of the educational role resulting from its complexity and from the administration's conflicting demands, pressures exercised by educational leadership, lack of professional growth opportunities, lack of resources, poor professional relations with colleagues, low pay, strained relations with students' parents, teacher's expectations of themselves and schools, and the lack of managerial support (p. 297).

Although these stressors are not present for every teacher, it is important to consider that teachers must be open to change when it comes to new program development regardless of the difficulties of their situation. Teachers can make or break a new program and can easily hinder and delay the activities of safety and security development significantly if they do not accept and support the change (Table 33.1).

It is also important to realize that a new safety program being developed at a manufacturing site, a church, or a school will be saturated with political issues and

TABLE 33.1
Initiating Change

Significant Change to School	Teacher Role in Program	District-Wide Buy-In Challenges
Adding external and internal security	The phrase "the safety and security of the staff and students at this school is paramount" needs to become the mantra of the teacher, not gossip or complaints about being spied on.	Parents will need to be "heard," so a school website for comments or a drop box or voice mail box of information that allows communication to the school regarding the changes would be helpful. Surveys can be conducted before the policy change to allow parent input. Asking the parent teacher organization/ association to discuss the additions of security guards in their groups would also be helpful for buy-in.
Adding gating and site entry restrictions	Use and support use of personal entry badging each day to keep traffic flowing through gates. Remind parents through home communications (invitations to classroom parties/holiday events, etc.) how to get onto the site easily. Learn about security threat rings and how protecting the exterior border of a property is the first step in protecting a site.	Aesthetics will be important in sites that are not in "rough neighborhoods." Schools should count and document trespassers on the property and have appropriate "no-trespassing" signage if gates are not protecting property. Changes for delivery trucks will also need to be made, and drop-off and pick-up times for students (or parking lots for high school) will become an issue that will need to be discussed.
Regular locker checks	Do not ignore the importance of regular locker checks and thoroughly explore lockers when you are assigned to do so. Under papers and library books could be drugs or weapons. Teach students policies on random locker checks. Teach students not to share lockers because if someone else puts something dangerous into their locker they will be held responsible.	Lockers are seen as personal property on a school site, but they are not. Parents may not agree with this statement, but because of the security risk, the safety of the students must be held above the personal belief that students need privacy in their lockers.

(continued)

TABLE 33.1 (Continued)
Initiating Change

Significant Change to School	Teacher Role in Program	District-Wide Buy-In Challenges
New overall security policy training	Be patient with the people enforcing the new safety rules and policies. Others will follow in your footsteps. Take part in activities designed to allow your participation whenever available.	Ownership of the programs is usually added to a person without any particular training. There is usually no single point of collection for all problems and complaints other than a personal e-mail in-box. Creation of programs for safety requires ongoing maintenance and regular review, updates, and ongoing refresher training; it is not a one-time event.
Background checks of volunteers	Request names and set up volunteers for parties and necessary activities in advance. Follow the policies and do not deviate from the requirements. Communicate to parents that you want them to come to your classroom but that a background check must be done before they can.	The cost of the checks will be questioned. What the school will do with information that is negative and sensitive will be a discussion point for school staff because the teacher will need to be informed that a parent cannot volunteer. A policy about what types of offenses are acceptable on a check will need to be discussed before the policy goes into effect.
Addition and upkeep of cameras or other technical computerized security systems to manage the building (building management systems)	Alert the maintenance staff if you see anything out of the ordinary (blinking lights on cameras or cameras pointed in the wrong direction) or if you hear any alarms going off. Learn what alarms mean and how to leave the building through the right evacuation routes so as not to cause additional difficulties for the maintenance staff.	Maintenance employees may not have the technical ability to run preventive maintenance programs on camera systems or building management systems and may need additional training. Parents may feel the privacy of their children is being encroached upon and will need to see policies relating to how the cameras will not be placed in private areas like locker rooms or restrooms. There will need to be strict policies about who can review what camera footage at what times and how it will be stored and catalogued if an issue is identified.

TABLE 33.1 (Continued)
Initiating Change

Significant Change to School	Teacher Role in Program	District-Wide Buy-In Challenges
Addition of a controversial figure to the administration staff supporting safety change	Support the administrative figure by learning about the new policies and by attending required and suggested training about safety. Do not negatively talk about the employee of the school to students or staff but rather listen and learn about their worries so you can help the school adjust the plans in the future as necessary to come up with the best possible safety plans.	Controversial figures may not last but sometimes are able to make significant changes quickly before they are dismissed. They may feel strongly about safety program development based on a tragedy, but they may not be able to find a middle ground. Sometimes after the tragedy occurs, buy-in for change is immediate but can slowly erode over a short period of time.
New badge system requiring staff and students to wear a badge each day	Wear your badge every day. Ask students in the hallway during passing times (middle or high school) where their badges are if they are not displayed appropriately. Use disciplinary procedures for students that are evading questions regarding their name when they do not have a badge or if students in your class do not wear their badges as required.	Parents may feel that wearing identification badges reduces their children to "numbers," so schools need to be prepared to talk about the importance of identification. Districts might use the badges for other purposes, such as loading funds for lunches or checking out books at the library. Teachers might be able to use their identification cards at the copy machine so paper can be charged back to their department or as an authentication at their computers as a log-on so that the badges become part of the normal day.

challenges to old ways of thinking. Dedication from the top down in the school will be required. The principals, secretarial staff, janitors, maintenance staff, food workers, volunteers, coaches, students, and especially the teachers will be impacted by new safety programs. Through passive, mandatory, mixed, and targeted communication strategies in the classroom, the teacher can choose how to play their vital role in the education of young people to carry out safety plans in the educational setting.

Integrating new security standards, procedures, and physical changes to any environment will be challenging. In a work environment such as a manufacturing site, a government building, or a bank, there are a group of adults involved in making decisions about how adults will deal with security changes. However, with K–12 safety and security integration plans, the student population is still under the direct care of their parents or guardians. Instead of working with adults at a site one-on-one regarding safety and security changes, the school district that is going through a makeover of their programs must also appease and appeal to the sense of the parents in addition to the staff and students. This is no small task in a regular setting, but in a school setting the stakes are even higher.

Change is growth, and growth often causes discomfort in the beginning. The start-up of a new technology like a parking lot gate system, turnstiles, or camera integration will be exciting, rewarding, and difficult. The new policies may cause stress and no small amount of grief for some staff, students, parents, and the community at one point or another. Yet the school will be safer when the final phases of the safety and security integration plans are completed and the continuous improvement phase is finally underway.

Policies will be created, and training will occur that will allow students and teachers to know what to do in emergency situations, such as a workplace violence emergency, which will reduce the number of fatalities and increase survival rates. Cohen, Allen, and Hyman (2006) include the following pointers:

> Provide teachers with professional development opportunities to increase their awareness of social and emotional skills, coordinate health promotion and violence prevention efforts, identify at risk students at the pre-K level and intervene early, eliminate bullying and actively address victim and bystander behaviors, promote tolerance to diversity, be positive role models for students and solve problems non-violently and respectfully, integrate interpersonal violence prevention into subject areas, conduct workshops on emotional abuse, and evaluate strengths and limitations of existing violence prevention as a springboard for planning.

When a phrase similar to "the safety and security of the staff and students at this school is paramount" becomes a true slogan of any school system, then communities will know that their students are being better protected from violence in educational systems in their own neighborhoods.

CASE STUDY

A new teacher was excited to attend a three-day conference in another state about her subject matter. When she got back to her school she was excited to apply many of the new techniques in her classroom, but could not wait to share with her coworkers information from one of the keynote speakers regarding a free webinar about designing safety and security programs in schools. She had a meeting with her mentor and told him her plans to invite all of the teachers to the free webinar, and he did not think it was necessary. He told her she could watch it on her own, but that everything at the school was safe and there were no problems, so inviting all the teachers would only make people worry.

- Should she share the webinar information according to her original plan? Why or why not?
- What other options would she have if she does not share the information regarding the webinar? What might happen to her if she does follow her original plan?

EXERCISES

1. Using the chart in Table 33.1 as a reference, list and discuss three additional challenges regarding safety and security changes in reference to the teacher and the district that may occur at a school:

Significant Change to School	Teacher Role in Program	District-Wide Buy-In Challenges

2. For the following list, identify an example of a K–12 safety topic that might be used by teachers in a classroom to support safety and security education in the school system. Discuss what age level the topic is geared toward, how the teacher could present the material to their class, and what students will be able to demonstrate at the end of the lesson (i.e., the objectives of the lesson).
   a. Passive safety communication strategy
   b. Mandatory safety communication strategy
   c. Mixed safety communication strategy
   d. Targeted safety communication strategy

REFERENCES

Aniskiewicz, R., and Wysong, E. (1990). Evaluating DARE: Drug education and the multiple meanings of success. *Policy Studies Review*, 9(4), 727–747.

Capizzi, A. M. (2009). Start the year off right: Designing and evaluating a supportive class-room management plan. *Focus on Exceptional Children*, 42(3), 1–12.

Cohen, J., Allen, J., and Hyman, L. (2006). Creating a safe, caring and respectful environment at home and in school. *Brown University Child & Adolescent Behavior Letter*, 22(12), 8.

Felix, E., Vernberg, E. M., Pfefferbaum, R. L., Gill, D. C., Schorr, J., Boudreaux, A., Garwitch, R. H., Galea, S., and Pfefferbaum, B. (2010). Schools in the shadow of terrorism: Psychosocial adjustment and interest in interventions following terror attacks. *Psychology in the Schools*, 47(6), 592–605.

Fergusson, D. M., Horwood, L. J., Beautrals, A. L., and Shannon, F. T. (1982). A controlled field trial of a poisoning prevention method. *Pediatrics*, 69(5), 515–520. Accessed March 17, 2012. http://www.sfu.ca/media-lab/archive/2007/387/Resources/Readings/MrYuk%20Field%20Trial.pdf.

Kann, L., Brener, N. D., and Wechsler, H. (2007). Overview and summary: School health policies and programs study 2006. *Journal of School Health*, 77(8), 385–397.

Mendler, A., and Mendler, B. (2010). What tough kids need from us. *Reclaiming Children & Youth*, 19(1), 27–31.

Murgatroyd, S. (2010). "Wicked problems" and the work of the school. *European Journal of Education*, 45(2), 259–279.

Nolen, J. L. (2003). Multiple intelligences in the classroom. *Education*, 124(1), 115–119.

Papastylianou, A., Kaila, M., and Polychronopoulos, M. (2009). Teachers' burnout, depression, role ambiguity and conflict. *Social Psychology of Education: An International Journal*, 12(3), 295–314.

Partnership for a Drug Free America. (n.d.). Brief history. Drugfree.org. Accessed March 17, 2012. http://www.drugfree.org/brief-history.

Pentz, M. (2000). Institutionalizing community-based prevention through policy change. *Journal of Community Psychology*, 28(3), 257–270.

34 Safety as a Principle of Leadership

Ronald Dotson

CONTENTS

Time: September 1990
Situation: Staff and Officer's Meeting; A Co. 4th CEB Charleston, West Virginia

On August 2, 1990, Iraqi forces crossed the border into Kuwait. I was at Camp Lejune, North Carolina, and was informed that I would not be returning to my reserve unit but instead would be sent to theatre with an active unit. Instead, I was sent back to my unit in Crosslanes, West Virginia. I had serious doubts that I would finish my degree at Marshall University, at least that semester. Thrust back to a reserve unit preparing for deployment to Southwest Asia was exciting and scary at the same time. It was early in September when I experienced a situation that would impact my character as a leader from that moment forward.

During a staff and officer's meeting, we were being briefed on a possible mission and an overview of the job that Task Force Ripper would be asked to perform in the effort to oust Iraqi forces from Kuwait. After being informed of how the Iraqi defensive positions had been laid out, we were told how the corps would be tasked with advancing through the massive obstacles of ditches, berms, wires, mines, artillery, and armor.

A senior staff sergeant and a popular platoon sergeant began questioning the plan of attacking fixed fortifications head on. The unit had trained hard and long for such scenarios and had always planned on diversion and flanking. So the staff sergeant continued, and his argumentative debate with the senior officers was well-grounded. He was concerned for the safety of all marines while I was busy thinking of how to prepare my heavy equipment operators to demolish the obstacles and complete the mission in front of us.

As the conversation continued, it became clear to me that three types of marine leaders were present in the room. The first was like the staff sergeant, concerned about the well-being of marines from a holistic point of view and questioning what did not make total sense from his immediate point of view. The second type was silent, trusting that their officers would not blindly jeopardize marines without good reason. The third type was like me, concentrating on forming a plan of training so that the mission would be successful. The other possibility of silence because of a blind sense of duty does not exist with marine leaders in such a meeting. Marines are disciplined, but not blind followers scared of taking initiative or presenting their views in a confidential meeting.

The common denominator was that all had the survival of their marines at heart. In no realm is leadership more important than military settings. Although personal risk is assumed and part of the entire enlistment decision, it is never turned a blind eye. At the heart of the relationship between leader and follower is the virtue of trust. Without it, the relationship fails.

I walked out of the meeting knowing that from that moment forward I would put my team members first without question and without regret. I learned that safety was the very foundation that allowed people to be motivated. Safety became a principle of leadership. Followers must be able to see a legitimate concern for their welfare, in terms of both safety and career, from their leadership.

HISTORY

Safety can be traced back to 2000 BC when the Samarians placed a building code in their written record (Goestch, 2002). The Samarians are known for the justice system that is based on the principle of "an eye for an eye." Their building code placed a "tit for tat" liability on the home builder to erect a strong and solid building. If the home were to collapse and take the life of a dweller, then the same positional member of the home builder's family would be put to death. Perhaps harsh by today's standard, the code revealed a clear concern for safety and perhaps reveals a problem in construction of the day. Anyone who has built a home can tell you that the home builder or general contractor becomes a leader in the process and the customer must have a level of trust that the structure will be safe and problem free.

During construction of his memorial temple, Ramses II instituted mandatory bathing, daily hygiene inspections, and quarantine procedures for the slaves that were used for the construction process (Goestch, 2002). The motivation was mainly a business decision aimed at timely construction, but the concern for the human resource was sincere. For such a revered leader, considered to be a god, to show concern for slaves was well ahead of its time, especially with the perspective of the lack of this concern in modern times.

Throughout history, safety only became a popular concern after tragedy. In 1802, Britain passed the Factory Act, but only in response to the tragic deaths of children (Hagan et al., 2001). It was early in the Industrial Revolution, and long hours, child labor, and poor working conditions were prevalent. In this case, the cotton mills of Manchester were at issue. An outbreak of fever spread quickly among the child laborers in the mills. After the pandemic subsided, legislation was passed limiting work hours for children and improving lighting and ventilation. Here in the United States, legislation was spotty from state to state and centered on working hours, lighting, and ventilation, and in 1867, Massachusetts passed the first laws on machine guarding. Enforcement did not receive a concentrated effort, and inconsistencies between jurisdictions existed. In 1869, the Bureau of Labor Statistics was formed in order to gather data on the issue of workplace injuries and illnesses and make its findings public (Hagan et al., 2001).

By 1905, workplace injuries had become a burden on communities. The Pittsburgh Study, a comprehensive look at workplace injuries in Alleghany County, revealed five hundred deaths for that year in that county alone (Bird et al., 2003). The economic and moral impact on communities allowed a move toward a modern workers' compensation system. The idea was not new. Germany had implemented a system of workers' compensation in 1882 and had much success with the program. However, the United States was not ready to adopt such a socialist idea just yet.

The injuries that were occurring in the American workplace were handled under common law as a tort. In this manner of a civil wrong, four elements had to be proven to the court in order for a favorable verdict for the worker. The worker had to show that the employer had a duty to protect, that the employer breached that duty, that the employer caused or had a causal relationship to the injury, and that there were actual damages (Larson and Larson, 2000). This was difficult for poor laborers to achieve. The cost of attorneys and of long, drawn-out legal processes were enough to prevent most from pursuing a claim. However, rooted in Western thought and in the American courts were three defenses that were overwhelming obstacles for many workers.

The first recognized defense was "assumption of the risk," which held that the worker, by accepting the job, had knowledge of the risk to them and therefore assumed responsibility for their own safety (Larson and Larson, 2000). This thought is still with us today. When you examine how post-traumatic stress disorders are handled under workers' compensation, this is revealed. Many argue that in the case of emergency first responders, they know from the moment of application that the job may involve witnessing terrible occurrences that will damage them psychologically and therefore should not be awarded compensation even when the mental barrier becomes such that normal functioning of their job is prohibited. Others counter this by explaining that you may acknowledge this possibility, but until it is actually experienced, you have little understanding of its personal impact. Regardless, in the early 1900s, assumption of the risk involved issues that today are readily placed on the employer. Take, for instance, working at height. Under the thought that the worker assumes the risk, personal fall protection would not be available for the worker unless the worker furnished it himself. Furthermore, the design and construction process may not facilitate the use of such protection.

The other defenses were contributory and comparative negligence. Contributory negligence held that if the worker contributed in any part to the risk of injury, then the employer was not at fault. This defense gave a little into what is called comparative negligence. In comparative negligence, the amount of contributory fault is determined, and it is used to determine award in compensation for the damage. Therefore, if a worker was deemed to have contributed to the risk at approximately 30%, the employer would be liable for the 70% remaining of the damages. Damages were and are today usually determined by the loss of income potential or in some cases by a set schedule or amount for the injury itself (Larson and Larson, 2002).

Under the common law handling of workplace injuries, workers did not win claims often. The economic impact rested upon a family that had little or no means of producing income to pay bills or continue mortgages. This increased the amount of youthful workers as well. Family dynamics of the time relied primarily on male members for income. Employers did not have a universally favorable situation either though. Legal costs were quite burdensome. A long, drawn-out legal process was impossible for most laborers to engage in, but it also strained business. Additionally, in the cases where they were found to be liable under a tort, the award was large and burdensome on business. Therefore, the modern workers' compensation system that was to evolve provided quick compensation to injured workers without regard to the degree of fault, and it was to be the only remedy available to workers. In other words, if the worker had workers' compensation available, a lawsuit against the employer was allowed (Larson and Larson, 2002).

It was in 1910 that Wisconsin passed a bill allowing workers' compensation. It was quickly challenged in court and found to be unconstitutional because business would not be afforded "due process." New York also passed similar legislation that was also ruled to be unconstitutional by the state supreme court. On the very day of its ruling against workers' compensation, the Triangle Shirtwaist Factory caught fire, killing 146 people, of which most were women and child laborers. The tragedy sparked massive public sentiment and paved the way for building codes and helped influence acceptance of workers' compensation legislation in 1917 (Wignot, 2011).

By 1915, the start of the investigational era, insurance carriers offering workers' compensation coverage began demanding investigations of workplace injuries. The goal was to reduce workplace accidents. The thought of the day was to blame the employee by looking for psychological causation. Ways to keep workers paying attention to their job centered on hard-line supervisors pushing employee production. Workers were challenged to the degree that they would not become bored, and complete attention would be required. Attempts were made to identify characteristics of accident-prone persons. However, this was not successful. Coaching of employees that were involved in accidents mainly centered on what the employee did wrong and what they could do to prevent the occurrence in the future. The investigational era ended in 1930. During this period many professional safety organizations such as American Society of Safety Engineers were formed. Many pamphlets and brochures were written and distributed. Education was considered the key to accident prevention, and in general, safety came to be accepted as a business management principle (Bird et al., 2003).

Later, in the second half of the twentieth century, the term "accident" came into question. The notion of not being a foreseeable event led research to compare the effectiveness of safety measures in workplace injuries and advancements in regards to public health. Herbert Heinrich first began the controversy over the term "accident" and the listed causes of accidents. Listed causes such as slips and falls were questioned by him as not being the cause at all. Instead these were types of incidents that lead to injury. Heinrich is considered the father of behavior-based safety initiatives and formed Heinrich's Law, a ratio of incident occurrence indicating that for every incident causing a major injury, 29 of minor injury would occur, as well as 330 incidents without injury. This was published in a 1931 book titled *Industrial Accident Prevention: A Scientific Approach* (Heinrich et al., 1980).

Dr. James Haddon was conducting a study for the U.S. Army in the 1950s related to head trauma. In 1963, he published his "energy-exchange theory" that basically classified how injuries occurred to the body. It identified two factors related to human injury. The first factor involved interference or energy exchange that interfered with whole body function. Examples included suffocation and drowning. The second factor involved an exchange of energy to the body that went above its local or whole body threshold. The development of this theory allowed the strategy of injury prevention to enter safety management. For Dr. Haddon, injury prevention was not about preventing the accident from occurring, but under realization that accidents will occur despite prevention effort, the key was to prevent the exchange of energy. Dr. Haddon became the first head of what is today called the National Highway Traffic Safety Administration. Formed under President Johnson, who appointed Dr. Haddon to lead the effort in 1966, measures such as the seatbelt requirement for manufactured automobiles were implemented. By 1974, automobile crash fatalities had been reduced by 28,000 (Bird et al., 2003).

Dr. Haddon and the success of initiatives with the automobile ushered in the function of engineering into safety management. Modern safety management includes the strategies of accident prevention and of injury prevention. Dr. Haddon developed ten strategies to prevent injury that are foundational keys to injury prevention strategy.

- Prevent the creation of the hazard in the first place
- Reduce the amount of hazard brought into being
- Prevent the release of the hazard
- Modify the rate or special distribution of the release of the hazard from its source
- Separate, in space or time, the hazard and that which is to be protected
- Separate the hazard and that which is to be protected by material barrier
- Modify basic qualities of the hazard
- Make that which is to be protected more resistant to the hazard
- Begin to counter the damage already done by the hazard
- Stabilize, repair, and rehabilitate the damaged object

Accident prevention efforts and injury prevention efforts used together have been shown to be effective safety management techniques. From the formation of the Occupational Safety and Health Administration in 1970 until 2000, data from the Bureau of Labor Statistics shows a 58% reduction in workplace fatalities (Bird et al., 2003). The overall trend has been one of decline in fatalities. The strategies of engineering and leadership combine to form what is meant by "safety" today.

Safety can be categorized as a function of engineering and as a function of leadership. As a function of engineering, it means that engineers of all specialties incorporate the design of new products and improvements of existing products around the human need for safety. As a function of leadership, safety concentrates on workplace risk management. It involves policy formation and enforcement, job hazard analysis, job task surveying, permitting and control, emergency response, auditing, training and educational efforts, insurance management, injury management, research, and other activities that reduce workplace injuries and preserve the human and environmental resources. Leadership and management skills are central in this aspect of safety.

THE LEVELS OF LEADERSHIP

The basic definition of leadership is influence. It sounds simple but is very complex. It really includes the process and relationship that exists between leader and follower. John Maxwell has very brilliantly captured the relationship of leader and follower in his model of the levels of leadership. In his model, there are five levels of leadership, as shown in Figure 34.1, and the levels are determined by the reason that the constituents allow the leader to influence them.

The positional level explains the notion that someone has respect for the authority of a boss or supervisor. In other words, the leader is in a position to give orders. Additionally, Maxwell suggests that anyone can be a leader at any level of an organization. He exemplifies this with the example of a waiter. If two people were visiting an unfamiliar restaurant and were seeking advice on what to order, they may ask the waiter for advice because of his knowledge and position. Permissive leadership is explained as when the followers want to follow the leader.

The next levels really distinguish the level of influence that the leader may be able exert on the constituents. In the production level constituents follow because of positive accomplishments that the leader has been able to achieve for the organization. The personal level reveals that the constituent follows because of positive things that the leader has done for the follower on a personal basis. Maxwell says that the highest level may be unattainable. At this level, constituents follow based solely on name and reputation. His example, a clearly unattainable one, is Jesus Christ (Maxwell, 1993).

The model shows a complex relationship between leader and follower. It may also show power. As the level of leadership increases from positional to personhood, the leader has more power within an organization. The leader may grow in influence

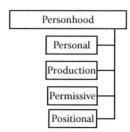

FIGURE 34.1   Levels of leadership.

over more and more of the constituents on the production or personal level, allowing for more and more support.

Military leaders realized long before industrialists that protection of the human resource was a function of leadership. Indiscriminate waste of willing fighters led to defeat. Strategic, tactical, and protective efforts, although aimed at victory, always included increasing the longevity of the soldier. Many technological and tactical advances played a role in safety as well as success. Take the design and use of Roman shields and swords. Not only did the tactics of an advancing wall and thrusting sword defeat enemies charging the ranks, but it succeeded by affording the soldier protection from a unified defensive wall rather than individual effort. Later, in medieval times, lords would commonly take care of fallen servants' families. The notion that leaders had a duty of safety and compensation to their servants is evident in the laws of Henry I. In referring to compensation, Arthur Larson, a recognized legal scholar, interprets the laws as declaring a duty to servants if the servant is on "mission of another"(Larson and Larson, 2000). Although soldiers often report that they perform for their buddy next to them, it is the safety of that comrade that motivates their action. The cause of the whole may indeed help motivate, but leaders without moral concern for their followers lose credibility, and the loss means that the cause then loses its motivational value (Ciulla, 2003).

As far back as the Babylonian Empire, a strong foundational leadership centering on justice moved society toward safety initiatives. From Hammurabi's Code addressing the liability of homebuilders in an "eye for an eye" justice system, the concern for safety was viewed as a just duty. Persons placed in leadership roles had a duty originating from a virtue of justice, to consider the safety of those around them. In modern studies of leadership, justice is considered a foundational virtue of leaders. Leaders must display a sense of justice that is viewed as moral by those that they lead. Safety is indeed one of those principles that come from the virtue of justice. Leaders must display a sincere effort for subordinate and peer safety or risk moral decay of their credibility.

Developed in the 1950s, Abraham Maslow's hierarchy of needs theorizes that humans are motivated by five needs. This theory perhaps best exemplifies the point of motivation and safety. Maslow's theory is to be viewed as a foundation model. Meaning that the pyramidal levels of need rely upon the preceding level for foundational support.

Foundationally on the bottom, the need for food, water, and shelter allow the need for physical and mental safety and security to develop. Social needs include acceptance and friendship. Esteem needs include respect, recognition, and achievement. Self-actualization is growth, development, and realizing potential (Robbins and DeCenzo, 1998). Considering this theory of motivation, once you get past basic human needs for survival such as food, air, water, and shelter, it is safety as a second tier foundation that allows the other human needs to be realized. It is again clear that if subordinates do not feel a sincere effort and minimal accomplishment of personal safety, then motivation is stifled. Leaders of an organization who do not fulfill this need for its members lose the ability to motivate membership, and the organization's potential for success is reduced.

Modern motivation was examined by Marcus Buckingham and Curt Coffman in their 1999 book, *First, Break All the Rules: What the World's Greatest Managers Do Differently*. Based upon the Gallup Organization's studies, Buckingham and Coffman draw some important lessons for managers. The book lays out four core tasks that managers must do in regards to personnel. They must recruit and hire the right person, must set clear expectations, motivate the person, and develop the person. Now, while leadership and management are distinctly different talents, they do go together. Managers without leadership ability and leaders without management skills seldom succeed alone. Attracting and motivating the right personnel is clearly leadership oriented. Setting expectations and allowing a developed career path are heavily reliant on management skills. Here a manager must follow the ideas of Frederick Taylor, the father of scientific management, in that he must break down the elements of the work into steps, necessary skills, and talent; scientifically select the personnel based on the work analysis; teach and train the worker; partner with the worker to analyze the work; and divide work and responsibility equally. In order to retain quality personnel, Buckingham and Coffman have identified six core questions that employees use to gauge job satisfaction. Central to their lesson is that suggestion that people do not leave organizations, they leave managers! The six core questions are as follows:

- Do I know what is expected of me at work?
- Do I have the equipment and materials to do my work properly?
- Do I have the opportunity to do what I do best every day?
- In the last seven days, have I received recognition?
- Does my supervisor care about me?
- Is there someone who encourages my development? (p. 34)

The lessons center on Maslow's highest need, self-realization. Workers must reach this level in order to reach their productive potential. If they do not or cannot reach this level, the organization risks losing them.

Safety plays a large role in satisfying these questions for the employee. Without a foundation of safety, self-realization cannot be properly obtained. Development of the employee does not adequately occur. Without a foundational concern for safety, any attempt to show concern for the worker on a personal basis will be viewed in time as an insincere ruse for their welfare.

Concern for the worker must begin with workplace safety and then extend to the person. Too often we do not view performing a job right as including safety compliance, but top companies all over the world are succeeding in establishing this in their work culture. Therefore, safety does play a role in satisfying this concern for the worker as well. Being expected to be safe without the proper training or equipment fails in satisfying the employee. Safety also affords a very effective means to ensure feedback and recognition. Tracking and displaying injury statistics is a very good means of showing concern, keeping employees focused on safety, and giving feedback and recognition.

Good leaders and managers know that safety is a principal of leadership. Without it, motivation and retention suffer. One of the most popular leadership principles is

"Lead by example." If safety were to be incorporated as a principle, it would read, "Provide your subordinates with the training, equipment, and policy in order to perform their job as safely as feasible." During the investigational era, the three Es of safety were first published. Safety management has generally adhered to education, equipment, and enforcement as a guiding principle. The National Realty and Construction Company case (479 F 2d 1257) established legal precedence in regards to the general duty clause, which states that companies have the duty to provide education and training, equipment, and enforcement of policy in regards to abatement of recognizable hazards that were likely to cause death or serious physical injury. Not only was this decision a reaffirmation that anyone in a leadership role had a duty to manage safety, but it further legally recognized the specific duties for safety management.

The safety of organizational members is clearly a function of leadership. Originating from a sense of justice and manifesting itself as a moral duty for those in leadership positions, safety will help the leader guide a successful organization. Often we tend to view safety as a cost-effective business duty, but it has a moral foundation that comes from a sense of justice. Not only should it be addressed as a best business practice but must be addressed because it is the right thing to do. In the mid-1800s in Massachusetts, the textile industry experienced a shortage of young workers because of the number of injured textile workers and the amount of available population. This was addressed with workplace legislation, but the problem was seemingly resolved only after an increase in Irish immigration caused by the potato famine in Ireland (Heinrich et al., 1980). It is such a shame that the dollar determined or impacted moral action in this situation.

FOUNDATIONS FOR SAFETY MANAGEMENT

Management involves four aspects that Henri Fayol first outlined in 1916: planning, organizing, controlling, and directing were found then to be the core of management duties. However, as Buckingham and Coffman have pointed out, management duties and skills involve much more detail when the core activities are broken down or expanded. The modern safety professional utilizes the techniques and skills that all business managers use. Scheduling, for example, is as critical in safety as in other areas of business. Safety specifically relies on several key management tools to accomplish its mission: job task surveys, job hazard analyses, permits, audits, statistical logs, and investigation reports. These tools are foundational in the mission of identifying risk, formulating countermeasures, implementing countermeasures, and reassessing the corrections for effectiveness. This has a reactive and proactive side to correcting and preventing workplace incidents.

In the 1920s, "accident prevention" was dominated by the domino sequence theory. At heart of this idea was a four-stage premise theorizing that injuries are caused by accidents, accidents are caused by unsafe acts of people or by exposure to improper mechanical conditions, these unsafe acts and conditions are a result of human faults, and that human faults are created by environment or inherited (Heinrich et al., 1980). In the application of this theory, the worker was the main concentration of effort and blame. It was the psychological focus, such as efforts to increase focus and compliance to rules and procedures, that dominated safety practice.

Later, Dr. Haddon introduced a concept centering not only on mechanical condition but also on the mechanical environment in his formulation of the energy-exchange theory and the strategies for injury prevention. Dr. Haddon's theories demonstrated a foundational concern that realized human psychological error was inevitable. Thus, the mechanical environment needed attention. Although machine guarding was not a new concept, his model for engineering controls in order to prevent human error from resulting in injury provided the missing link for safety management. His work had been inspired by his recognition that early safety management practices had not been as effective at injury prevention as public medicine had been in the early twentieth century (Bird et al., 2003).

The domino theory of injuries listed five steps that lead to injury. In was first the environmental and social climate and ancestry that allowed the second step of human error to develop. This error in turn led to unsafe acts or mechanical and physical hazards. These acts or hazards then allowed an accident to occur, and then some accidents produced injury. Undesirable human traits such as nervousness were either inherited or created and exacerbated by their environment. These traits created human faults that then allowed unsafe acts such as not wearing protective gloves, or even engineered oversight of the need for machine guarding (Heinrich et al., 1980). Dr. Haddon was removing the fifth step of injury occurrence. Today's efforts of ergonomic control also attempt to remove the possibility of an error as well.

Until the second half of the twentieth century, management efforts centered solely on eliminating the third sequential step of unsafe acts. Heinrich's work was one of the first in progression to have several models of accident causation. Frank Bird began looking at management factors contributing to accident causation. Models that considered the systematic approach to accident causation became known as "management models" (Heinrich et al., 1980). From early on in the twentieth century, the theory of accident proneness has been explored. The theory assumed that everyone has an equal opportunity for accidents to all; however, statistics show that most have no accidents, some have few accidents, and few have multiple accidents. This led to the inconclusive finding that certain traits are associated with accidents.

Other behavior type models have also become popular. The life-changing units theory surmised that accident potential is situational and that those persons exposed to certain stresses or events, such as the death of a spouse, were more open to accidents. Peterson's motivation and reward satisfaction model theorized that the ability to perform the work and level of motivation impacted accident liability. In this model, factors such as happiness, job advancement, and other intrinsic and extrinsic reward impacted motivation. Human factor models also developed. Dr. Russell Ferrell from the University of Arizona formulated the Ferrell theory. This model theorized three overall situations that lead to human error that are in a causal chain leading to an accident:

- Overload as a mismatch of human capacity and the load it is subjected to in a motivational arousal state
- Incorrect response by a person because of an incompatibility to which the person is subjected
- Improper activity because of ignorance of the situation or as a deliberate taking of risk

Later deliberate taking of risk was further explored by Peterson to account for why deliberate risk was being taken. He surmised that a person takes deliberate risk in two situations:

- The person has a perceived low potential for an accident.
- Because it is more logical to do so because of peer pressure, priorities, or an unconscious desire to be injured (Heinrich et al., 1980).

Accident causation models are used by safety managers and engineers today as foundational guides to incident investigation, system design, and product design, to prevent accidents from occurring. Today, human-centered design is a practiced principle that designs operating and management systems and products with human considerations in mind. The practice relies on accident causation models to guide this in practice. In short, tasks are divided between human and machine based upon identified strengths of each. Mixed with injury prevention techniques developed by Dr. William Haddon, safety has been effective since at least the late 1960s.

Today safety management involves aspects aimed at all five dominos and utilizes the various models for accident causation as a foundation for best management practices and guiding incident analysis. Training and education are being expanded to counter environmental influences that contribute to workplace incidents. Foundationally, education involves more than a picture of safety as a profit maximizer, but, as suggested by this entire chapter, must use safety as a foundational leadership practice. Of course, educational and training efforts attempt to reduce human error either in design or practice, but establishing the thought of safety as equal in importance to any other technical step or the teaching of safety as a principle of leadership attempts to overcome environmental and social influence of behavior traits. For example, establishing rules for design such as human-centered design principles attempts to eliminate human error or the second domino. Establishing safety as a cultural trait then attempts to eliminate the first domino. During the investigational era, training and education were considered the key aspects in accident prevention, but the training centered on the technical skill aspect and the identification of undesired traits. It was not aimed at changing or influencing immoral leadership practices.

Safety management involves two main strategies: accident prevention and injury prevention. A third strategy has also developed to eliminate the accident from occurring, but not by eliminating human error. Redundant type controls and computer programming of machine controls have created a new strategy for accident elimination. Here the commission of an error may be recognized by computer controls through sensors and may stop the process before the incident occurs but after the human error created an unsafe act or mechanical condition.

IDENTIFYING HAZARDS

The first step for safety management is to identify hazards. This is done with various methods, but first and foremost incident tracking must be developed into a system to identify trends within a specific company or facility. This allows comparison with national industry data. The two may not always match. Knowing what types of incidents are occurring and the causes is imperative. It has been suggested that the two categories of incidents are with cost and without cost. However, is any incident truly without cost? A more proper way to divide categories of incidents for a safety professional is with injury and without injury to the human resource, as shown in Table 34.1. Various subdivisions can then be inserted.

With this line of organization, the human worker is the central concern. Therefore, a near miss involves a near injury and may or may not involve property damage. An example could be a worker that walks into the path of a forklift, tractor, or school bus but is not hit and no property damage occurs. Recordable incidents include injuries that would be required to be listed on the OSHA 300 log. Reportable incidents would then be those that may go on the log but would also require direct reporting above the log to a regulatory agency.

More specific typing is required. Injuries can usually be typed in a manner that matches the reports of the workers' compensation carrier. Examples include lacerations, bruises, amputations, and burns. The tracking forms must also consider other metrics that could be central in identifying trends and determining causation. These usually include day, time, supervisor, process or machine-operated activity before accident, and many others. On this form, the data should actually be metrics used to track and compare incidents within the facility or company. Other contributing factors can be listed on the investigation form.

Beyond incident tracking, safety professionals also commonly use job task surveying and job hazard analysis to identify risks. A job task survey is a listing of jobs that the position will be tasked to perform in the organization, will be evaluated on, or perform routinely in order to accomplish formal tasks. For example, if you were listing the job tasks of an elementary school teacher, would the survey list classroom decoration? Whether you formally list it as a job requirement, it becomes standard practice for teachers to decorate and arrange their rooms in a way that facilitates learning. The job hazard analysis then breaks the identified tasks into major steps and identifies the hazards and the correction to the hazard required of each step. This systematic way of analyzing the job duties helps the safety manager avoid missing relevant hazards that may cause or are leading to injury and therefore loss.

TABLE 34.1
Injury Comparison

With Injury	Without Injury
First-aid incident	Near miss
Recordable incident	Property damage
Environmental incident	Environmental incident
Reportable to an agency	Reportable to an agency

COUNTERING THE HAZARDS

The safety professional must formulate a corrective action to abate the identified hazards. This can be arrived at from an educational aspect where best management practices are established by standard or law. Safety managers learn the acceptable and best ways to fix hazards based on successful practices from others, research, or training and education of standards, but more foundationally they must understand causation.

Frank Bird introduced his model of root cause analysis in his book, *Practical Loss Control Leadership* (Bird et al., 2003). The model identifies three levels of causation that must be addressed for proper countermeasure formation. Understanding and identifying factors of the incident under all three levels is necessary to address the incident at a root level and prevent future occurrence.

The first level is immediate. Under the notion of immediate cause, unsafe acts and conditions are classic examples of factors that immediately or directly allowed for the accident to happen. This could be as simple as a worker deciding not to wear cut-resistant gloves while handling a sharp object. It could also be a danger area or point that was left unguarded, an unsafe condition mandated by standards.

The root cause is the condition that allowed the immediate cause to develop. These factors can be divided into personal factors and job-related factors. Personal factors include lack of skill, lack of training, physical limitation, and mental issues such as nervousness or boredom. Job factors include inadequate engineering, maintenance, normal wear, or maybe abuse of the machine.

Bird introduced a third level that is widely accepted. These management factors would be the additional conditions that allowed the root cause to develop that management could be or should be addressing. Examples can include a lack of policy and breakdowns structurally, resourcefully, or in systemic procedures. Basically, the question must be asked as to what the company was not doing that allowed a root cause to develop and what measures could be taken to prevent the occurrence. This manner of corrective action planning ensures that the reactive investigation can be used proactively to learn from past mistakes.

In 1892, the first safety committee was formed in a steel mill to produce a countermeasure for a flywheel that had broken apart and struck a worker. Investigations and countermeasure planning can be approached from this aspect as well. The investigator must be able to form a committee of sorts to cover the possible angles of the investigation and determine all levels of cause. Good investigators involve experts in order not to miss possibilities.

CASE STUDY

The principal of East High School has aggressively embraced servant leadership, a relatively new leadership style that attempts to "up-end the pyramid." This means that rather than a top-down hierarchical model, a new model is presented where higher members in the organizational structure are present to serve those at lower levels by listening and implementing ideas they have to make the school function more smoothly. This model presents a bottom-up philosophy where teachers and staff will have a strong voice in how the school moves forward.

- How can this leadership model facilitate the development of safety as a key component in the organizational culture of the school?
- What pitfalls might exist with such a leadership style?

EXERCISES

1. How does leadership affect teacher behavior and duties?
2. What do you think are the most important leadership traits for a teacher to possess?
3. How would you look for these when screening and interviewing for new teachers?

REFERENCES

Bird, F., Jr., Germain, G. L., and Clark, M. D. (2003). *Practical Loss Control Leadership* (3rd ed.). Det Norske Veritas, Duluth, GA.

Buckingham, M., and Coffman, D. (1999). *First, Break All the Rules: What the World's Greatest Managers Do Differently*. The Free Press, New York.

Ciulla. (2003). *The Ethics of Leadership*. Wadsworth/Thomson Learning, Belmont, CA.

Goestch, D. L. (2002). *Occupational Safety and Health for Technologists, Engineers, and Managers* (5th ed.). Prentice Hall, Upper Saddle River, NJ.

Hagan, P. E., Montgomery, J. F., and O'Reilly, J. T. (2001). *Accident Prevention Manual for Business and Industry Administration and Programs* (12th ed.). National Safety Council, Itasca, IL.

Heinrich, H.W. Peterson, D., and Roos, N. (1980). *Industrial Accident Prevention: A Safety Management Approach* (5th ed.). McGraw-Hill, New York.

Larson, L. K., and Larson, A. (2000). *Worker's Compensation Law: Cases, Materials, and Text* (3rd ed.). Lexis, New York.

Maxwell, J. C. (1993). *Developing the Leader Within You*. Thomas Nelson, Nashville, TN.

Robbins, S. P., and DeCenzo, D. A. (1998). *Fundamentals of Management: Essentials Concepts and Applications* (2nd ed.). Prentice Hall, Upper Saddle River, NJ.

Wignot, J. (2011). *Triangle Fire: The Tragedy That Forever Changed Labor and Industry* (DVD). Public Broadcasting Service, Arlington, VA.

35 School Safety Management

Ronald Dotson

CONTENTS

The phrase "school safety" can cause a degree of confusion in that though the word "safety" is used, it is actually used to refer to issues of school security. School safety as a comprehensive discipline involves the protection of students, school associates, and visitors or guests on school grounds. Student safety is central in the topic of school safety to teachers and administrators. A review of websites will produce little about injury prevention and injury effects on education in regard to teacher and associate loss. However, it will produce an abundance of information on bullying, playground injuries to students, and assaults. The tendency is to talk about keeping students safe. This is a noble idea and of great importance. In order to actually take care of students first, it may be necessary to place teachers first in some categories. Workplace safety is one of those categories. Wagner and Simpson (2009) assert that safety is a foundational principle of the moral architecture of a school. If so, developing a culture of safety must be more than putting metal detectors on entrances, fencing school grounds, escorting children to the bus, controlling visitors, and using a school nurse.

Schools play such a vital role in developing our virtuosic behaviors. Schools can be places that indoctrinate students into conformity or develop a moral sense of behavior based on traits. Wagner and Simpson (2009) assert that a school with a higher degree of morals will be a safer school. Teaching our youth to act right because it is right is much better than indoctrinating them into blindly following rules. In essence, we may be decreasing their faith in societal institutions and forcing them to take back the power to make the rules.

I picture this theory playing out with bullying in the following manner. A bullied person is taught not to physically defend him- or herself when physically bullied, or the student might face the same punishment as the bully, which could be expulsion

415

from school. Therefore, the student becomes a victim of bullying, but also a victim of the zero tolerance rules that were crafted in order to protect them.

This contradiction in practice and policy can also be seen in regard to personal safety. Safety is indeed a principle of leadership. It is also a personal duty for an individual. The human resource is an increasingly critical component to organizational success. In today's business world, lean manufacturing guarantees that human loss impacts production immediately, and reliance on key workers can rival the old days of skilled laborers building individual products. Today, we would incur overtime cost, training cost, and possible penalties for missing shipment quotas. Companies simply do not have an abundance of workers standing by in order to take up slack at a moment's notice. The loss of special skills and knowledge can also be devastating. Take, for example, an engineer who is in charge of the design process for a product for which the company is seeking a contractual award. His omission could adversely impact the process.

Therefore, not only are companies today worried about safety at work but at home as well. It is unrealistic to expect this key person to refrain from farming, do it yourself projects at home, or other activities that have risk. At a minimum, training and education efforts have become common for away-from-work exposures. For example, defensive driving courses for employees that drive company vehicles are becoming more and more common. These courses sometimes extend to family members. In education, there are three aspects that drive safety to be a core value. First, safety can be cost effective. With fiscal responsibility becoming an ever-growing issue, the workers' compensation and general liability premiums and their management may have a significant impact on a district. Second, educators must consider that when teachers miss class time, it can adversely impact student achievement. Third, educators have a duty to teach our youth to consider human preservation in order to make the leadership principle and business management principle of cost effectiveness merge in the organizations of our communities.

Because personal safety is vital to cost effectiveness and mission success, it must be approached from a virtuosic view. Have you ever stopped to consider how much early practices learned in school affect safety behavior on a job site or in a factory? Let us look at an observed practice of traffic duty by teachers. A teacher stands on the sidewalk adjacent to a painted crosswalk leading from the parking lot to the main school entrance. The point is to prevent a vehicle from impacting a student, parent, or teacher that is coming into the school. However, the teacher does not wear high visibility vests, carry a stop paddle, and no signage is visible to pedestrians or drivers. The practice is to step out in front of a car to let the pedestrian cross. It translates into the pedestrians crossing the traffic way without stopping. This practice actually violates the safe practices of crossing a road whether it is at school or elsewhere. The proper procedure is to stop as a pedestrian and wait until traffic notices you. In a crosswalk, proper procedure would be stopping and waiting until the teacher has stopped the vehicle without exposing himself or herself to injury and signals you to cross. Improper modeling not only teaches children and adults to disregard proper road-crossing procedures but more importantly tells them that personal safety is not important. This can increase the frequency of taking a personal risk when the youth now becomes an adult.

Student involvement in occupational safety at schools is foundational for teaching safety as a principle of leadership. Successful efforts in safety management always involve the employees and rely on frontline supervision. In a school setting, this is even more important. It is generally accepted that students will mimic the behavior of the adults around them. When educators take personal risk or fail to address occupational safety issues, students learn precedence from actions and inactions. For example, a teacher that uses a chair to reach an item on a book cabinet or decorate a room is teaching a child to disregard personal safety in order to quickly accomplish an immediate goal. Any leadership principle must be executed correctly. Results and goal achievement are important, but personal safety can never be compromised, or the lesson becomes one of taking a shortcut. When the shortcut then results in failing to meet a goal, the cost negates the achievement. This loss may occur in the future from when the shortcut was successfully taken. Learned actions become reinforced when successful but rarely are fully abandoned after failure.

The first step in safety management is to identify the hazards that are present. It is important to note that the sooner the hazard is addressed and abated, the lower the solution will be in terms of cost. Safety professionals use a job hazard analysis (JHA) to identify the risks of the task that a person will be required to perform. This can be accomplished by observing the person performing the task. One overlooked job task for a teacher is the decorating of the classroom. The weeks before classes resume from summer break are filled with classroom preparation duties for teachers. In fact, if a teacher did not have a room that was inviting and conductive to an educational setting, the principal may take action or make notes of discrepancy during a performance review. Overlooking this activity can lead to a failure to address occupational issues such as heavy lifting and working at a height.

The second step is to identify the hazards that go along with the steps of each job task. The process identifies the job steps, the hazards associated with each step, and the abatement strategies for each hazard. This formal analysis helps to insure that hazards are not overlooked. This is a foundational technique for safety management. An example of a very basic analysis instrument may look like Table 35.1.

TABLE 35.1
Sample Job Hazard Analysis

XYZ High School
JOB HAZARD ANALYSIS

Job Classification: _____

Department/Job Site: _____

Surveyor: _____

Date: _____ Review Date: _____

Task	Critical Steps	Hazards	Countermeasures

The job requirements of a teacher also include peripheral duties that might be overlooked without performing a formal JHA. It is beneficial to determine the job duties of the position first by surveying and observing the position. Once the critical steps are identified, there are four sources that must be used to properly research the abatement of hazards. The Occupational Safety and Health Administration (OSHA) regulations or standards are the minimum starting point. Other federal and state regulatory agencies such as the Department of Transportation or the Environmental Protection Agency provide standards that cover safety for their relevant areas. National consensus standards produced by organization such as the National Fire Protection Association and the American National Standards Institute also provide detailed safety requirements. Some of these standards are "incorporated by reference" into regulations and then become the law. Others are voluntary, unless a certifying body requires such for certification. There are also "best management practices" (BMPs) that are commonly practiced in order to meet requirements. These BMPs come from training, experience, and networking with other safety professionals. School policies are a source for abatement as well. Policies may reflect school administration preferences that are not specific to requirements.

Developing the countermeasures or solutions for the hazards presented relies on a team approach. A crossfunctional team is formed to address the root causes of an incident or solve problems as early as possible in the process of producing a new process or modifying an existing process. The team consists of various levels of the organization. This includes the faculty or staff member. In a school setting, this means that the teacher, parents in some situations, and possibly students would be involved in addressing the hazard. As with any organizational change, identifying the key advocates or leaders from all levels is important. Those that participate will better follow and implement the abatement. This is a bottom-up approach. It avoids teachers taking the view that this is just another initiative or administrative burden placed on them. This is not to say that upper administration's support is not important. It is a reminder that all levels must be included in a collaborative style of leadership.

In safety management, initiatives are not always driven from the top down. Instead, it becomes necessary to allow faculty and staff to assist in problem solving and producing solutions. After all, faculty and staff are the real experts when examining how a job is performed. Several principles of safety leadership are adhered to when it comes to worker motivation. One of these is that workers tend to comply with procedures and processes when they have participated in their formation. Other key principles include utilizing the key advocates with work groups in order to gain support for safety initiatives, and finally, the idea that workers can accept change if it appears in small increments (Bird et al., 2003). These rules of safety leadership are effective and foundational to having a positive safety culture. Having compliance from mere strict enforcement does not create a true culture because the rules will be followed only when someone is thought to be looking.

Active employee safety committees are partnerships between the company and the employees with the goal of creating and maintaining a safe and healthful workplace. Active employee safety committees mean that employees become advocates for their represented group, perform safety audits, engage in safety training, participate in investigations, and receive employee questions and complaints. Their role is an important piece of a successful safety culture. Schools utilize committee groups like any other organization. Although committees may be organized in different manners, many schools leave out the student. Key advocates with the students exist as well. Understand more of why and how policy and practice impact the student perspective, and perhaps user-friendly initiatives will be more successful.

BUDGETING

Safety management requires budgeting to be integrated into management philosophy. Management by objectives (MBO) is an effective method for keeping all levels of the organization educated on clear and measurable goals. By setting goals for safety based upon loss experience or rating the types of incidents by frequency and cost, safety can impact the bottom line of the organization. Of course, under MBO philosophy, each goal is achieved by one or more objectives that are documented and designed to be measurable.

With proper documentation and tracking of organizational metrics, goals can be easily identified. Then further investigation may involve a cross-functional team to solve issues comprehensively. Realistic numbers may be ascertained from trials or by sharing information with other organizations that have attempted similar strategies. The objectives then written for each goal can be examined from a cost analysis. The cost for implementing the strategy for each objective can be reasonably estimated from quotes given by vendors for their products, from contractor quotes, or from in-house costs of implementation. The realistic goal reduction can be estimated from real numbers centering on the average direct cost of each type of incident the strategy is aimed at preventing. The project stages are given a cost of investment for each goal, and then an estimated return on investment can be calculated. This is a line-by-line return of investment organized from the goals and objectives of the safety program efforts for the next fiscal year or over a period of time.

Safety costs should be factored into a product the same as any other cost. This is true for education as well. The days of looking at last year's safety expenditures and increasing them based upon inflation or hit-and-miss programs are over. Setting annual and long-term measurable goals allows safety to be compared with any other aspect of doing business. Safety is not an incidental cost of doing business.

FILLING SCHOOL VOIDS WITH SAFETY

Today educators are facing critical issues because of the increase in diversity awareness and the understanding of its impact. Issues such as safety that were not viewed as diversity dilemmas in past decades are now being explored from a moral ground of inclusion and inherent rights to education for students in topics such as bullying. However, discussions regarding school safety tend to exclude the workplace safety needs of teachers and staff. The concentration is on security.

Safety is a fundamental of leadership and can be traced back to written building codes from the Babylonian Empire around 2000 BC that were based on the principle of "an eye for an eye" (Goetsch, 2008, p. 4). Society recognized the moral responsibility that a home builder had to his customers over four thousand years ago. The leadership trait of justice cannot be fulfilled without leaders having credible concern for the well-being of the people around them. The whole philosophy surrounding modern workers' compensation had its beginnings with the tradition of a lord waiving taxes and moving the family of one of his fallen soldiers into the area of his home, in order to assure their well-being (Larson and Larson, 2000).

Safety has a moral foundation in what is considered just and fair, and this is evident throughout the history of safety from all over the world and from varying cultures. It must therefore be a goal of any legitimate and credible leader to keep his subordinates and peers as safe as reasonably possible in the completion of work done at the direction and responsibility of the leader. This must be true for educators and educational service workers as well. Misguided perceptions that teaching duties are low risk and somehow less deserving of risk management efforts that can be seen at the very heart of most manufacturing jobs is a failure of educational leadership. Examining the picture of occupational injuries in education is eye-opening and reveals a void in the foundation of educational leadership.

In order for any organization to establish all facets of safety as a form of culture, safety must be foundational at all levels of the organization, shared responsibility and accountability practices must be in place, and thought of safe practices must be second nature with its members. Procedures and practices that do not utilize safe practices or encourage safety shortcuts or instill bad safety habits then reduce the positive effects of safety on all levels. If it is acceptable to overlook safety in one process, then it must be okay to take another shortcut. In other words, it is just not that important. Children are impressionable, and if a precedent is accepted early on, then it becomes more difficult to change behavior as an adult.

School safety must also include the prevention of loss among district employees. While groundskeepers, maintenance personnel, and bus drivers may be the first positions thought to be the "most dangerous" jobs in a school, teacher safety produces some eye-opening statistics that propel the issue to a void that impacts day-to-day operations. A review of injury statistics from the Bureau of Labor Statistics shows that of all general industry employers, 3.6 out of 100 workers received an injury requiring medical attention above first aid in 2009. This includes manufacturing facilities like steel mills, slitters, automotive stampers, automotive assembly, and warehousing. While educational services fall just below this at 3.2 (national level reported under North American Industrial Classification System code 61), many states have rates that would be considered unacceptable for administrative job categories in manufacturing (U.S. Bureau of Labor Statistics, 2010).

Let us first examine the background information behind the numbers from the Bureau of Labor Statistics. The North American Industrial Classification System is used to benchmark injury rates for fair comparison among industries. The 61 designation is for education workers. It is further broken down with subsequent numbers added after the initial 61 designation to reflect the following:

- 6111: Elementary and secondary schools
- 6112: Junior colleges
- 6113: Colleges, universities, and professional schools

- 6114: Business schools and computer and management training
- 6115: Technical and trade schools
- 6116: Other schools and instruction
- 6117: Educational support services

Employees of these establishments report their loss experience under this classification for benchmarking (U.S. Bureau of Labor Statistics, 2010).

The injury rate is calculated from samples of educational service workers from all over the country. The actual rate is calculated the same for all categories. The number of reported injuries that meet the criteria for placement on the OSHA 300 log is multiplied by 200,000, or the typical hours worked without overtime for one hundred workers in a given year, and then is divided by the number of hours actually worked. This translates then to an incident rate per one hundred employees. An injury tracked on the 300 log must meet the following criteria:

- It must be work-related.
- It must be a new injury or significant aggravation of an existing condition.
- It must:
  - Require days off from work
  - Require job transfer
  - Require job restriction
  - Result in loss of consciousness
  - Result in death
  - Result in a serious condition diagnosed by a physician
  - Require treatment beyond first aid

First aid has many facets to its definition:

- Use of heat or cold
- Use of massage, not from a therapist or other health-care practitioner
- Temporary use of splints
- Use of wound coverings such as bandages, gauze, etc.
- Nonsurgical removal of splinters or foreign bodies in the eye and skin

Nevada, for example, reported 4,400 incidents among its 97,400 employees surveyed in educational services (North American Industrial Classification System code 61) for a rate of 4.52 injuries per one hundred educational service workers. Vermont reported 5,200 such injuries among 9,500 employees in the survey. Vermont's rate then is 54.74 injuries per 100 educational service employees. Kentucky reported 2,600 injuries among 15,800 employees for a rate of 16.46 injuries per one hundred educational service workers. The trend in the last five years has shown an increase in injuries (U.S. Bureau of Labor Statistics, 2010). By U.S. region, the comparison is shown in Figure 35.1.

The data collected from the Kentucky School Board Association from 2005 through 2009 show that out of ninety-six districts in Kentucky, teachers as a job class had a much higher amount of injuries than the job classifications that one might think

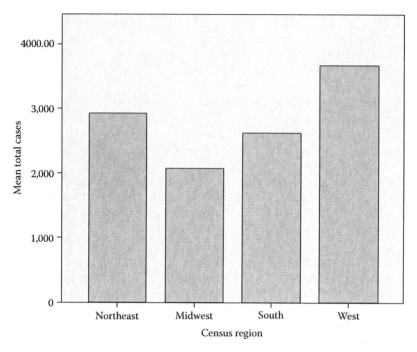

FIGURE 35.1 Number of reported injuries.

are most dangerous. As shown in Figure 35.2, teachers had 4,709 incidents compared with 1,676 for custodians, 2,312 in food services, 462 among maintenance workers, and 876 among bus drivers. When the type of injuries reported among educational workers is reviewed, it produces an inexcusable picture: 4,824 of all incidents were falls, 2,647 were back injuries, 866 were falls from a different level, 2,542 were from being struck by an object, and 1,279 were bites. During an interview with Joe Isaacs, former loss control expert with the Kentucky School Board Association and now with the Kentucky League of Cities, falls from a different level occur mainly because of teachers standing on chairs, overexertion injuries are mainly teacher back injuries, and bites occur from student assaults on the teachers and assistants (Isaacs, 2010).

The dollar amount is the statistic that makes this void in educational management a deep crater impacting day-to-day operations (Figures 35.3 and 35.4). Falls accounted for $11,241,952 in loss from 2005 through 2009. Overexertion injuries, which we identified as mainly back strains among teachers, amounted to $10,338,849 in lost revenue. The total cost for this same period equaled $33,913,211.00. Although this was cost paid by the carrier of the insurance, in this case the Kentucky School Board Association, the cost is then passed along to districts in the form of an increased premium (Isaacs, 2010).

A potential explanation for the disparity between teachers, maintenance workers, and bus drivers could be related to the amount of safety training and injury prevention strategies practiced. For example, bus drivers have a licensing and commercial driving safety management system that they and the district must follow. In addition to this federal and state effort, bus drivers receive training and evaluation that

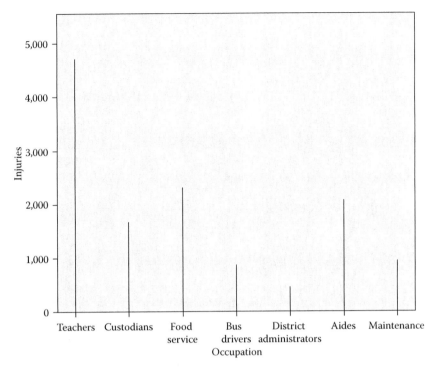

FIGURE 35.2 Injuries per occupation.

specifically target safety and skill. A good example of this is the driving simulators that travel from district to district. Maintenance workers are also exposed to routine safety training and safety management issues. Examples include locking out and controlling hazardous energy during maintenance and repair operations. The simple

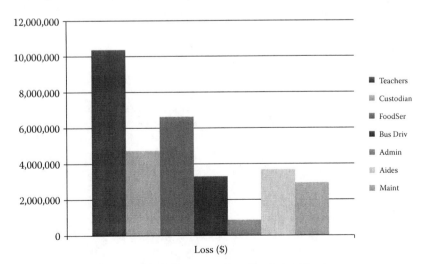

FIGURE 35.3 Loss measured in dollars per occupational classification.

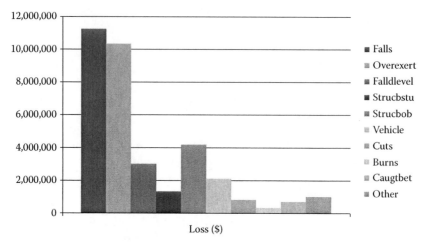

FIGURE 35.4   Amount of loss by type of injury.

locking of a circuit breaker with a locking device prevents many electrocutions. The use of fall protection while working from heights is another example. Teachers may be overlooked. Last spring while attending an educational course on the position of the school principal that was designed for teachers seeking a master's degree, safety was not identified as a priority for the job of principal by the teachers. When asked to identify the job duties of a principal, not one of fourteen teachers of various levels of experience from new substitute to twenty-year tenured faculty identified safety as a job duty. This may serve as an indicator that teacher safety is overlooked. A classic pitfall was exemplified in the course on school principals where eight principals in attendance from Kentucky schools mentioned that student safety was their first job concern. If school safety was such a concern to principals for students, why did the teachers from the same districts not mention this as a job priority? Future studies can answer this question and many others (Biggins, 2010).

CONCLUSION

It is clear from examining surface-level statistics that workplace safety for educational service workers impacts day-to-day operations of school districts. It is clear that budgets can be impacted positively by more effective risk management activities. We can also explore how safety impacts student achievement. Lengthy absences of a teacher because of personal injury after the students get acclimated to the teacher often result in multiple substitute teachers and an interruption in classroom dynamics. More foundationally, we should explore how moral architecture impacts school safety with educational service workers as well as students. Wagner and Simpson (2009) point out that those schools with higher levels of moral architecture may be safer because children learn to choose to do what is right because it is fair. This is in sharp contrast to schools where behavior is merely mandated. If Wagner and Simpson are correct in their assertion that higher levels of moral architecture result in safer schools and if safety is indeed a moral principle for leadership, districts must begin to focus on workplace safety as well as student safety. Safety can help fill voids instead of being a void itself.

CASE STUDY

In the past fiscal year of operations, the central office adopted a plan for cutting costs by not replacing some higher-level administrators who were retiring and allowing others to take on their management duties. The expected savings were to be in the range of $200,000 to $300,000 per year. However, because of an increase in workers' compensation premiums from work-related injuries that included a high-profile teacher on disability, the expected savings are now reduced significantly. The district's insurance carrier offered a loss prevention specialist to audit the district and suggest improvements. One of the suggested improvements centered on performing official job hazard analysis audits for all personnel.

- How would leadership in workplace safety influence this situation?
- How would you build a business case for safety within the school system?

EXERCISES

1. Perform a job hazard analysis for an elementary school teacher. You might consider the criteria that teachers are judged on by their principal as well as what the official job description lists as teacher duties. You should include surveys or interviews with teachers and observations of job performance in order to have a comprehensive view of actual and peripheral duties.
2. Is there a method where students can be involved in identifying and reporting slip, trip, and fall hazards?
3. Identify the various hazards that a grounds worker or maintenance worker might encounter.

REFERENCES

Biggins, R. (2010). *Notes from Educational Course EAD 801.* Eastern Kentucky University, Richmond, KY.

Bird, F., Jr., Germain, G. L., and Clark, M. D. (2003). *Practical Loss Control Leadership* (3rd ed.). Det Norske Veritas, Duluth, GA.

Goetsch, D. (2008). *Occupational Safety and Health for Technologists, Engineers, and Managers.* Pearson Prentice Hall, Upper Saddle River, NJ.

Isaacs, J. (2010). School liability issues. 2010 Kentucky School Board Association Annual Meeting. Louisville, KY.

Larson, L., and Larson, A. (2000). *Workers' Compensation Law: Cases, Materials, and Text* (3rd ed.). Lexis Publishing, New York.

U.S. Bureau of Labor Statistics. (2010). Injury/illness data table. Accessed May 2, 2012. http://www.bls.gov/news.release/archives/osh_10212010.pdf.

Wagner, P., and Simpson, D. (2009). *Ethical Decision Making in School Administration: Leadership as Moral Architecture.* SAGE Publications, Thousand Oaks, CA.

Index